ENGINEERING PHYSICS - II

ENGINEERING PHYSICS - II

For

B.E./B.Tech.

STRICTLY AS PER THE LATEST SYLLABUS PRESCRIBED BY G.B. TECHNICAL UNIVERSITY, LUCKNOW

By

Dr. A.K. KATIYAR
Asst. Professor, Department of Applied Science
Institute of Engineering and Technology
Lucknow (U.P.)

NARINDER KUMAR
M.Sc. PES (I)
Formerly, Senior Lecturer,
Department of Physics
S.D. Govt. College, Ludhiana
Punjab

UNIVERSITY SCIENCE PRESS

(An Imprint of Laxmi Publications Pvt. Ltd.)

BANGALORE • CHENNAI • COCHIN • GUWAHATI • HYDERABAD JALANDHAR • KOLKATA • LUCKNOW • MUMBAI • PATNA RANCHI • NEW DELHI

Published by :

UNIVERSITY SCIENCE PRESS

(*An Imprint of Laxmi Publications Pvt. Ltd.*)
113, Golden House, Daryaganj,
New Delhi-110002

Phone : 011-43 53 25 00
Fax : 011-43 53 25 28

www.laxmipublications.com
info@laxmipublications.com

Price : ₹ **195.00** *Only.*

First Edition : 2009
Second Edition : 2013

OFFICES

✆ **Bangalore**	080-26 75 69 30	✆ **Chennai**	044-24 34 47 26
✆ **Cochin**	0484-237 70 04, 405 13 03	✆ **Guwahati**	0361-251 36 69, 251 38 81
✆ **Hyderabad**	040-24 65 23 33	✆ **Jalandhar**	0181-222 12 72
✆ **Kolkata**	033-22 27 43 84	✆ **Lucknow**	0522-220 99 16
✆ **Mumbai**	022-24 91 54 15, 24 92 78 69	✆ **Patna**	0612-230 00 97
✆ **Ranchi**	0651-221 47 64		

UEP-9348-195-ENGG PHYSICS-II (UP)-KUM

C—6038/012/012

Typeset at : Kalyani Computer Services, New Delhi.

Printed at : Ajit Printers, Delhi.

Contents

PREFACE TO THE SECOND EDITION

This second edition of Engineering Physics-II book is designed strictly as per the new syllabus of UP Technical University. This book not only covers entire UPTU syllabus, but also gives complete information about the subject. The matter of the book is presented in such a simple and logistic manner that even a student belongs to poor background of subject can understand the subject easily. Authors feel that by reading this book, particularly engineering students, for whom this book is prepared can make themselves perfect in the field covered in the book.

The chapters of this book cover the subject matters of Wave Mechanics, X-ray Diffraction, Dielectrics, Magnetic Properties of Materials, Electromagnetic, Superconductivity and Nano-technology. Each chapter starts with basic knowledge and covers all information to be provided to readers. Topics are explained with knowledgeable examples and relevant numerical problems for better understanding. Looking the present pattern of question paper of the University, a large number of multiple choice questions with answers are given at the end of each chapter. This may also be treated by the students as a strong self evaluation exercise for examination point of view. A large number of solved and unsolved problems are backbone of the book.

Though we made every effort for the excellent presentation and error free book, even then some misprint/errors might be crept in. We shall be thankful for pointing out them and giving suggestions for further improvement.

—Authors

PREFACE TO THE FIRST EDITION

This book is prepared as per new course introduced in second semester for engineering students of UP Technical University. It will intend to satisfy the entire requirements of the students preparing for engineering physics-II. Some emerging topics like superconductivity and nanotechnology provide basic informative knowledge to readers. This book is the result of a long teaching experience of authors in this field. Hence presentation of the matter is thoroughly lucid and comprehensive. Way of expression is so simple that a student poor in background of the subject can also be benefited.

The salient features of the book are as under:

- The subject matter of each and every topic is quite rich in content.
- The accuracy of the subject matter as well as the accuracy of printing matter has been assured completely.
- A large number of multiple choice questions have been provided. This may be treated by the students as a strong self evaluation exercise for examination point of view.
- A large number of solved and unsolved problems are backbone of the book.

Although every effort has been made for the excellent presentation of the work, even we are open to suggestions for future incorporation.

—Authors

SYLLABUS

ENGINEERING PHYSICS - II

EAS-202

Unit-1 : Wave Mechanics and X-ray Diffraction

Wave-particle duality, de-Broglie matter waves, Phase and Group velocities, Davisson-Germer experiment, Heisenberg uncertainty principle and its applications. Wave function and its significance, Schrödinger's wave equation — particle in one dimensional box.

Diffraction of X-rays by crystal planes, Bragg's spectrometer, Compton's effect.

10 Hrs.

Unit-II : Dielectric and Magnetic Properties of Materials

Dielectric constant and Polarization of dielectric materials, Types of Polarization (Polarizability). Equation of internal fields in liquid and solid (One-Dimensional), Claussius-Mossotti Equation, Ferro and Piezo electricity (Qualitative), Frequency dependence of dielectric constant, Dielectric Losses, Important applications of dielectric material, Langevin's theory for dia and paramagnetic material, Phenomena of hysteresis and its applications.

Ultrasonic: Generation, detection and application of ultrasonics. **08 Hrs.**

Unit-III : Electromagnetics

Displacement Current, Maxwell's Equations (Integral and Differential Forms). Equation of continuity, EM-Wave equation and its propagation characteristics in free space and in conducting media, Poynting theorem and Poynting vectors. **06 Hrs.**

Unit-IV : Superconductivity and Science and Technology of Nanomaterials

Temperature dependence of resistivity in superconducting materials, Effect of magnetic field (Meissner effect), Type I and Type II superconductors, Temperature dependence of critical field, BCS theory (Qualitative), High temperature superconductors. Characteristics of superconductors in superconducting state, Applications of Superconductors.

Introduction to Nanomaterials—Basic principle of nanoscience and technology, creation and use of buckyballs, structure, properties and uses of Carbon nanotubes, Applications of nanotechnology. **06 Hrs.**

UNIT I

WAVE MECHANICS & X-RAY DIFFRACTION

This unit contains the following chapters:

1. Wave Mechanics
2. X-rays

1

Wave Mechanics

1.01 WAVE-PARTICLE DUALITY

(*i*) Dual Nature of Radiation

Different theories have been put forward at different times regarding the real nature of radiation and light. A successful theory would make the least possible assumptions, explain the existing facts and should be able to explain various other phenomena which are discovered later. On this test, no theory of light is really complete and successful.

(*a*) **Corpuscular theory.** Historically speaking, in the later half of the seventeenth century, Newton gave the '*corpuscular theory*' of light, according to which, light consists of tiny particles or corpuscles which are shot out by the luminous body. This theory quite satisfactorily explained rectilinear propagation, reflection, refraction etc. But it did not tackle the question as to how and why a candle light and a powerful arc light should send the same corpuscles of the same energy with the same velocity.

Further, on refraction, this theory suggests that light should travel faster in the denser medium, and slower in the rarer medium. This later on proved to be just the reverse. We now know that light travels *slower* in the denser medium than in a rarer medium. And again reflection was explained on the basis of repulsion of a light corpuscle by the reflecting surface while refraction was explained as due to attraction by the refracting medium. Thus, a medium could either attract or repel the corpuscle of light. But we all know that glass partially reflects and partially refracts light. How could glass then attract as well as repel the light corpuscle simultaneously ? To add to this confusion, Newton himself discovered that a thin oil film on water shows various colours. This phenomenon could not be explained on this theory. To cap it all, new phenomena of interference, diffraction, polarisation etc., could never be explained on the basis of corpuscular theory. So, Newton's corpuscular theory was discarded. After this, light began to be regarded as a wave phenomenon.

(*b*) **Huygens' theory.** A new theory by Christian Huygens was put forward. This was called the *wave theory*. An illuminated body sends out a disturbance in the form of a wave in the space. Huygens assumed the existence of a hypothetical medium ether, through which waves could pass. Later on, Maxwell modified the concept by stating that light consists of *electromagnetic waves* requiring no material medium. All the same, Maxwell's theory of light could explain rectilinear propagation, reflection, refraction etc., in addition to phenomena like interference, diffraction, polarisation etc.

(*c*) **Photon picture of light.** The wave theory held a firm ground till spectroscopy and photoelectricity came into picture. Lines of the spectrum are possible only when an electron jumps from a higher energy level to a lower energy level, releasing the difference of energy in the form of radiation of a definite frequency in 'packets' or discrete units called *quanta* (Its singular is quantum.).

Likewise a definite quantum of light radiation, called photon, is falling on a surface to release an electron. Thus, these and similar phenomena could *never* be explained on the basis of wave theory. The only way we can explain these phenomena is by assuming that light really consists of packets of energy called 'photons' having a definite frequency. So, we revert to the particle (corpuscle or quantum) aspect of the nature of radiation or light. This caused a lot of confusion, whether radiation has a particle nature or a wave nature.

(*d*) **Dual nature of light.** The confusion is resolved by the assumption that light has in fact *a dual nature*. In certain experiments, it behaves *only* as particles and in certain others, it behaves *only* as waves. One thing is certain that no experiment has ever been found in which light simultaneously behaves like both particles and waves. It is *either* this *or* that; and never this *and* that aspect. This is known as the Bohr's law of complimentarity of radiation.

> ■ ***In the later part of the 19th century and in the beginning of 20th century, it was realised that black body radiation and the photoelectric effect can be understood only on the basis of particle model of light.***
>
> ■ ***Some experiments require light to be a wave while others require light to be a particle. This led to the acceptance of dual (particle and wave) nature of light.***

At present, we reconcile ourselves to this dual aspect by asserting that when the radiation is emitted, it is in quanta. When it is transmitted, it is in waves; and finally when it is absorbed, it is again in quanta.

(*ii*) **Dual Nature of Matter/Matter Waves**

It is an established fact that radiation has a dual aspect behaving either as waves or as particles under suitable circumstances. A particle having a finite mass has been supposed to be associated with a definite amount of energy, given by Einstein in his mass-energy relationship, $E = mc^2$, where E is the energy associated with a particle of mass *m*, *c* being the velocity of light. On this basis, a wave has a particle aspect.

> ■ ***In analogy with the behaviour of light, de Broglie proposed that all material particles should also show dual behaviour. The particle character of matter, e.g., electron, had been established from a number of experiments and its wave character was confirmed by diffraction and interference experiments. Many modern instruments like electron microscope are based on the wave nature of electrons. Neutrons, protons, hydrogen atoms and even C_{60} fullerene molecules have also been shown to possess wave character.***
>
> ■ ***We have seen that light and other electromagnetic radiations have a dual nature i.e., wave and particle nature. Both these characters are instrumental in explaining the characteristics of these radiations. Inspired by this, Louis de Broglie, a French physicist in 1924 advanced a new idea by stating that all moving material particles (microscopic as well as macroscopic) have dual nature i.e., wave and particle nature. However, the wave associated with matter is known as matter wave with characteristics different from the electromagnetic wave. He co-related the two characters in the form of an equation known as de-Broglie Equation after his name.***
>
> $$\lambda = \frac{h}{mv}$$
>
> ***Here λ denotes the wave nature of the moving material object while mv (momentum) represents its particle nature.***

In 1924, it was argued by the French physicist Louis de Broglie that if radiation has a dual aspect, why not think of particles of matter also as having this dual aspect ? Nature loves symmetry and if mass-energy symmetry is present for waves, then the same symmetry should be available for material particles. In simple words, what we know as material particles (like electrons, protons, neutrons etc.) should also behave like waves of a definite λ, under suitable conditions. Thus, particles may also behave as waves. This is *the dual nature of matter*. The waves which are associated with matter are called **matter waves** or **de-Broglie waves** after the name of the French scientist who gave us this idea. Experiments performed later on actually proved the duality of matter. We know that electrons are particles having a discrete mass and moving with a definite velocity. Sometimes, these behave like waves and produce diffraction patterns as was shown by Davisson and Germer; as also by Thomson. Thus, the concept of de-Broglie waves is not merely the flight of fancy of a scientist but actually in existence ! So we accept this duality of matter, as we accept the duality of radiation. It has to be noted that *Bohr's law of complimentarity* holds good here. This means that in no experiment, matter exists *both* as a particle and as a wave simultaneously. It is either one or the other aspect and not this and that aspect together ! The two aspects (wave and particle) are complimentary to each other.

1.02 De-BROGLIE WAVE EQUATION

According to Einstein relation of mass-energy equivalence, $E = mc^2$, where m is the mass of the particle and c is the velocity of light. Further, according to Planck's quantum theory, $E = h\nu$, where ν is the frequency of radiation and 'h' is Planck's constant. Combining the two, we get

$$E = h\nu = mc^2$$

But $c = \nu\lambda$ for waves. Thus, for a quantum of light (photon), we get

$$E = h \cdot \frac{c}{\lambda} = mc^2; \text{ giving us } \lambda = \frac{h}{mc}$$

The product of mass and velocity is the momentum p.

$$\therefore \qquad \boxed{\lambda = \frac{h}{p}}$$

Thus, a quantum (particle) having a momentum p is associated with a wavelength λ.

This relation was for a photon and it proved true experimentally. It now occurred to de Broglie that what is true for a photon, may also be true for a material particle of mass m moving with velocity v (not c).

$$\therefore \qquad \boxed{\lambda = \frac{h}{mv}} \qquad (\because \quad p = mv)$$

This means that a material particle of mass m moving with a velocity v, can be considered as a wave of length λ, given by the above relation. This is *de Broglie wave equation*. It firmly establishes the aspect of unity of matter ; since the momentum p can be associated only with a particle, while λ is associated only with a wave.

Discussion

(i) $\lambda \propto \frac{1}{v}$ *i.e.*, if $v = 0$, $\lambda = \infty$.

Thus, the matter waves are associated with material particles only if they are in motion. Moreover, the wavelength of a particle should decrease as its velocity increases.

(ii) $\lambda \propto \frac{1}{m}$. Smaller the mass of the particle, higher is the wavelength associated with it.

For a given velocity, heavier particles should have shorter wavelength than lighter particles. In the case of macroscopic objects such as a bullet or a ball, the de Broglie wavelength is so small that it cannot be measured directly by any known means.

(iii) $\lambda \propto \frac{1}{p}$. Larger the momentum of the particle, shorter is the wavelength.

(iv) Wavelength associated with a material particle is independent of the charge of the particle.

KNOWLEDGE PLUS

- $$\lambda \propto \frac{1}{p} \quad \text{or} \quad \text{Wave nature} \propto \frac{1}{\text{Particle nature}}$$

  Thus, we conclude that the wave and particle nature of the moving material objects are inversely proportional to each other.
- The experimental confirmation of the de-Broglie relation was obtained when Davisson and Germer observed that a beam of electrons is diffracted. The wavelengths of the electrons determined by the diffraction experiments were found to be in agreement with the values predicted by de Broglie relation.

Louis de-Broglie (1892-1987)

De Broglie, a French physicist, studied history as an undergraduate in the early 1910's. His interest turned to science as a result of his assignment to radio communications in World War I. He received his Dr. Sc. from the University of Paris in 1924. He was professor of theoretical physics at the University of Paris from 1932 until his retirement in 1962. He was awarded the Nobel Prize in Physics in 1929.

1.03 SIGNIFICANCE OF De-BROGLIE RELATIONSHIP

Although the dual nature of matter is applicable to all material objects but it is significant for microscopic bodies only. For large bodies, the wavelengths of the associated waves are very small and cannot be measured by any of the available methods. Therefore, practically these bodies are said to have no wavelengths. Thus, any material body in motion can have wavelength but it is measurable or significant only for microscopic bodies such as electron, proton, atom or molecule. This may be illustrated as follows :

The wavelength of an electron with mass 9.11×10^{-31} kg and moving with the velocity of 10^6 m s^{-1} is 7.28×10^{-10} m as shown below :

$$\lambda = \frac{h}{mv} = \frac{6.63 \times 10^{-34} \text{ kg m}^2 \text{ s}^{-1}}{(9.11 \times 10^{-31} \text{ kg}) \times (10^6 \text{ m s}^{-1})}$$

$$= 7.28 \times 10^{-10} \text{ m} \qquad [\text{J} = \text{kg m}^2 \text{ s}^{-2}]$$

This wavelength associated with the moving electron is of the same order of magnitude as of X-rays which can be easily measured.

On the other hand, the wavelength associated with a ball, for example, weighing 10 g and moving with the same velocity as that of electron (10^6 m s^{-1}) is only 6.63×10^{-38} Å as shown below :

$$\lambda = \frac{h}{mv} = \frac{6.63 \times 10^{-34} \text{ kg m}^2 \text{ s}^{-1}}{(10 \times 10^{-3} \text{ kg}) \times (10^6 \text{ m s}^{-1})} = 6.63 \times 10^{-38} \text{ m}$$

This wavelength is shorter than any electromagnetic radiation and cannot be measured by any known method. Therefore, it is difficult to grasp the idea of waves associated with the moving ball. In other words, we can say that the ball does not have waves associated with it. Similarly, the other large bodies are said to be associated with waves but their wavelengths are too small to be measured by any conceivable method.

1.04 LIMITATION OF De-BROGLIE EQUATION

It has been stated earlier that the de-Broglie equation or relationship must cover every moving object but in actual practice it is true only for microscopic particles such as electrons, protons, neutrons, atoms, ions etc. It has no relevance for moving semimicro and macro objects where particle character is very large and the wave character is negligible. In order to clarify this view point, let us apply this relation to a moving microparticle (say electron) and a moving macroparticle (say a ball weighing 100 g or 0.1 kg) both travelling with the same speed *e.g.* 10^5 m s^{-1}. Let us calculate the wavelength in both the cases.

For electron :

$$\lambda = \frac{h}{mv} = \frac{(6.626 \times 10^{-34} \text{ kg m}^2 \text{ s}^{-1})}{(9.11 \times 10^{-31} \text{ kg}) \times (10^5 \text{ m s}^{-1})}$$

$$= 7.27 \times 10^{-9} \text{ m}$$

For ball :

$$\lambda = \frac{h}{mv} = \frac{(6.626 \times 10^{-34} \text{ kg m}^2 \text{ s}^{-1})}{(0.1 \text{ kg}) \times (10^5 \text{ m s}^{-1})}$$

$$= 6.626 \times 10^{-38} \text{ m}$$

The wavelength of the electron wave is within the range of the electromagnetic waves and can be measured. However, the wavelength of the wave associated with the ball is very small as compared to any of the known electromagnetic waves. Therefore, it is not possible to measure the same. Thus, practically an electron has a wave character while no such character is associated with a ball which signifies a moving macroscopic or macro particle. The same is also true for the moving semimicro particles although their mass is small as compared to the macro particles.

> ***Practically, de-Broglie relation can be applied only to the moving microscopic particles including electrons, protons, atoms, small molecules etc. while it has no relevance for the moving semimicro or macroparticles.***

1.05 DERIVATION OF BOHR'S POSTULATE OF QUANTISATION OF ANGULAR MOMENTUM FROM De-BROGLIE RELATION

This postulate gives the quantum condition that only such orbits are permitted in which the angular momentum of the electron is an integral multiple of $h/2\pi$. In mathematical language, for a permitted stationary orbit, the condition

$$mvr = \frac{nh}{2\pi}$$

must be satisfied. Here, n is an integer known as the principal quantum number. De Broglie gave the theory that such permitted orbits can exist only when the orbit's circumference equals an integral multiple of the wavelength of the electron. There is a stationary wave system for the de Broglie waves for an electron in that orbit and the orbit must contain a whole number of electron waves (Fig. 1.1). Thus, in the diagram, we have a permitted stationary orbit for an electron where $n = 3$. The whole circumference of the orbit is equal to three complete waves of wavelength λ given by the de Broglie relation. Thus in a permitted orbit, we must have

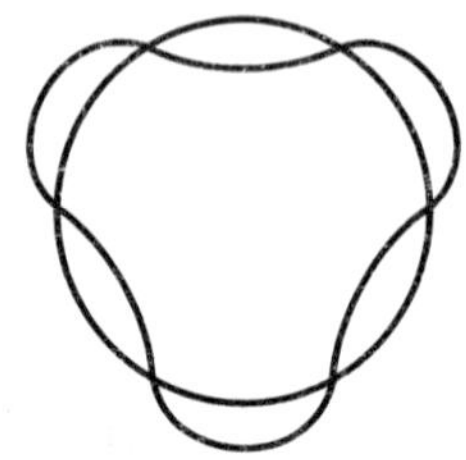

Fig. 1.1

$$2\pi r = n\lambda$$

de-Broglie wavelength, $$\lambda = \frac{h}{mv}$$

Equating $$2\pi r = n\frac{h}{mv} \quad \text{or} \quad m\,vr = n\frac{h}{2\pi}$$

This is the same condition as given by Bohr's quantum postulate. So, we note that the concept of de Broglie waves is fully reconciled with the observed quantum conditions of the Bohr's model of hydrogen atom. Later experiments by Thomson and also by Davisson and Germer have conclusively established the truth of the existence of de Broglie waves in nature.

Niels Bohr, a Danish physicist received his Ph.D. from the University of Copenhagen in 1911. He then spent a year with J.J. Thomson and Ernest Rutherford in England. In 1913, he returned to Copenhagen where he remained for the rest of his life. In 1920 he was named Director of the Institute of theoretical Physics. After first World War, Bohr worked energetically for peaceful uses of atomic energy. He received the first Atoms for Peace award in 1957. Bohr was awarded the Nobel Prize in Physics in 1922.

Niels Bohr
(1885-1962)

1.06 De-BROGLIE WAVELENGTH IN TERMS OF ABSOLUTE TEMPERATURE

We know that $$\lambda = \frac{h}{mv}$$

or $$\lambda = \frac{h}{p}$$

Now $$E_k = \frac{1}{2}mv^2 = \frac{m^2v^2}{2m} = \frac{p^2}{2m}$$

or $$p = \sqrt{2m\,E_k}$$

$$\therefore \quad \lambda = \frac{h}{\sqrt{2m\,E_k}}$$

In thermal equilibrium state, at temperature T, the kinetic energy associated with a particle is of the order of $\frac{3}{2}k_B T$, where k_B is Boltzmann's constant.

$$\therefore \quad \lambda = \frac{h}{\sqrt{2m\frac{3}{2}k_B T}} \quad \text{or} \quad \lambda = \frac{h}{\sqrt{3m\,k_B T}}$$

1.07 De-BROGLIE WAVELENGTH OF AN ELECTRON WAVE (NON-RELATIVISTIC CASE)

Consider an electron accelerated through a potential difference of V volt. According to work-energy principle, work done (eV joule) on the electron must be equal to the gain of kinetic energy $\left(\frac{1}{2}mv^2\right)$ of the electron.

$$\therefore \quad eV = \frac{1}{2}mv^2 \quad \text{or} \quad v = \sqrt{\frac{2eV}{m}}$$

Now, de-Broglie wavelength associated with moving electron is given by :

$$\lambda = \frac{h}{mv} = \frac{h}{m\sqrt{\frac{2eV}{m}}} = \frac{h}{\sqrt{2meV}}$$

Putting $h = 6.62 \times 10^{-34}$ Js, $m = 9.1 \times 10^{-31}$ kg and $e = 1.6 \times 10^{-19}$ C, we get

$$\lambda = \frac{6.62 \times 10^{-34}}{\sqrt{2 \times 9.1 \times 10^{-31} \times 1.6 \times 10^{-19} \times V}}\ \text{m}$$

or $$\lambda = \frac{12.27 \times 10^{-10}}{\sqrt{V}}\ \text{m} \quad \text{or} \quad \lambda = \frac{12.27}{\sqrt{V}}\ \text{Å}$$

1.08 DISTINCTION BETWEEN WAVE AND PARTICLE

A particle always occupies a well-defined and specific position in space which cannot be simultaneously occupied by another particle. If there are more than one particles in a given region of space, then their sum is equal to the number of individual particles. The sum can neither be more nor less.

On the other hand, a wave is spread out or delocalised in space. Unlike particles, two or more waves can co-exist in the same region. Two or more waves can even superimpose to form a resultant wave. The resultant wave can be larger or smaller than the individual waves.

The important differences between wave and particle are given below :

S.No.	*Wave*	*Particle*
1.	A wave is delocalised or spread out in space.	A particle occupies a well defined or localised position in space.
2.	Two or more waves can exist simultaneously (or co-exist) in space. In other words, they can interfere with each other.	Two particles cannot occupy the same position in space. In other words, the particles do not interfere with each other.
3.	If two or more waves are present in the same region, then due to interference, the resultant wave can be either smaller or bigger than the individual waves.	If two or more particles are present in the same region in space, their sum is equal to the number of the individual particles.
4.	Waves cannot be counted.	Particles can be counted.

1.09 DETECTION OF MATTER WAVES

The waves associated with material particles are called matter waves.

According to de-Broglie, the matter waves are generated by both microscopic and macroscopic bodies. However, the de-Broglie wavelength of a matter wave associated with a macroscopic body is so small that it cannot be detected by any instrument. This explains as to why the wave nature of matter is not observed in our daily life. The following illustrations are useful in understanding this concept.

Illustration 1. Let us calculate the de-Broglie wavelength λ of earth, taking the mass of earth to be 6×10^{24} kg and orbital velocity to be 3×10^4 m s^{-1}.

$$\lambda = \frac{h}{mv} = \frac{6.625 \times 10^{-34}}{6 \times 10^{24} \times 3 \times 10^4} = 3.7 \times 10^{-63} \text{ m}$$

Clearly, λ is so small that it cannot be detected by any measuring instrument.

Illustration 2. Consider a stone of mass 0.1 kg moving at a speed of 1 m s^{-1}. Its de-Broglie wavelength is given by

$$\lambda = \frac{h}{mv} = \frac{6.625 \times 10^{-34}}{0.1 \times 1} \text{ m} = 6.625 \times 10^{-33} \text{ m}$$

This wavelength is also very small and cannot be detected by any measuring instrument.

Illustration 3. Consider an electron moving at a speed of 10^5 m s^{-1}. If its mass is assumed to be 10^{-30} kg, then the de-Broglie wavelength is given by

$$\lambda = \frac{6.625 \times 10^{-34}}{10^{-30} \times 10^5} \text{ m} = 6.625 \times 10^{-9} \text{ m}$$

This value is within the limit of measurement. It is of the order of wavelength of X-rays which can be detected.

1.10 PROPERTIES OF MATTER WAVES

(i) The de-Broglie wavelength of a wave associated with a moving heavy particle is less than the wavelength associated with a lighter particle.

(ii) The de-Broglie wavelength of a wave associated with a fast moving particle is less than the wavelength associated with a slow moving particle.

(iii) The de-Broglie wavelength of a wave associated with a particle at rest is infinite. In-fact, the wave becomes indeterminate. Clearly, the concept of matter waves is valid only when the material particles are in motion.

(iv) The de-Broglie wavelength does not depend upon the charge of the particle. The matter waves are generated by moving charged particles as well as by moving neutral particles.

(v) The velocity of a matter wave is greater than the velocity of electromagnetic wave.

(vi) The velocity of matter wave depends upon the velocity of the material particle.

(vii) The wave and particle aspects of matter never appear simultaneously in the same experiment.

1.11 DISTINCTION BETWEEN ELECTROMAGNETIC WAVES AND MATTER WAVES

If we throw a pebble (small piece of stone) in a pond, water waves can be seen in the pond which surround the pebble. These are called matter waves and are quite different from the electromagnetic waves in the form of light, heat, X-rays etc. The electron waves are also matter waves and quite often called de-Broglie waves.

The important differences between electromagnetic waves and matter waves are given below :

Electromagnetic waves	*Matter waves*
1. Electromagnetic waves are associated with electric and magnetic fields which are perpendicular to each other and also to the direction of propagation of radiation.	Matter waves are generally not associated with electric and magnetic fields. Matter waves obtained from charged particles are associated with electric and magnetic fields.
2. Electromagnetic waves can be radiated into space or emitted.	Matter waves are neither radiated into space nor emitted by the particles. These are simply associated with the particles.
3. They do not require any medium for propagation *i.e.,* these waves can pass through vacuum.	They require medium for propagation *i.e.,* these waves cannot pass through vacuum.
4. All electromagnetic waves travel with the same velocity.	Matter waves travel with different velocities.
5. The velocity of all electromagnetic waves is equal to that of light (3×10^8 m/s).	The velocity of matter waves is generally less than that of light.
6. The wavelengths of electromagnetic radiations are much large and are given by the relation, $$\lambda = \frac{c}{\nu}$$ where ν is the frequency.	The matter waves have shorter wavelengths given by de-Broglie equation. $$\lambda = \frac{h}{mv}$$
7. Wavelengths cover a full range of spectrum.	These have very short wavelengths.

1.12 PROOF OF De-BROGLIE WAVE EQUATION

We can derive the expression for the wavelength of the wave associated with a material particle by using the general equation of a standing wave and Lorentz transformation equations in relativity.

Assume that a material particle like electron or proton is equivalent to a standing wave system in the region of space occupied by the particle.

At any instant t, the value of ψ at a point (x, y, z) is given by :

$$\psi = \psi_0 \sin \omega t$$

or
$$\psi = \psi_0 \sin (2\pi \nu t) \quad ...(i)$$

Here ψ_0 is the amplitude of the wave at a point (x, y, z) and ν is the frequency of the particle as observed by an observer at rest relative to the particle under consideration.

Equation (i) represents the standing wave associated with the particle.

Consider a particle moving with velocity v along the positive direction of x-axis. Using inverse Lorentz transformation,

$$\psi = \psi_0 \sin \left[\frac{2\pi\nu \left(t' + \frac{vx'}{c^2} \right)}{\sqrt{1 - \frac{v^2}{c^2}}} \right] \quad ...(ii)$$

The standard wave equation is given by :

$$\psi = \psi_0 \sin \left[2\pi\nu' \left(t' + \frac{x'}{u'} \right) \right] \quad ...(iii)$$

Here ψ_0 is the amplitude, ν' is the frequency and u' is the phase velocity of the wave along x-axis.

Comparing (ii) and (iii), we get

$$\nu' = \frac{\nu}{\sqrt{1 - \frac{v^2}{c^2}}} \quad \text{and} \quad u' = \frac{c^2}{v}$$

According to mass-energy relation,

$$E = mc^2$$

Also,
$$E = h\nu'$$

$\therefore$
$$h\nu' = mc^2 \quad \text{or} \quad \nu' = \frac{mc^2}{h}$$

The wavelength of the wave associated with a moving particle is given by

$$\lambda = \frac{\text{phase velocity}}{\text{frequency}}$$

$$\lambda = \frac{u'}{\nu'} \quad \text{or} \quad \lambda = \frac{c^2 / v}{mc^2 / h}$$

or
$$\lambda = \frac{h}{mv} \quad \text{or} \quad \lambda = \frac{h}{p}$$

1.13 IS ELECTRON A PARTICLE OR A WAVE

A question comes to our mind *"Is the electron really a particle or is it a wave ?" and "what does it look like ?"* Upto the year 1924, the electron was exclusively regarded as a particle. However, in 1924, de-Broglie suggested that the electron behaves both as a material particle and as a wave. Just like light, some experimental facts can be explained by assuming electrons as tiny particles while some other facts can be explained only on the basis of wave character of the electrons. Thus, the electron behaves as a particle as well as a wave.

The electrons are so small that they cannot be seen even with the help of a powerful microscope. But there is no doubt regarding their existence, rather they are the essential constituents of all forms of matter. Thus, they can be imagined to look like very tiny dots executing wave like motion and moving with speed 3×10^8 m s^{-1}.

1.14 EVIDENCE TO SUPPORT WAVE AND PARTICLE NATURE OF THE ELECTRON

With the help of de-Broglie relationship, we have studied that an electron has a dual nature *i.e.*, wave as well as particle nature. This was indeed a remarkable breakthrough in the structure of the atom. Evidences have been cited to support both these characters associated with the electron.

Evidence in favour of particle nature. An electron has all the characteristics of a particle *i.e.*, it has a definite mass, charge, energy and also momentum. In case an electron is made to strike against a screen coated with zinc sulphide (ZnS), a bright spot known as scintillation is produced at the point of impact. It has been observed that only one scintillation point is noticed per electron. This clearly shows that the striking electron is localised like a particle and not spread out like a wave.

Evidence in favour of wave nature. The wave nature of the electrons was supported by the following experiments :

(*i*) Davisson and Germer diffraction experiment.

(*ii*) G.P. Thomson experiment.

1.15 DAVISSON AND GERMER EXPERIMENT

In the year 1927, C.P. Davisson and L.H. Germer performed a diffraction experiment on the electrons. This was the first experimental proof of the wave nature of material particles.

Principle

This experiment is based on the scattering of a beam of electrons by nickel crystal. The de-Broglie wavelength for an electron is less than 1 Å. This is too small to be used with ordinary diffraction gratings that we use in the case of light. A crystal lattice in which the atomic distances between layers are of the order of 1 Å is ideal for the purpose of studying the diffraction of electron waves.

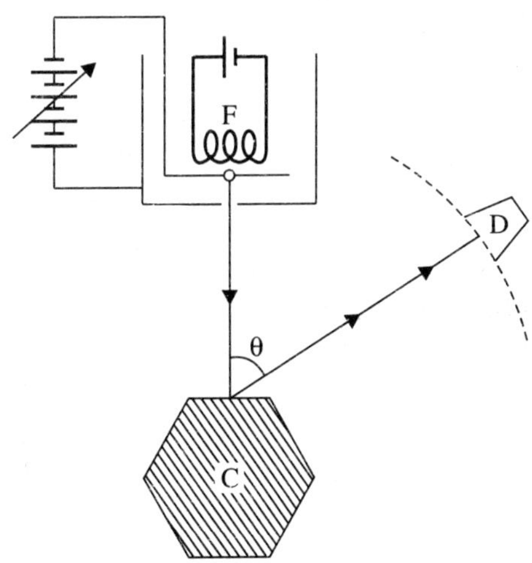

Fig. 1.2. Davisson and Germer experiment.

Experimental set-up

The experimental set-up is given in a sketch form in Fig. 1.2. A nickel crystal C is taken. Electrons are made incident on it as a narrow beam. The incident electrons are produced by an electron gun G. The gun consists of a tungsten filament F which is connected to a low tension battery. Electrons emitted from the filament F are made to pass through pin-holes under suitable accelerating potentials. The accelerating potential is provided with the help of a high-tension battery. Ordinarily, we get an electron beam of energy 50 eV from the gun.

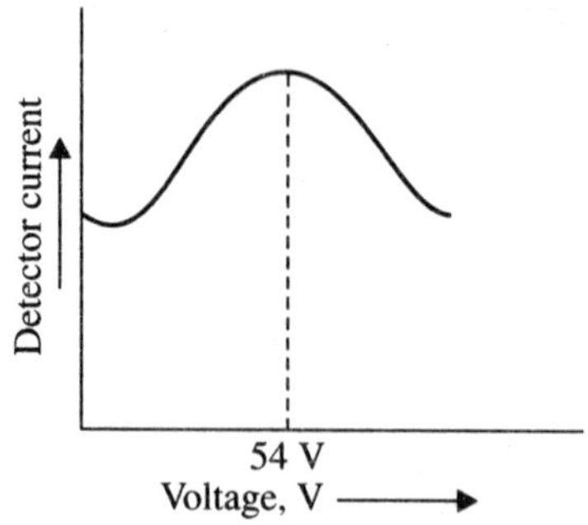

Fig. 1.3. Variation of detector current with voltage.

The incident beam of electrons is made to fall normally on the crystal. The beam is diffracted by the crystal and received at an angle θ by a detector D. The intensity of the diffracted electrons is measured by the detector as a function of angle θ. In the original experiment, the detector was set at an angle of 50° to the direction of the incident beam. The scattered electron current for different values of voltage V was noted. The graph between the voltage and the detector current is as shown in Fig. 1.3. The existence of peak in the graph can be explained as due to constructive interference of waves scattered from atoms in different planes of the crystal. The peak occurs at 54 V. The observed phenomenon is similar to the diffraction of X-rays. So, it establishes the *wave nature of electrons*.

It has been observed that the scattered electron current (or detector current) is maximum when the following relation given by Bragg for X-ray diffraction is satisfied.

$$2d \sin\theta = n\lambda$$

where n is the order of diffraction, d is the atomic spacing between successive crystal planes and θ is the angle at which strong reflection takes place.

The value of d, as determined by X-ray reflection by nickel crystal, comes out to be 0.91 Å. The value of θ comes out to be 65°.

For $\quad n = 1, \lambda = 2d \sin\theta$

$$= 2 \times 0.91 \times \sin 65° = 1.65 \text{ Å} \quad ...(1)$$

We know that

$$\lambda = \frac{h}{\sqrt{2me\,V}}$$

In the given experiment, $\quad V = 54$ volt

$$\therefore \quad \lambda = \frac{6.63 \times 10^{-34}}{(2 \times 9 \times 10^{-31} \times 1.6 \times 10^{-19} \times 54)^{1/2}} \text{ m} = 1.67 \text{ Å} \quad ...(2)$$

It is clear from equations (1) and (2) that the wavelength determined by two different methods comes out to be the same. This confirms that the electrons are diffracted in the same way as de-Broglie waves.

KNOWLEDGE PLUS

Following figures show the diffraction rings received on screen or photographic plate.

Diffraction is a well known phenomenon and is quite common in X-rays. A comparison of the diffraction rings for the electrons and X-rays revealed that the two are exactly similar. Since wave character of the X-rays is already established, the experiment performed by Davisson and Germer supported the wave character of the electrons. The exact pattern of the diffraction

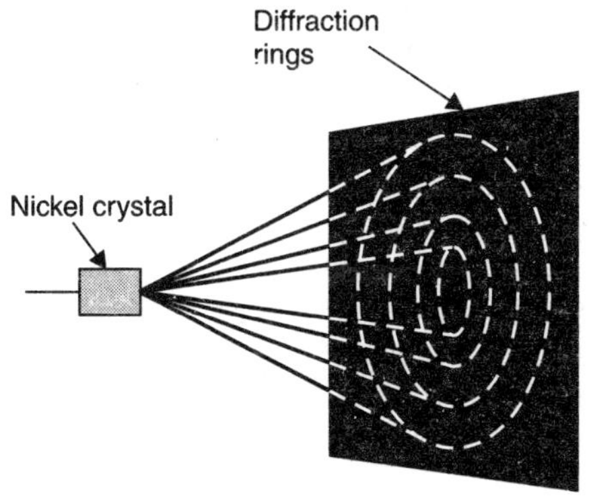

Fig. 1.4

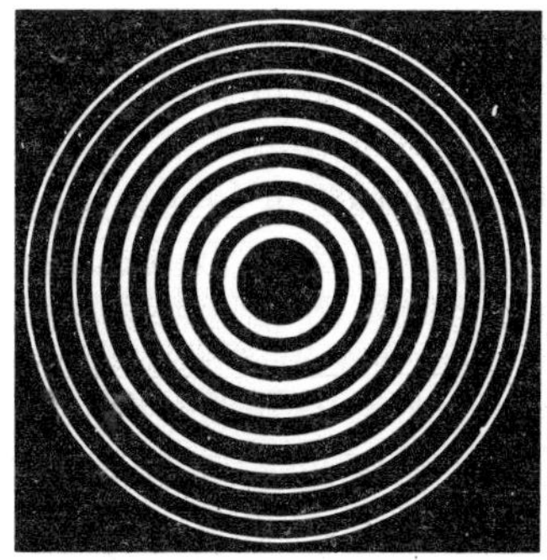

Fig. 1.5

rings has been shown separately in the above figure. Moreover, the wavelengths of the electron waves as determined by diffraction experiment were found to be in order with the values as calculated theoretically with the help of de-Broglie relationship.

1.16 THOMSON'S EXPERIMENT ON ELECTRON DIFFRACTION

In 1928, G.P. Thomson performed an experiment by replacing nickel crystal by a metal foil. He got almost the same type of diffraction patterns for the electron beam as in Davisson and Germer experiment. This further supported the wave character of the electrons.

Thomson obtained diffraction pattern by passing an electron beam through a thin metal foil. The experimental arrangement employed by him is sketched in Fig. 1.6.

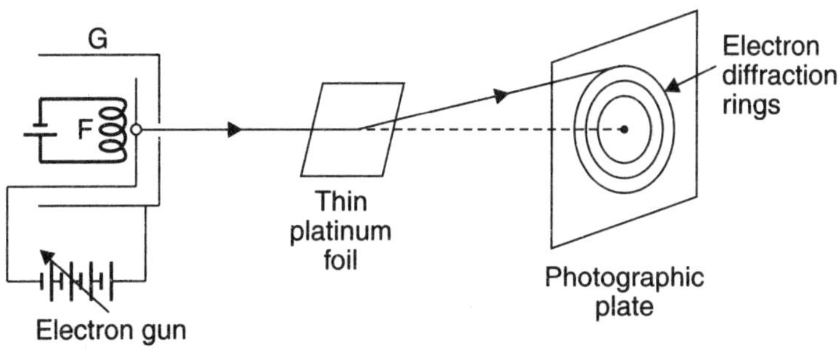

Fig. 1.6. Thomson's experiment.

A beam of electrons obtained from an electron gun is made to fall normally on a thin platinum foil. The foil is nearly 10^{-8} m thick. The foil can also be of aluminium or gold. A photographic plate is placed behind the foil. Diffraction pattern in the form of concentric rings is obtained on the photographic plate. These rings are clearly due to the diffraction of waves (associated with electrons). This is because of the randomly oriented crystals in the foil. Knowing the geometry of the apparatus, the voltage used to accelerate the electron beam and the diameter of the electron diffraction rings, the wavelength associated with the beam of electrons can be determined. It will be noted that the experimentally determined value of wavelength will be very close to the theoretically calculated (from de-Broglie equation) value of wavelength. Thus, de-Broglie hypothesis of matter waves stands verified experimentally.

If, in the above experimental arrangement, we use X-rays in place of the electron beam and an amorphous solid in place of a foil, we will get diffraction rings similar to the one obtained with electron beam.

KNOWLEDGE PLUS

It is interesting to note that the diffraction phenomenon was noticed not only in case of electrons but also for some other microscopic particles like protons, neutrons and small atoms like hydrogen etc. This confirms that all of them have a wave character. The wave-like character of the electrons is used in electron-microscopes and is quite helpful in measuring the size of the very small particles.

1.17 De-BROGLIE WAVES ASSOCIATED WITH ATOMS AND MOLECULES

It is not necessary to stick only to electrons to establish the truth of de-Broglie waves. In fact, de-Broglie's hypothesis is valid for all moving mass particles.

Eastermann and Stern were the first to show diffraction of atoms and molecules by crystals. They repeated Davisson and Germer experiment with helium as well as hydrogen molecules using crystals of lithium fluoride and sodium chloride. The beam of helium molecules from a high temperature source was made incident on a crystal of lithium fluoride in vacuum. The beam got diffracted at various angles. The maximum intensity was obtained at angle θ given by

$$2\,d \sin\theta = n\lambda$$

The de-Broglie wavelength of helium molecule was calculated. An excellent agreement was noted between the observed and calculated values.

A similar experiment was performed by Johnson. A beam of hydrogen atoms was scattered by a crystal of lithium fluoride. The scattered beam was made incident on a plate coated with molybdenum oxide. When struck by hydrogen, the oxide got reduced to molybdenum. Thus, a diffraction pattern was obtained. It was found to be in accordance with de-Broglie hypothesis.

We can conclude from the above discussion that just as radiation has dual nature, particles of matter also possess a dual nature. Of course, both the aspects (particle and wave) are never exhibited simultaneously. Depending upon the experiment, it is either one aspect or another. Neither a photon nor a particle of matter show both wave and particle behaviour simultaneously in the same experiment. This is called *Bohr's principle of complimentarity*.

1.18 ELEMENTARY IDEA OF ELECTRON MICROSCOPE (PRACTICAL APPLICATION OF De-BROGLIE WAVES)

The wavelength of de-Broglie waves associated with an electron is given by

$$\lambda = \frac{h}{\sqrt{2m\,eV}}$$

Substituting the values of h, m and e, we get $\lambda = \dfrac{12.3}{\sqrt{V}}$ Å where V is the voltage through which an electron is accelerated. If an electron has an energy of 300 eV, *i.e.*, it is accelerated through a potential difference of 300 volt, then

$$\lambda = \frac{12.3}{\sqrt{300}} \text{ Å} = 0.71 \text{ Å}$$

Resolving power is the ability of an instrument to show as two different lines which are normally so close that these appear as one to the naked eye. Smaller the wavelength, greater is the resolving power. An ordinary optical microscope uses visible radiation of wavelength 6000 Å. If somehow we could use electrons and accelerate them so that we get de-Broglie waves of small wavelength, then the resolving power and also the magnifying power of the instrument will be greatly increased. Thus, an electron microscope has been devised and put to practical use. This is one of the best proofs of the truth of the existence of de-Broglie waves. Also, it is one of the best applications of matter waves.

Accelerating the electron through a high potential difference will further reduce the wavelength and increase the utility of the instrument. However, a proper focussing of these beams is not that easy. In ordinary optical microscope, the focussing is done with the help of suitable lenses. In an electron microscope, the beam is focussed by specially adjusted electric and/or magnetic fields. The operational sketch of an electron microscope along with an optical microscope for the sake of comparison is shown in Fig. 1.7.

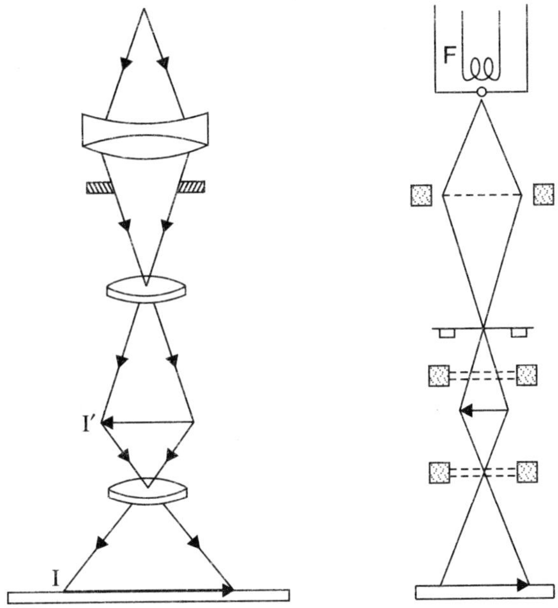

(i) Electron microscope

(ii) Optical microscope

Fig. 1.7

The electron beam emitted by the filament is accelerated through a large potential difference in an electron gun. The electron beam emerging out of the gun is made to pass through the centre of an electromagnet. The electrons get deflected into a parallel beam which is made to fall on the object to be magnified. It is essential that the object should be thin so that electrons may pass through it. More electrons will be transmitted through the 'transparents' of the object while fewer number of electrons will pass through comparatively denser portions. The second electromagnet causes the beam to diverge thereby producing enlarged image of the object. A third electromagnet can be used to focus a still larger image. This image can also be obtained on a photographic plate for permanent record.

In order to get sharp focussing, the intensity of magnetic fields produced by electromagnets is adjusted. In order to examine in detail some parts of the object, a fine micro-movement can be given to the object.

The whole arrangement is mounted in a highly evacuated glass tube.

An electron microscope can magnify an object 100,000 times with sufficiently adequate resolving power.

Uses. An electron microscope is an extremely useful tool for scientific research. In industry, it is used to study the structure of fibres, paper and plastics. In medical science, it is used to study viruses and the structural details of Bacteria.

1.19 A NOTE ON WAVE PACKETS OR WAVE GROUP

A moving material particle is not equivalent to a single wave. It is equivalent to a wave packet or wave group.

A wave packet is a concentrated bunch of waves. It is made up of a number of waves slightly differing in velocity and wavelength. Their phase and amplitude are such that they interfere constructively over a small region of space somewhere within which the particle can be located. Outside this space, they interfere destructively so that the amplitude is reduced to zero. Thus, a wave packet has a beginning and an end.

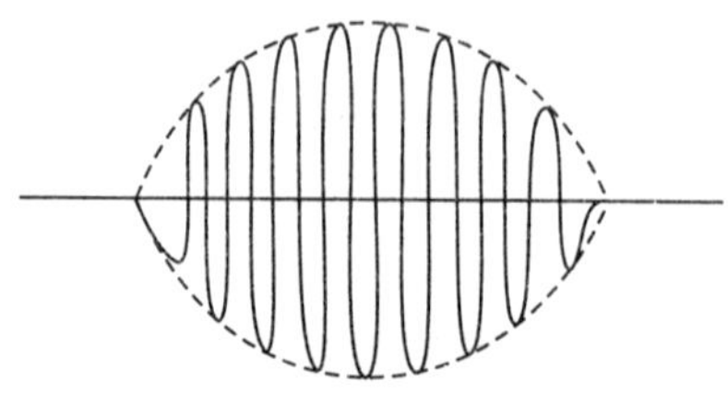

Fig. 1.8

The wave pattern consists of points of maximum amplitude and points of minimum amplitude. Between any two consecutive minima, there is a position of maximum amplitude.

The velocity with which a wave packet moves is called group velocity.

The average velocity of the individual monochromatic wave with which a wave packet is constructed is called wave velocity or phase velocity.

1.20 EXPRESSION FOR GROUP VELOCITY OF A WAVE PACKET

Consider a wave packet consisting of two waves which differ slightly in angular frequencies and phase velocities. Assume that the two waves have equal amplitude.

Let the two waves be represented by the equations

$$y_1 = a \sin(\omega_1 t - k_1 x)$$

and

$$y_2 = a \sin(\omega_2 t - k_2 x)$$

Here, ω_1 and ω_2 are the angular frequencies, k_1 and k_2 are the propagation constants and a is the amplitude.

Using principle of superposition, we get

$$y = y_1 + y_2$$

or

$$y = a \sin(\omega_1 t - k_1 x) + a \sin(\omega_2 t - k_2 x)$$

or

$$y = 2a \sin \frac{(\omega_1 t - k_1 x) + (\omega_2 t - k_2 x)}{2} \cos \frac{(\omega_1 t - k_1 x) - (\omega_2 t - k_2 x)}{2}$$

or

$$y = 2a \cos\left[\frac{(\omega_1 - \omega_2)t}{2} - \frac{(k_1 - k_2)x}{2}\right] \sin\left[\frac{(\omega_1 + \omega_2)t}{2} - \frac{(k_1 + k_2)x}{2}\right] \quad ...(i)$$

or

$$y = \text{A} \sin\left[\frac{(\omega_1 + \omega_2)t}{2} - \frac{(k_1 + x_2)x}{2}\right]$$

Here,

$$\text{A} = 2a \cos\left[\frac{(\omega_1 - \omega_2)t}{2} - \frac{(k_1 - k_2)x}{2}\right]$$

A is the amplitude of the wave packet. It is modulated both in space and time by a very slowly varying envelope of frequency $\frac{\omega_1 - \omega_2}{2}$ and propagation constant $\frac{k_1 - k_2}{2}$. The maximum value of A is $2a$.

Equation (*i*) can now be rewritten as follows :

$$y = 2a \cos\left(\frac{\Delta\omega}{2}t - \frac{\Delta k}{2}x\right)\sin(\omega t - kx)$$

This equation represents a wave of angular frequency ω and wave number k that is superimposed by a modulation of angular frequency $\frac{\Delta\omega}{2}$ and wave number $\frac{\Delta k}{2}$. The modulation produces successive wave groups as shown in Fig. 1.9.

ω and k represent the mean values of angular frequencies and propagation constants of the superposing component waves.

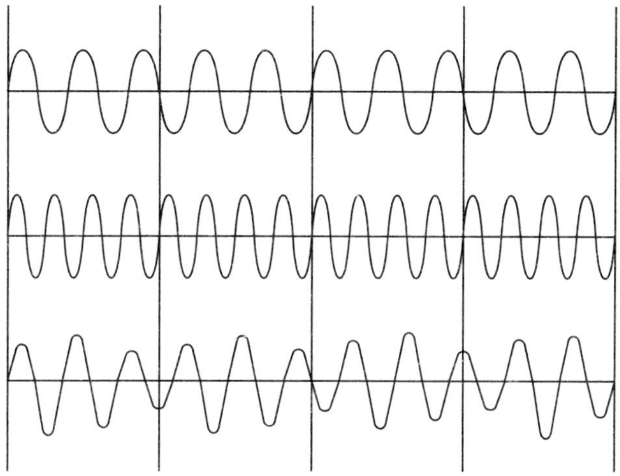

Fig. 1.9

The amplitude of the wave packet is $2a \cos\left(\frac{\Delta\omega}{2}t - \frac{\Delta k}{2}x\right)$. Its phase is $(\omega t - kx)$.

In a wave, there is a progressive change of phase from one particle to the next. So, we can consider different planes, each of constant phase. *The velocity of propagation of the planes of constant phase through medium is known as wave velocity or phase velocity.*

For a plane of constant phase,

$\omega t - kx = \text{constant}$

Differentiating, $\frac{d}{dt}(\omega t) - \frac{d}{dt}(kx) = 0$

or $\omega - k\frac{dx}{dt} = 0$ or $-k\frac{dx}{dt} = -\omega$

or $\frac{dx}{dt} = \frac{\omega}{k}$ or $\boxed{\text{Phase velocity, } v_p = \frac{\omega}{k}}$

Again, the amplitude of the wave packet is constant.

$$\therefore \quad \frac{\Delta\omega}{2}t - \frac{\Delta k}{2}x = \text{constant}$$

or $$\frac{\Delta k}{2}x = \frac{\Delta\omega}{2}t \quad \text{or} \quad x = \frac{\Delta\omega}{\Delta k}t$$

or $$\frac{dx}{dt} = \frac{\Delta\omega}{\Delta k} \quad \text{or} \quad \text{Group velocity, } v_g = \frac{\Delta\omega}{\Delta k}$$

In the limiting case, $$\boxed{v_g = \frac{d\omega}{dk}}$$

1.21 VELOCITY OF De-BROGLIE WAVES (WAVE VELOCITY OR PHASE VELOCITY OR De-BROGLIE WAVE VELOCITY)

According to de-Broglie, a wave of wavelength λ is associated with a particle of mass m moving with velocity v.

$$\lambda = \frac{h}{mv} \quad \text{...}(i)$$

Now, $$E = h\nu \quad \text{or} \quad \nu = \frac{E}{h}$$

Using Einstein's mass-energy relation,

$$\nu = \frac{mc^2}{h} \quad \text{...}(ii)$$

The de-Broglie wave velocity v_p is given by

$$v_p = \nu\lambda$$

or $$v_p = \left(\frac{mc^2}{h}\right)\left(\frac{h}{mv}\right) \quad \text{[Using equations (i) and (ii)]}$$

or $$v_p = \frac{c^2}{v}.$$

The particle velocity v is always less than the velocity c of light. So, *the de-Broglie wave velocity v_p must be greater than the velocity c of light.* This would directly mean that the de-Broglie wave associated with a particle travels much faster than the particle itself. The wave would thus leave the particle far behind. This result contradicts the wave concept of particle. Moreover, this brings us in conflict with the theory of relativity. Thus, we conclude that a moving particle is not associated with a single wave train. In other words, a monochromatic de-Broglie wave cannot transport a particle or carry energy. ***The contradiction was resolved by Schrodinger by postulating that a material particle in motion is equivalent to a wave packet (group of waves) rather than a single wave.***

1.22 RELATION BETWEEN GROUP VELOCITY AND PHASE VELOCITY

Phase velocity, $$v_p = \frac{\omega}{k} \quad \text{or} \quad \omega = v_p k \quad \text{...}(i)$$

Group velocity, $$v_g = \frac{d\omega}{dk}$$

or $$v_g = \frac{d\omega}{d\left(\frac{2\pi}{\lambda}\right)} \qquad \left[\because\ k = \frac{2\pi}{\lambda}\right]$$

or $$v_g = \frac{d\omega}{-\frac{2\pi}{\lambda^2}d\lambda} \quad \text{or} \quad v_g = -\frac{\lambda^2}{2\pi}\frac{d\omega}{d\lambda}$$

or $$v_g = -\frac{\lambda^2}{2\pi}\frac{d}{d\lambda}(v_p k) \qquad [\text{Using} = n(i)]$$

or $$v_g = -\frac{\lambda^2}{2\pi}\frac{d}{d\lambda}\left(\frac{2\pi v_p}{\lambda}\right) \quad \text{or} \quad v_g = -\lambda^2 \frac{d}{d\lambda}\left(\frac{v_p}{\lambda}\right)$$

or $$v_g = -\lambda^2\left[v_p\left(-\frac{1}{\lambda^2}\right) + \frac{1}{\lambda}\frac{dv_p}{d\lambda}\right] \quad \text{or} \quad \boxed{v_g = v_p - \lambda\frac{dv_p}{d\lambda}} \qquad \ldots(ii)$$

This is the relation between group velocity and phase (or wave) velocity in a *dispersive medium. It is clear from equation (*ii*) that, in a dispersive medium, the group velocity is less than the phase velocity.

In a **non-dispersive medium, v_p is constant.

$$\therefore \qquad \frac{dv_p}{d\lambda} = 0$$

$\therefore$ from eqn. (*ii*), $$\boxed{v_g = v_p}$$

Conclusion. *In a non-dispersive medium or free space, the group velocity of a wave packet is equal to phase or wave velocity.*

1.23 RELATION BETWEEN GROUP VELOCITY AND PARTICLE VELOCITY

Group velocity, $$v_g = \frac{d\omega}{dk} = \frac{\frac{d\omega}{dv}}{\frac{dk}{dv}} \qquad \ldots(i)$$

Now $$\omega = 2\pi\nu = \frac{2\pi E}{h} \qquad \left[\because\ E = h\nu,\ \nu = \frac{E}{h}\right]$$

$$= \frac{2\pi mc^2}{h} \qquad [\because\ E = mc^2]$$

$$= \frac{2\pi c^2}{h} m = \frac{2\pi c^2}{h}\frac{m_0}{\sqrt{1-\frac{v^2}{c^2}}} = \frac{2\pi m_0 c^2}{h\sqrt{1-\frac{v^2}{c^2}}}$$

$$= \frac{2\pi m_0 c^2}{h}\left[1-\frac{v^2}{c^2}\right]^{-1/2}$$

---

*A dispersive medium is that medium in which the wave velocity is frequency-dependent.

**A medium in which the wave velocity is independent of frequency is called non-dispersive medium.

Differentiating both sides w.r.t. v, we get

$$\frac{d\omega}{dv} = \frac{2\pi m_0 c^2}{h}\left(-\frac{1}{2}\right)\left(1-\frac{v^2}{c^2}\right)^{-3/2}\left(-\frac{1}{c^2}\times 2v\right)$$

or
$$\frac{d\omega}{dv} = \frac{2\pi m_0 v}{h\left(1-\frac{v^2}{c^2}\right)^{3/2}} \qquad ...(ii)$$

Again, propagation constant or wave number, k

$$= \frac{2\pi}{\lambda} = \frac{2\pi m v}{h} \qquad \left(\because \lambda = \frac{h}{mv}\right)$$

$$= \frac{2\pi v}{h}\frac{m_0}{\sqrt{1-\frac{v^2}{c^2}}} = \frac{2\pi m_0}{h}\left[\frac{v}{\left(1-\frac{v^2}{c^2}\right)^{1/2}}\right]$$

Differentiating both sides w.r.t. v, we get

$$\frac{dk}{dv} = \frac{2\pi m_0}{h}\frac{d}{dv}\left\{v\left[1-\frac{v^2}{c^2}\right]^{-1/2}\right\}$$

$$= \frac{2\pi m_0}{h}\left\{v\left(-\frac{1}{2}\right)\left(1-\frac{v^2}{c^2}\right)^{-3/2}\left(-\frac{2v}{c^2}\right)+\left(1-\frac{v^2}{c^2}\right)^{-1/2}\right\}$$

$$= \frac{2\pi m_0}{h}\left(1-\frac{v^2}{c^2}\right)^{-3/2}\left[\frac{v^2}{c^2}+1-\frac{v^2}{c^2}\right]$$

$$\frac{dk}{dv} = \frac{2\pi m_0}{h\left(1-\frac{v^2}{c^2}\right)^{3/2}} \qquad ...(iii)$$

From equations (i), (ii) and (iii), we get

$$v_g = \frac{\dfrac{2\pi m_0 v}{h\left(1-\frac{v^2}{c^2}\right)^{3/2}}}{\dfrac{2\pi m_0}{h\left(1-\frac{v^2}{c^2}\right)^{3/2}}} \quad \text{or} \quad \boxed{v_g = v}$$

Conclusions. ***(i)*** A wave group or a wave packet associated with a moving particle travels with the velocity of the moving particle.

(ii) A moving particle is equivalent to a wave packet or a group of waves.

1.24 RELATION BETWEEN v_g AND v_p FOR A NON-RELATIVISTIC FREE PARTICLE

$$E = h\nu$$

Also,
$$E = \frac{1}{2} m v_g^2$$

$$\therefore \qquad h\nu = \frac{1}{2} m v_g^2 \quad \text{or} \quad \nu = \frac{m v_g^2}{2h}$$

Again,

$$\lambda = \frac{h}{m v_g}$$

But

$$v_p = \nu\lambda$$

Putting the values of ν and λ, we get

$$v_p = \frac{m v_g^2}{2h} \times \frac{h}{m v_g} \quad \text{or} \quad \boxed{v_p = \frac{v_g}{2}}$$

NUMERICAL EXAMPLES BASED ON De-BROGLIE WAVES

■ **Example 1.01.** *Calculate the mass of a photon of sodium light having wavelength 5894 Å and velocity 3×10^8 m s^{-1} ($h = 6.6 \times 10^{-34}$ kg m^2 s^{-1}).*

Solution. According to de-Broglie wavelength,

$$\lambda = \frac{h}{mv} \quad \text{or} \quad m = \frac{h}{\lambda v}$$

$\lambda = 5894$ Å $= 5894 \times 10^{-10}$ m,

$v = 3.0 \times 10^8$ m s^{-1},

$h = 6.6 \times 10^{-34}$ kg m^2 s^{-1}

$$m = \frac{6.6 \times 10^{-34} \text{ kg m}^2 \text{ s}^{-1}}{(5894 \times 10^{-10} \text{m}) \times (3.0 \times 10^8 \text{ m s}^{-1})}$$

$= \mathbf{3.74 \times 10^{-36}}$ **kg.**

■ **Example 1.02.** *Find the de-Broglie wavelength for an electron of energy V electron volt.*

[UPTU, B.Tech. I, Q.Bank 2001]

Solution. $\lambda = \frac{h}{mv} = \frac{h}{p} = \frac{h}{\sqrt{2mE}}$

Substituting values, we get

$$\lambda = \frac{6.62 \times 10^{-34}}{\sqrt{2 \times 9.1 \times 10^{-31} \times 1.6 \times 10^{-19}\,\text{V}}}$$

$$= \frac{\mathbf{12.27}}{\sqrt{\mathbf{V}}} \text{ Å.}$$

■ **Example 1.03.** *Calculate the mass of photon of sodium light having wavelength 5894 Å and velocity 3×10^8 m s^{-1} ($h = 6.626 \times 10^{-34}$ kg m^2 s^{-1}).*

Solution. According to de-Broglie equation,

$$\lambda = \frac{h}{mv}$$

$\lambda = 5894$ Å $= 5894 \times 10^{-10}$ m,

$v = 3 \times 10^8$ m s^{-1},

$h = 6.626 \times 10^{-34}$ kg m^2 s^{-1}.

Mass of the photon,

$$m = \frac{h}{\lambda v} = \frac{(6.626 \times 10^{-34} \text{ kg m}^2 \text{ s}^{-1})}{(5894 \times 10^{-10} \text{m}) \times (3 \times 10^8 \text{ m s}^{-1})}$$

$= \mathbf{3.75 \times 10^{-36}}$ **kg.**

■ **Example 1.04.** *Calculate the de-Broglie wavelength associated with a proton moving with a velocity equal to $\frac{1}{20}$ of the velocity of light.*

[UPTU, B.Tech. I, Q.Bank 2001]

Solution. $\lambda = \frac{h}{mv}$

$$= \frac{6.62 \times 10^{-34}}{1.67 \times 10^{-27} \times \frac{1}{20} \times 3 \times 10^8} \text{ m}$$

$= \mathbf{2.643 \times 10^{-14}}$ **m.**

■ **Example 1.05.** *The sodium flame test has a characteristic yellow colour due to emission of wavelength 589 nm. What is the mass equivalence of one photon of this wavelength ?*

Solution. According to de-Broglie equation

$$\lambda = \frac{h}{mv}$$

Wavelength of photon

$= 589$ mm $= 589 \times 10^{-9}$ m

Velocity of photon

$= 3 \times 10^8$ m s^{-1}

$$m = \frac{h}{\lambda v} = \frac{(6.626 \times 10^{-34} \text{ kg m}^2 \text{ s}^{-1})}{(589 \times 10^{-9} \text{m}) \times (3 \times 10^8 \text{ m s}^{-1})}$$

$= \mathbf{3.75 \times 10^{-36}}$ **kg.**

■ **Example 1.06.** *What is de-Broglie wavelength of an electron accelerated from rest through a potential difference of 100 V ?*

[UPTU, B.Tech. I, Q. Bank, 2001]

Solution. $\lambda = \frac{h}{mv} = \frac{h}{p} = \frac{h}{\sqrt{2m E_k}}$

$$= \frac{h}{\sqrt{2\,meV}}$$

$$= \frac{6.62 \times 10^{-34}}{\sqrt{2 \times 9.1 \times 10^{-31} \times 1.6 \times 10^{-19} \times 100}} \text{ m}$$

$$= \frac{6.62}{\sqrt{29.12}} \times 10^{-34+24} \text{ m}$$

$$= \frac{6.62}{5.396} \times 10^{-10} \text{ m} = \mathbf{1.227\ Å.}$$

■ **Example 1.07.** *Show that the wavelength of a 150 g rubber ball travelling at a velocity of 50 m s^{-1} is short enough to be determined.*

Solution. According to de-Broglie equation,

$$\lambda = \frac{h}{mv}$$

$h = 6.626 \times 10^{-34}$ kg m^2 s^{-1},

m = 150 g = 0.150 kg,

v = 50 m s^{-1}

∴ Wavelength, λ

$$= \frac{(6.626 \times 10^{-34} \text{ kg m}^2 \text{ s}^{-1})}{(0.15 \text{ kg}) \times (50 \text{ m s}^{-1})}$$

$= \mathbf{8.83 \times 10^{-35}}$ **m.**

Rubber ball is a macro particle and its wavelength is so small that it is not possible to determine its value.

■ **Example 1.08.** *A tennis ball of mass 6.0 × 10^{-2} kg is moving with a speed of 62 m s^{-1}. Calculate the wavelength associated with this moving tennis ball. Will the movement of this ball exhibit a wave character ? Explain (h = 6.63 × 10^{-34} kg m^2 s^{-1}).*

Solution. According to de-Broglie equation :

$$\lambda = \frac{h}{mv}$$

$m = 6.0 \times 10^{-2}$ kg,

v = 62 m s^{-1}, $h = 6.63 \times 10^{-34}$ kg m^2 s^{-1}

$$\therefore \quad \lambda = \frac{6.63 \times 10^{-34} \text{ kg m}^2 \text{ s}^{-1}}{6.0 \times 10^{-2} \text{ kg} \times 62 \text{ m s}^{-1}}$$

$= \mathbf{1.8 \times 10^{-34}}$ **m.**

This wavelength is too small to be measured and, therefore, the wave nature of such a ball cannot be detected. Thus, the ball will not have wave character and will describe a fixed path.

■ **Example 1.09.** *How fast is an electron moving if it has a wavelength equal to the distance it travels in one second ?*

Solution. Wavelength of electron (λ) = Velocity of electron (v)

But Wavelength $(\lambda) = \frac{h}{mv}$

$$\therefore \quad v = \frac{h}{mv} \quad \text{or} \quad v^2 = \frac{h}{m}$$

$$v = \left(\frac{h}{m}\right)^{1/2} = \left(\frac{6.626 \times 10^{-34} \text{ kg m}^2 \text{ s}^{-1}}{9.1 \times 10^{-31} \text{ kg}}\right)^{1/2}$$

$= (7.28 \times 10^{-4})^{1/2} = \mathbf{2.7 \times 10^{-2}}$ **m s^{-1}.**

■ **Example 1.10.** *Calculate the wavelength of 1000 kg rocket moving with a velocity of 3000 km per hour. (h = 6.626 × 10^{-34} J s)*

Solution. According to de-Broglie equation

$$\lambda = \frac{h}{mv}$$

Mass of rocket = 1000 kg

Velocity of rocket = 3000 km h^{-1}

$= \frac{3000 \times 10^3}{60 \times 60}$ m s^{-1} = 0.083 × 10^4 m s^{-1}

Planck's constant, h

= 6.626 × 10^{-34} J s

= 6.626 × 10^{-34} kg m^2 s^{-1}

$$\therefore \quad \lambda = \frac{(6.626 \times 10^{-34} \text{ kg m}^2 \text{ s}^{-1})}{(1000 \text{ kg}) \times (0.083 \times 10^4 \text{ m s}^{-1})}$$

$= \mathbf{7.98 \times 10^{-40}}$ **m.**

■ **Example 1.11.** *Calculate the de-Broglie wavelength of an electron accelerated from rest to a potential difference V, where V = 50 volt.*

[UPTU, B.Tech. I, Sem 2001]

Solution.

$$\lambda = \frac{h}{mv} = \frac{h}{p} = \frac{h}{\sqrt{2m E_k}}$$

$$= \frac{h}{\sqrt{2\,meV}}$$

$$= \frac{6.62 \times 10^{-34}}{\sqrt{2 \times 9.1 \times 10^{-34} \times 1.6 \times 10^{-19} \times 50}} \text{ m}$$

$= \mathbf{1.737\ Å}$

■ **Example 1.12.** *The mass of an electron is 9.1×10^{-31} kg. If its K.E. is 3.0×10^{-25} J, calculate its wavelength.*

Solution. Since K.E. $= \frac{1}{2} mv^2$

$$v = \left(\frac{2 \text{ K.E.}}{m}\right)^{1/2}$$

$$= \left(\frac{2 \times 3.0 \times 10^{-25} \text{ kg m}^2 \text{ s}^{-2}}{9.1 \times 10^{-31} \text{ kg}}\right)^{1/2}$$

$= 812 \text{ m s}^{-1}$

$$\lambda = \frac{h}{mv}$$

$$= \frac{6.626 \times 10^{-34} \text{ J s}}{(9.1 \times 10^{-31} \text{ kg})(812 \text{ m s}^{-1})}$$

$= 8967 \times 10^{-10} \text{ m} = \mathbf{8967\ Å.}$

■ **Example 1.13.** *Can a photon and an electron of the same momentum have the same wavelengths ? Compare their wavelengths if the two have the same energy.*

[UPTU, B.Tech. I, Q. Bank, 2001]

Solution. Momentum of electron,

$$p_e = \frac{h}{\lambda_e}$$

Momentum of photon, $p_{ph} = \frac{h}{\lambda_{ph}}$

Now, $p_e = p_{ph}$ (given)

$\therefore \quad \boldsymbol{\lambda_e = \lambda_{ph}}$

$$\text{Now, } \lambda_e = \frac{h}{\sqrt{2mE}} \quad \ldots(i)$$

and $$\lambda_{ph} = \frac{h}{p}$$

$$E = h\nu = \frac{hc}{\lambda} = pc \quad \text{or} \quad p = \frac{E}{c}$$

$$\therefore \quad \lambda_{ph} = \frac{h}{E/c}$$

or $$\lambda_{ph} = \frac{hc}{E} \quad \ldots(ii)$$

Dividing (*ii*) by (*i*), we get

$$\frac{\lambda_{ph}}{\lambda_e} = \frac{hc}{E}\frac{\sqrt{2mE}}{h}$$

$$= \frac{\sqrt{\mathbf{2mc^2}}}{\mathbf{E}}.$$

■ **Example 1.14.** *Calculate the kinetic energy of moving electron which has a wavelength of 4.8 pm (mass of electron = 9.11×10^{-31} kg, $h = 6.63 \times 10^{-34}$ J s).*

Solution. According to de-Broglie equation,

$$\lambda = \frac{h}{mv}$$

$h = 6.63 \times 10^{-34}$ J s,

$\lambda = 4.8 \text{ pm} = 4.8 \times 10^{-12}$ m,

$m = 9.11 \times 10^{-31}$ kg

$$4.8 \times 10^{-12} \text{ m} = \frac{6.63 \times 10^{-34} \text{ kg m}^2 \text{ s}^{-1}}{9.11 \times 10^{-31} \text{ kg} \times v}$$

or $$v = \frac{6.63 \times 10^{-34} \text{ kg m}^2 \text{ s}^{-1}}{(4.8 \times 10^{-12} \text{ m}) \times (9.11 \times 10^{-31} \text{ kg})}$$

$= 1.516 \times 10^8 \text{ m s}^{-1}$

$$\therefore \quad \text{Kinetic energy} = \frac{1}{2} mv^2$$

$$= \frac{1}{2} \times (9.11 \times 10^{-31} \text{ kg}) \times (1.516 \times 10^8 \text{ m s}^{-1})^2$$

$= 1.05 \times 10^{-14} \text{ kg m}^2 \text{ s}^{-2}$

$= \mathbf{1.05 \times 10^{-14}\ J.}$

■ **Example 1.15.** *Calculate de-Broglie wavelength of an electron (mass = 9.1×10^{-31} kg) moving at 1% speed of light ($h = 6.63 \times 10^{-34}$ kg m^2 s^{-1}).*

Solution. We know that

$$\lambda = \frac{h}{mv}$$

$m = 9.1 \times 10^{-31}$ kg,

$h = 6.63 \times 10^{-34}$ kg m^2 s^{-1},

v = 1% of speed of light

$$= \frac{1 \times 3.0 \times 10^8}{100} \text{ m s}^{-1}$$

$= 3.0 \times 10^6 \text{ m s}^{-1}$

($\because$ speed of light $= 3.0 \times 10^8$ m s^{-1})

$$\lambda = \frac{6.63 \times 10^{-34} \text{ kg m}^2 \text{ s}^{-1}}{(9.1 \times 10^{-31} \text{ kg}) \times (3.0 \times 10^6 \text{ m s}^{-1})}$$

$= \mathbf{2.43 \times 10^{-10}\ m.}$

■ **Example 1.16.** *Calculate the wavelength of an electron that has been accelerated in a particle accelerator through a potential difference of 100 million volt.*

Given : $1 eV = 1.6 \times 10^{-19} J, m_e = 9.1 \times 10^{-31} kg, h = 6.6 \times 10^{-34} J s.$

Solution.

$$\text{K.E. of electron} = \frac{1}{2} mv^2$$

$$= 100 \times 10^6 \text{ eV} = 10^8 \times 1.6 \times 10^{-19} \text{ J}$$

$$= 1.6 \times 10^{-11} \text{ J} = 1.6 \times 10^{-11} \text{ kg m}^2 \text{ s}^{-2}$$

$$v^2 = \frac{2 \times \text{K.E.}}{m}$$

$$= \frac{2 \times 1.6 \times 10^{-11} \text{ kg m}^2 \text{ s}^{-2}}{9.1 \times 10^{-31} \text{ kg}}$$

$$= 0.35 \times 10^{20} \text{ m}^2 \text{ s}^{-2}$$

Now, $v = (0.35 \times 10^{20} \text{ m}^2 \text{ s}^{-2})^{1/2}$

$$= \mathbf{5.92 \times 10^9 \text{ m s}^{-1}}.$$

According to de-Broglie equation,

$$\lambda = \frac{h}{mv}$$

$$= \frac{6.6 \times 10^{-34} \text{ kg m}^2 \text{ s}^{-1}}{9.1 \times 10^{-31} \text{ kg} \times 5.92 \times 10^9 \text{ m s}^{-1}}$$

$$= \mathbf{1.22 \times 10^{-11} \text{ m}}.$$

■ **Example 1.17.** *Calculate the wavelength of de-Broglie waves associated with a proton of kinetic energy 500 eV.*

Given : $m_p = 1.67 \times 10^{-27} kg, h = 6.63 \times 10^{-34} J s.$

Solution. Kinetic energy, E = 500 eV

$$= 500 \times 1.6 \times 10^{-19} \text{ J}$$

$$= 500 \times 1.6 \times 10^{-19} \text{ kg m}^2 \text{ s}^{-2}$$

Planck's constant, $h = 6.63 \times 10^{-34}$ J s

$$= 6.63 \times 10^{-34} \text{ kg m}^2 \text{ s}^{-1}$$

Mass of proton, $m_p = 1.67 \times 10^{-27}$ kg

Kinetic energy, E

$$= \frac{1}{2} mv^2$$

$$2 m\text{E} = \frac{1}{2} mv^2 \times 2m = m^2v^2$$

$$mv = (2 m\text{E})^{1/2}$$

According to de-Broglie equation

$$\lambda = \frac{h}{mv}$$

$$= \frac{6.63 \times 10^{-34} \text{ kg m}^2 \text{ s}^{-1}}{[2 \times (1.67 \times 10^{-27} \text{ kg}) \times (500 \times 1.6 \times 10^{-19} \text{ kg m}^2 \text{ s}^2)]^{1/2}}$$

$$= \mathbf{1.28 \times 10^{-12} \text{ m}}.$$

■ **Example 1.18.** *Find out the number of waves made by a Bohr electron in one complete revolution in its 3rd orbit.*

Solution. In general, the number of waves in any orbit is

$$\text{Number of waves} = \frac{\text{Circumference of orbit}}{\text{Wavelength}}$$

$$= \frac{2\pi r}{\lambda}$$

But $\lambda = \dfrac{h}{mv}$

$$\text{No. of waves} = \frac{2\pi r}{h/mv} = \frac{2\pi r . mv}{h}$$

or

$$= \frac{2\pi(mvr)}{h}$$

The angular momentum of Bohr's 3rd orbit is

$$mvr = \frac{3h}{2\pi}$$

$$\therefore \quad \text{Number of waves} = \frac{2\pi}{h} \times \frac{3h}{2\pi} = 3$$

So, the number of waves in Bohr's third orbit is **3.**

■ **Example 1.19.** *Calculate the de-Broglie wavelength of neutron of energy 28.8 eV. Given :* $h = 6.62 \times 10^{-34}$ *J s,* $m = 1.67 \times 10^{-27}$ *kg.*

[UPTU, B. Tech. I, Q. Bank, 2001]

Solution. Rest mass energy of neutron

$$= m_nc^2 = 1.67 \times 10^{-27} \times 3 \times 10^8 \times 3 \times 10^8 \text{ J}$$

$$= 1.503 \times 10^{-10} \text{ J}$$

$$= \frac{1.503 \times 10^{-10}}{1.6 \times 10^{-13}} \text{ MeV}$$

$$= 939.375 \text{ MeV}$$

The kinetic energy of the neutron under consideration is very small as compared to the rest mass energy. So, we would ignore relativistic considerations.

Now, $\lambda = \frac{h}{\sqrt{2mE}}$

Substituting values,

$$\lambda = \frac{6.62 \times 10^{-34}}{\sqrt{2 \times 1.67 \times 10^{-27} \times 28.8 \times 1.6 \times 10^{-19}}} \text{ m}$$

$= \mathbf{0.0534 \text{ Å}}$.

■ Example 1.20. *Calculate the de-Broglie wavelength of an electron that has been accelerated by a potential difference of 1 kV.*

Solution. The kinetic energy of the electron under the potential difference of 1 kV or 1000 V is given by the relation :

$$\frac{1}{2} mv^2 = eV$$

$e = 1.602 \times 10^{-19}$ C, $v = 1000$ V,

$m = 9.1 \times 10^{-31}$ kg

$$\therefore \quad \frac{1}{2} \times 9.1 \times 10^{-31} \times v^2$$

$$= 1.602 \times 10^{-19} \text{ C} \times 1000 \text{ V}$$

$$\therefore \quad v^2 = \frac{2 \times 1.602 \times 10^{-19} \times 1000}{9.1 \times 10^{-31}}$$

$$v^2 = 3.52 \times 10^{14} \text{ m}^2 \text{ s}^{-2}$$

or $\quad v = 1.88 \times 10^7 \text{ m s}^{-1}$

Now, $\lambda = \frac{h}{mv}$

$$= \frac{6.63 \times 10^{-34} \text{ J s}}{(9.1 \times 10^{-31} \text{ kg})(1.88 \times 10^7 \text{ m s}^{-1})}$$

$= \mathbf{3.87 \times 10^{-11} \text{ m}}$.

■ Example 1.21. *Calculate the momentum of a particle which has a de-Broglie wavelength of 0.1 nm. ($h = 6.63 \times 10^{-34}$ kg m^2 s^{-1})*

Solution. According to de-Broglie relationship,

$$\lambda = \frac{h}{mv} \quad \text{or} \quad mv = \frac{h}{\lambda}$$

$h = 6.63 \times 10^{-34}$ kg m^2 s^{-1},

$\lambda = 0.1$ nm $= 0.1 \times 10^{-9}$ m $= 1 \times 10^{-10}$ m

$$mv = \frac{6.63 \times 10^{-34} \text{ kg m}^2 \text{ s}^{-1}}{1 \times 10^{-10} \text{ m}}$$

$= \mathbf{6.63 \times 10^{-24} \text{ kg m s}^{-1}}$.

■ Example 1.22. *When would the wavelength associated with an electron become equal to the wavelength associated with proton ? (Mass of electron = 9.1095×10^{-28} g, mass of proton = 1.6725×10^{-24} g).*

Solution. According to de-Broglie equation,

$$\lambda = \frac{h}{mv}$$

For electron, $\lambda_e = \frac{h}{m_e v_e}$

For proton, $\lambda_p = \frac{h}{m_p v_p}$

But $\quad \lambda_e = \lambda_p$

$$\therefore \quad \frac{h}{m_e v_e} = \frac{h}{m_p v_p} \quad \text{or} \quad m_e v_e = m_p v_p$$

or $\quad v_e = \frac{m_p v_p}{m_e}$

$$= \frac{1.6725 \times 10^{-24}}{9.1095 \times 10^{-28}} \times v_p = 1836\, v_p$$

Velocity of electron = 1836 × velocity of proton.

■ Example 1.23. *What will be the kinetic energy of an electron if its de-Broglie wavelength equals the wavelength of the yellow line of sodium (5896 Å). The rest mass of electron is $m_0 = 9.1 \times 10^{-31}$ kg and $h = 6.63 \times 10^{-34}$ J s.*

[UPTU, B.Tech. I, Q. Bank, 2001]

Solution. If we ignore relativistic considerations, then $m = m_0$, $\lambda = \frac{h}{m_0 v}$ and kinetic energy

$$E_k = \frac{1}{2} m_0 v^2$$

Now, $E_k = \frac{1}{2} m_0 \left[\frac{h}{m_0 \lambda}\right]^2$

$$= \frac{h^2}{2\, m_0 \lambda^2}$$

Substituting values,

$$E_k = \frac{(6.63 \times 10^{-34})^2}{2 \times 9.1 \times 10^{-31} (5896 \times 10^{-10})^2} \text{ J}$$

$$= 6.95 \times 10^{-25} \text{ J} = \frac{6.95 \times 10^{-25}}{1.6 \times 10^{-19}} \text{ eV}$$

$= \mathbf{4.34 \times 10^{-6} \text{ eV}}$.

■ **Example 1.24.** *An electron is moving with a kinetic energy of 2.275 × 10^{-25} J. Calculate its de-Broglie wavelength. (Mass of electron = 9.1 × 10^{-31} kg, = 6.6 × 10^{-34} J s)*

Solution. Kinetic energy of electron,

$$\frac{1}{2}mv^2 = 2.275 \times 10^{-25} \text{ J}$$

or $$= 2.275 \times 10^{-25} \text{ kg m}^2 \text{ s}^{-2}$$

or $$v^2 = \frac{2 \times 2.275 \times 10^{-25}}{\text{m}}$$

$$m = 9.1 \times 10^{-31} \text{ kg}$$

$$\therefore \quad v^2 = \frac{2 \times 2.275 \times 10^{-25} \text{ kg m}^2 \text{ s}^{-2}}{9.1 \times 10^{-31} \text{ kg}}$$

$$= 0.5 \times 10^6 \text{ m}^2 \text{ s}^{-2}$$

$$v = 0.707 \times 10^3 \text{ m s}^{-1}$$

Now, $\lambda = \frac{h}{mv}$

$$\lambda = \frac{6.6 \times 10^{-34} \text{ kg m}^2 \text{ s}^{-1}}{(9.1 \times 10^{-31} \text{ kg}) \times (0.707 \times 10^3 \text{ m s}^{-1})}$$

$$= 1.026 \times 10^{-6} \text{ m} = \mathbf{1026 \text{ nm.}}$$

■ **Example 1.25.** *A beam of helium atoms moves with a velocity of 2.0 × 10^3 m s^{-1}. Find the wavelength of particles constituting the beam. (h = 6.626 × 10^{-34} J s)*

Solution. Mass of helium atom

$$= \frac{4}{6.023 \times 10^{23}}$$

$$= 6.64 \times 10^{-24} \text{ g} = 6.64 \times 10^{-27} \text{ kg}$$

According to de-Broglie equation,

$$\lambda = \frac{h}{mv}$$

$$h = 6.626 \times 10^{-34} \text{ J s},$$

$$m = 6.64 \times 10^{-27} \text{ kg},$$

$$v = 2.0 \times 10^3 \text{ m s}^{-1}$$

$$\therefore \quad \lambda = \frac{6.626 \times 10^{-34} \text{ kg m}^2 \text{ s}^{-1}}{(6.64 \times 10^{-27} \text{ kg})(2.0 \times 10^3 \text{ m s}^{-1})}$$

$$= 4.99 \times 10^{-11} \text{ m}$$

$$\therefore \quad \lambda = \mathbf{49.9 \text{ pm.}}$$

■ **Example 1.26.** *Show that the de-Broglie wavelength for a material particle of rest mass m_0 and charge q accelerated from rest through a potential difference of V volt relativistically is given by*

$$\lambda = \frac{h}{\left\{2m_0qV\left(1 + \frac{qV}{m_0c^2}\right)\right\}^{1/2}}$$

[UPTU, B.Tech. I, Q. Bank, 2001]

Solution. Since the velocity is relativistic therefore we use relativistic formula for E.

$$E^2 = p^2c^2 + m_0^2c^4$$

or $$p^2c^2 = E^2 - m_0^2c^4$$

But $$E = E_k + m_0c^2 = qV + m_0c^2$$

$$\therefore \quad p^2c^2 = (qV + m_0c^2)^2 - m_0^2c^4$$

or $$p^2 = 2m_0qV + \frac{q^2V^2}{c^2}$$

or $$p = \sqrt{2m_0qV\left[1 + \frac{qV}{2m_0c^2}\right]}$$

Now, $\lambda = \frac{h}{p}$

$$= \frac{\mathbf{h}}{\sqrt{\mathbf{2m_0qV}\left[\mathbf{1 + \frac{qV}{2m_0c^2}}\right]}}.$$

■ **Example 1.27.** *Obtain the de-Broglie wavelength of a neutron of kinetic energy 150 eV. An electron beam of this energy is suitable for crystal diffraction experiments. Would a neutron beam of the same energy be equally suitable ? Explain. Given : mass of neutron = 1.675 × 10^{-27} kg, h = 6.63 × 10^{-34} Js.*

Solution. $\lambda = \frac{h}{p} = \frac{h}{\sqrt{2mE}}$

$$= \frac{6.63 \times 10^{-34}}{\sqrt{2 \times 1.675 \times 10^{-27} \times 150 \times 1.6 \times 10^{-19}}} \text{ m}$$

$$= \frac{6.63 \times 10^{-34}}{\sqrt{804} \times 10^{-23}} \text{ m} = \mathbf{2.34 \times 10^{-12} \text{ m.}}$$

The inter-atomic spacing is nearly a hundred times greater. So, a neutron beam of 150 eV energy is not suitable for diffraction experiments.

■ **Example 1.28.** *Two particles A and B are in motion. If the wavelength associated with particle A is 5 × 10^{-8} m, calculate the wavelength of the particle B if its momentum is half that of A.*

Solution. According to de-Broglie equation,

$$\lambda = \frac{h}{mv} = \frac{h}{p}$$

For particle A, $p_A = \frac{h}{\lambda_A}$

For particle B, $p_B = \frac{h}{\lambda_B}$

Momentum of particle B, $p_B = \frac{1}{2} p_A$

$$p_B = \frac{1}{2} p_A = \frac{h}{\lambda_B} \quad \text{or} \quad p_A = \frac{2h}{\lambda_B} \qquad ...(i)$$

$$p_A = \frac{h}{\lambda_A} \qquad ...(ii)$$

Equating (*i*) and (*ii*),

$$\frac{2h}{\lambda_B} = \frac{h}{\lambda_A} \quad \text{or} \quad \lambda_B = 2\lambda_A$$

$$\therefore \quad \lambda_B = 2 \times 5 \times 10^{-8} \text{ m}$$
$$= 10 \times 10^{-8} \text{ m} = \mathbf{10^{-7} \text{ m.}}$$

Example 1.29. *What should be the ratio of velocities of CH_4 and O_2 molecules so that they are associated with de-Broglie waves of equal wave lengths ?*

Solution. According to de-Broglie equation,

$$\lambda = \frac{h}{mv}$$

For methane (CH_4), $\lambda_{CH_4} = \frac{h}{m_{CH_4} \times v_{CH_4}}$

For oxygen (O_2), $\lambda_{O_2} = \frac{h}{m_{O_2} \times v_{O_2}}$

Since the wavelength of CH_4 and O_2 are to be equal,

$$\therefore \quad \lambda_{CH_4} = \lambda_{O_2}$$

or $$\frac{h}{m_{CH_4} \times v_{CH_4}} = \frac{h}{m_{O_2} \times v_{O_2}}$$

$$m_{CH_4} \times v_{CH_4} = m_{O_2} \times v_{O_2}$$

or $$\frac{v_{CH_4}}{v_{O_2}} = \frac{m_{O_2}}{m_{CH_4}} = \frac{32}{16} = \mathbf{2.}$$

Example 1.30. *The kinetic energy of an electron is 4.55×10^{-25} J. The mass of electron is 9.1×10^{-31} kg. Calculate velocity, momentum and the wavelength of the electron.*

Solution.

Kinetic energy $= \frac{1}{2} mv^2 = 4.55 \times 10^{-25}$ J

$$= 4.55 \times 10^{-25} \text{ kg m}^2 \text{ s}^{-2}$$

or $$v^2 = \frac{2 \times \text{KE}}{m}$$

$$= \frac{2 \times 4.55 \times 10^{-25} \text{ kg m}^2 \text{ s}^{-2}}{9.1 \times 10^{-31} \text{ kg}}$$

$$= 10^6 \text{ m}^2 \text{ s}^{-2}$$

or Velocity, $v = (10^6 \text{ m}^2 \text{ s}^{-2})^{1/2} = \mathbf{10^3 \text{ m s}^{-1}}$

Momentum of electron $= mv$

$$= (9.1 \times 10^{-31} \text{ kg}) \times (10^3 \text{ m s}^{-1})$$
$$= \mathbf{9.1 \times 10^{-28} \text{ kg m s}^{-1}}$$

According to de-Broglie equation,

$$\lambda = \frac{h}{mv}$$

$$= \frac{6.626 \times 10^{-34} \text{ kg m}^2 \text{ s}^{-1}}{(9.1 \times 10^{-31} \text{ kg}) \times (10^3 \text{ m s}^{-1})}$$

$$= 0.728 \times 10^{-6} \text{ m} = \mathbf{7.28 \times 10^{-7} \text{ m.}}$$

Example 1.31. *A proton and an electron have same de-Broglie wavelength. Which of them moves faster and which possesses more kinetic energy ? Justify your answer.*

Solution. Let the subscripts 1 and 2 be used for proton and electron respectively.

$$E_1 = \frac{1}{2} m_1 v_1^2 \quad \text{or} \quad m_1 v_1^2 = 2E_1$$

or $$m_1^2 v_1^2 = 2E_1 m_1$$

or $$m_1 v_1 = \sqrt{2E_1 m_1}$$

Now, $$\lambda = \frac{h}{m_1 v_1} = \frac{h}{\sqrt{2E_1 m_1}}$$

or $$E_1 = \frac{h^2}{2m_1 \lambda^2} \qquad ...(i)$$

$$E_2 = \frac{1}{2} m_2 v_2^2 \quad \text{or} \quad m_2 v_2^2 = 2E_2$$

or $$m_2^2 v_2^2 = 2E_2 m_2$$

or $$m_2 v_2 = \sqrt{2E_2 m_2}$$

$$\lambda = \frac{h}{m_2 v_2} = \frac{h}{\sqrt{2E_2 m_2}}$$

or $$E_2 = \frac{h^2}{2m_2 \lambda^2}$$

Dividing, $$\frac{E_1}{E_2} = \frac{m_2}{m_1}$$

But $m_2 < m_1$ $\therefore$ $E_1 < E_2$

So, **electron has more K.E.** than the proton.

$$\text{Now, } \frac{E_1}{E_2} = \frac{\frac{1}{2}m_1v_1^2}{\frac{1}{2}m_2v_2^2} = \frac{m_1v_1^2}{m_2v_2^2}$$

or $$\frac{v_1}{v_2} = \sqrt{\frac{E_1m_2}{E_2m_1}}$$

But $E_1m_2 < E_2m_1$ $\therefore$ $v_1 < v_2$

or $v_2 > v_1$

So, **electron has greater velocity** than the proton.

■ Example 1.32. *The de-Broglie wavelength of a particle of kinetic energy K is λ. What would be the wavelength of the particle if its kinetic energy were $\frac{K}{4}$?*

Solution. $$\lambda = \frac{h}{\sqrt{2mK}}$$

$$\lambda' = \frac{h}{\sqrt{2m\frac{K}{4}}} = 2\frac{h}{\sqrt{2mK}}$$

or $\lambda' = 2\lambda$.

■ Example. 1.33. *A neutron is an uncharged particle of mass 1.67 × 10⁻²⁷ kg. Calculate the de-Broglie wavelength of the neutron moving with a velocity such that kinetic energy is 0.04 eV.*

(Given : h = 6.6 × 10⁻³⁴ J s)

Solution. $$\lambda = \frac{h}{\sqrt{2mE_k}}$$

$$= \frac{6.6 \times 10^{-34}}{[2 \times 1.67 \times 10^{-27} \times 0.04 \times 1.6 \times 10^{-19}]^{1/2}} \text{ m}$$

$$= \frac{6.6 \times 10^{-34}}{0.46 \times 10^{-23}} \text{ m} = 14.35 \times 10^{-11} \text{ m}$$

$= 1.435 \times 10^{-10}$ m = **1.435 Å.**

■ Example 1.34. *Calculate the de-Broglie wavelength for electrons and protons if their speed is 10^5 m s^{-1}.*

Solution. (*i*) Mass of electron = 9×10^{-31} kg

de-Broglie wavelength,

$$\lambda_e = \frac{h}{m_ev} \quad \text{or} \quad \lambda_e = \frac{6.63 \times 10^{-34}}{9 \times 10^{-31} \times 10^5} \text{ m}$$

= **7.4 × 10⁻⁹ m**

(*ii*) Mass of proton = 1.67×10^{-27} kg

$$\lambda_p = \frac{h}{m_pv} = \frac{6.63 \times 10^{-34}}{1.67 \times 10^{-27} \times 10^5} \text{ m}$$

= **3.97 × 10⁻¹² m.**

■ Example 1.35. *Obtain the de-Broglie wavelength of a neutron of kinetic energy 150 eV. (Given : mass of neutron = 1.675 × 10⁻²⁷ kg, h = 6.62 × 10⁻³⁴ Js.)*

Solution. $E = 150 \times 1.6 \times 10^{-19}$ J

$= 2.4 \times 10^{-17}$ J,

$m = 1.675 \times 10^{-27}$ kg

$$\lambda = \frac{h}{mv} = \frac{h}{\sqrt{2mE}}$$

$$= \frac{6.62 \times 10^{-34}}{\sqrt{2 \times 1.675 \times 10^{-27} \times 2.4 \times 10^{-17}}} \text{ m}$$

= **2.335 × 10⁻¹² m.**

■ Example 1.36. *An electron and a photon have the same de-Broglie wavelength of 10⁻¹⁰ m. Which of the two has greater kinetic energy ?*

Solution. *For the case of electron*

$$E = \frac{1}{2}mv^2 \quad \text{or} \quad v = \sqrt{\frac{2E}{m}}$$

$$\text{Also, } \lambda = \frac{h}{mv} = \frac{h}{m}\sqrt{\frac{m}{2E}} = \frac{h}{\sqrt{2mE}}$$

or $$\sqrt{2mE} = \frac{h}{\lambda}$$

or $$E = \frac{h^2}{2m\lambda^2} \quad ...(i)$$

For the case of a photon

$$E' = \frac{hc}{\lambda} \quad ...(ii)$$

Dividing (*ii*) by (*i*), we get

$$\frac{E'}{E} = \frac{hc}{\lambda} \times \frac{2m\lambda^2}{h^2} \quad \text{or} \quad \frac{E'}{E} = \frac{2m\lambda c}{h}$$

Substituting values,

$$\frac{E'}{E} = \frac{2 \times 9.1 \times 10^{-31} \times 10^{-10} \times 3 \times 10^8}{6.6 \times 10^{-34}}$$

$= 82.7$

Clearly, $E' > E$

So, the photon possesses more kinetic energy than that of electron.

■ Example 1.37. *An electron and a photon have got the same de-Broglie wavelength. Which of the two has greater 'total energy' ?*

Solution. *For the case of electron*

$$E = mc^2$$

Also, $\lambda = \frac{h}{mv}$, $m = \frac{h}{\lambda v}$

$$\therefore \quad E = \frac{hc^2}{\lambda v} \qquad ...(i)$$

For the case of photon

$$E' = h\nu = \frac{hc}{\lambda} \qquad ...(ii)$$

Dividing (*i*) by (*ii*), we get

$$\frac{E}{E'} = \frac{hc^2}{\lambda v} \times \frac{\lambda}{hc} = \frac{c}{v}$$

But $c > v$

($\because$ An electron cannot travel with the speed of light.)

$$\therefore \quad \frac{E}{E'} > 1 \quad \text{or} \quad E > E'$$

So, the 'total energy' of an electron is greater than the 'total energy' of a photon.

■ **Example 1.38.** *An electron microscope uses electrons accelerated by a voltage of 50 kV. Determine the de-Broglie wavelength associated with the electrons. If other factors such as numerical aperture etc. are taken to be roughly the same, how does the resolving power of an electron microscope compare with that of an optical microscope which uses yellow light ? (Given : wavelength of yellow light = 5990 Å.)*

Solution. de-Broglie wavelength,

$$\lambda = \frac{h}{\sqrt{2mE}}$$

$h = 6.62 \times 10^{-34}$ Js, $m = 9.1 \times 10^{-31}$ kg,

$E = 50 \text{ keV} = 50 \times 10^3 \times 1.6 \times 10^{-19}$ J

$= 8 \times 10^{-15}$ J

$$\lambda = \frac{6.62 \times 10^{-34}}{\sqrt{2 \times 9.1 \times 10^{-31} \times 8 \times 10^{-15}}} \text{ m}$$

$$= \frac{6.62}{\sqrt{145.6}} \times 10^{-11} \text{ m} = \frac{6.62}{12.07} \times 10^{-11} \text{ m}$$

$= 5.48 \times 10^{-12}$ m

Wavelength of yellow light,

$\lambda' = 5990 \times 10^{-10} \text{ m} = 5.99 \times 10^{-7}$ m

Now, $\frac{\lambda}{\lambda'} \approx 10^{-5}$

Since resolving power is inversely proportional to wavelength therefore the resolving power of electron microscope is 10^5 times larger than the resolving power of optical microscope.

EXERCISES

1. Calculate the wavelength and momentum of an α-particle (mass = 6.6×10^{-27} kg) moving with a speed of 10^5 cm s^{-1}. [**Ans.** $\lambda = 1.0 \times 10^{-10}$ m, momentum = 6.6×10^{-24} kg m s^{-1}]
2. What is the wavelength of an electron if its kinetic energy is 1.375×10^{-25} J ? (mass of electron = 9.1×10^{-31} kg, $h = 6.6. \times 10^{-34}$ kg m^2 s^{-1}) [**Ans.** 1.32×10^{-6} m]
3. Calculate the ratio between the wavelength of an electron and a proton if the proton is moving with half the velocity of the electrons (mass of proton = 1.67×10^{-24} g and mass of electron = 9.11×10^{-28} g) [**Ans.** λ (electron) / λ (proton) = 916]
4. The kinetic energy of a sub-atomic particle is 5.85×10^{-25} J. Calculate the frequency of the particle wave. (Planck's constant, $h = 6.63 \times 10^{-34}$ kg m^2 s^{-1}) [**Ans.** 1.76×10^9 s^{-1}]
5. Calculate the de-Broglie wavelength of a helium atom at 27°C moving with a velocity of 2.40×10^2 m s^{-1} (atomic mass of helium atom = 4 a.m.u.) [**Ans.** 0.416 nm]
6. Calculate the kinetic energy of a moving electron which has a wavelength of 4.8 pm. (mass of electron = 9.11×10^{-31} kg, $h = 6.63 \times 10^{-34}$ J s) [**Ans.** 10.52×10^{-15} kg m^2 s^{-2}]
7. Find the wavelength of carbon dioxide molecule moving at a velocity of 440 m s^{-1}. [**Ans.** 2.06×10^{-11} m]
8. What must be the velocity of a beam of electrons if they are to display a de-Broglie wavelength of 100 Å ? [**Ans.** 7.25×10^4 m s^{-1}]

9. An electron beam emerges from an accelerator with kinetic energy 100 eV. What is its de-Broglie wavelength ?
($m = 9.1 \times 10^{-31}$ kg, $h = 6.6 \times 10^{-34}$ Js, 1 eV $= 1.6 \times 10^{-19}$ J) [**Ans.** 1.2×10^{-10} m]
10. An electron having mass 9.1×10^{-31} kg is moving with a velocity of 10^3 m s^{-1}. Calculate its wavelength and kinetic energy ($h = 6.63 \times 10^{-34}$ J s). [**Ans.** 7.3×10^{-7} m, 4.55×10^{-25} J]
11. Calculate the de-Broglie wavelength of an electron moving at 1.5% speed of light. [**Ans.** 1.62×10^{-10} m]
12. Calculate the wavelength of 100 kg rocket with a velocity of 300 km per hour. [**Ans.** 7.95×10^{-38} m]
13. The kinetic energy of a sub-atomic particle is 5.65×10^{-25} J. Calculate the frequency of the particle wave. [**Ans.** 1.705×10^9 s^{-1}]
14. A beam of helium atoms moves with a velocity of 2×10^3 m s^{-1}. Find the wavelength associated with helium atoms. [**Ans.** 4.98×10^{-11} m]
15. Calculate the wavelength associated with an electron moving with a velocity of 10^3 m s^{-1}. [**Ans.** 7.28×10^{-7} m]
16. An electron is moving with a kinetic energy of 2.275×10^{-25} J. Calculate its de-Broglie wavelength. [**Ans.** 1.04×10^{-6} m s^{-1}]
17. Calculate the kinetic energy of O_2 molecule which has wavelength of 2.5 pm. [**Ans.** 6.61×10^{-19} J]
18. Calculate the momentum of a particle which has a de-Broglie wavelength of 0.1 nm. [**Ans.** 6.626×10^{-24} kg m s^{-1}]
19. Calculate the mass of a photon of wavelength of 3.6 Å. [**Ans.** 6.135×10^{-33} kg]
20. Calculate the momentum of a particle which has de-Broglie wavelength of 2.5×10^{-10} m. [**Ans.** 2.65×10^{-24} kg m s^{-1}]
21. Calculate the velocity of the electron in Bohr's first orbit of hydrogen atom. [**Ans.** 2.18×10^6 m s^{-1}]
22. Find the number of quanta of radiations of frequency 4.67×10^{13} s^{-1} that must be absorbed in order to melt 5 g of ice. The energy required to melt 1 g of ice is 333 J. [**Ans.** 5.38×10^{22}]
23. What is the de-Broglie wavelength of an electron moving with a velocity of 500 km s^{-1} ? Given : mass of electron $= 9 \times 10^{-31}$ kg and $h = 6.6 \times 10^{-34}$ J s. [**Ans.** 1.45 nm]
24. Calculate de-Broglie wavelength for electrons moving with speed of 10^5 m s^{-1}. Given : $h = 6.6 \times 10^{-34}$ Js and mass of electron $= 9 \times 10^{-31}$ kg. [**Ans.** 73.33 Å]
25. What is the de-Broglie wavelength of a 2 MeV proton ? Given : mass of proton $= 1.67 \times 10^{-27}$ kg. [**Ans.** 2.025×10^{-14} m]
26. What is the de-Broglie wavelength of a 0.04 eV neutron ? Given : mass of neutron $= 1.67 \times 10^{-27}$ kg. [**Ans.** 1.4 Å]
27. Find the de-Broglie wavelength of an electron moving with a velocity 1.5×10^8 m s^{-1}. Given : rest mass of electron $= 9.1 \times 10^{-31}$ kg. [**Ans.** 4.2×10^{-12} m]
28. (*a*) What will be the de-Broglie wavelength of electron having kinetic energy of 500 eV ? [**Ans.** 0.55 Å]
(*b*) Calculate the wavelength of de-Broglie waves associated with a proton of kinetic energy 500 eV. [**Ans.** 0.013 Å]
(*c*) Calculate the de-Broglie wavelength of a beam of electrons having kinetic energy of 100 eV. [**Ans.** 1.23 Å]
29. Calculate the wavelength of matter waves associated with a particle of mass 1 gram moving with a velocity of 2000 km s^{-1}. [**Ans.** 3.31×10^{-37} m]
30. What voltage must be applied to an electron source to produce electron of wavelength 0.4 Å ?
Given : $e = 1.6 \times 10^{-19}$ C and $m = 9 \times 10^{-31}$ kg. [**Ans.** 954.1 volt]
31. What voltage must be applied to an electron microscope to produce electron of wavelength 0.6 Å ? [**Ans.** 602.2 volt]

32. State de-Broglie hypothesis on the wave nature of matter. Obtain the de-Broglie wavelength of an electron of kinetic energy 100 eV. Mass of electron = 9.1×10^{-31} kg, $e = 1.6 \times 10^{-19}$ C, $h = 6.63 \times 10^{-34}$ Js. [**Ans.** 1.2×10^{-10} m]

33. Calculate the de-Broglie wavelength of a 3 kg object moving with a speed of 2 m s^{-1}. Given : $h = 6.62 \times 10^{-34}$ Js. [**Ans.** 1.1×10^{-34} m]

34. Calculate the de-Broglie wavelength of an electron of energy 400 eV. Given : $h = 6.6 \times 10^{-34}$ Js, $m = 9.1 \times 10^{-31}$ kg. [**Ans.** 0.61 Å]

35. For what kinetic energy of a neutron will the associated de-Broglie wavelength be 1.40×10^{-10} m ? Mass of a neutron = 1.675×10^{-27} kg, $h = 6.63 \times 10^{-34}$ Js. [**Ans.** 4.15×10^{-2} eV]

1.25 HEISENBERG'S UNCERTAINTY PRINCIPLE

According to classical mechanics, a moving particle has a definite momentum and occupies a definite position in space. It is possible to determine accurately both its position and momentum. In our daily life, we observe that all moving objects have well-defined paths or trajectories. The path or trajectory of an object can be determined by knowing its position and velocity at various instants of time.

In the year 1927, a German physicist Werner Heisenberg pointed out that we can never measure accurately both the position and momentum of a microscopic particle as small as an electron. Consequently, it is not possible to talk of the trajectory of an electron. In simple words, the classical mechanics does not describe satisfactorily the behaviour of the particles of atomic dimensions.

On the basis of dual nature of matter,*Werner Heisenberg, a German physicist in 1927, put forward an important principle known as *the principle of indeterminacy* or *the uncertainty principle.*

Statement of uncertainty principle.

> ***It is not possible to measure simultaneously both the position and velocity (or momentum) of a microscopic particle with absolute accuracy or certainty.***

Mathematically, uncertainty principle may be put as : "*the product of uncertainty in position and momentum is always constant.*" It may be expressed as :

$$\Delta x \times \Delta p \geq \frac{h}{4\pi}$$

where Δx = uncertainty in position and Δp = uncertainty in momentum.

The sign $\geq$ means that the product of Δx and Δp can be either greater than or equal to $h/4\pi$. It can never be less than $h/4\pi$. The sign of equality refers to minimum uncertainty and is equal to $h/4\pi$.

The constancy of the product of uncertainties means that :

(i) If Δx *is small, i.e.,* the position of the particle is measured accurately, Δp *would be large, i.e.,* there would be large uncertainty in its momentum.

(ii) On the other hand, if Δp *is small i.e.,* the momentum of the particle is measured more accurately, Δx *would be large i.e.,* there would be large uncertainty with regard to the position of the particle.

In other words, *if the position of a particle is measured accurately, there will be more error in the measurement of momentum. Conversely, if momentum is measured more accurately, the position will not be accurately known.*

---

*He was awarded the 1932 Nobel Prize in Physics.

We know that $p = mv$.

$$\therefore \qquad \Delta p = m\Delta v \qquad (\because \ m \text{ is constant.})$$

So, $\Delta x \times \Delta p \geq \frac{h}{4\pi}$ may be written as :

$$\Delta x \times m(\Delta v) \geq \frac{h}{4\pi} \quad \text{or} \quad \Delta x \times \Delta v \geq \frac{h}{4\pi m}$$

So, the position and velocity of an object cannot be simultaneously known with accuracy.

The uncertainty principle can be best understood with the help of an example. Suppose you are asked to measure the thickness of a sheet of paper with an unmarked metrestick. Obviously, the result obtained would be extremely inaccurate and meaningless. In order to obtain any accuracy, you would have to use an instrument graduated in units smaller than the thickness of a sheet of paper. Analogously, in order to determine the position of an electron, we must use a "metrestick" calibrated in units smaller than the dimensions of an electron. To observe an electron, we can illuminate it with "light" or electromagnetic radiation. The "light" used must have a wavelength smaller than the dimensions of an electron. But photon of such "light" would possess very high energy. The high momentum photons of such light $\left(p = \frac{h}{\lambda}\right)$ would change the energy of electrons by collision. In this process, we, no doubt, would be able to calculate the position of the electron, but we would know very little about the velocity of the electron after the collision.

Werner Heisenberg (1901–1976)

Werner Heisenberg (1901–1976) received his Ph. D. in physics from the University of Munich in 1923. He then spent a year working with Max Born at Göttingen and three years with Niels Bohr in Copenhagen. He was professor of physics at the University of Leipzig from 1927 to 1941. During World War II, Heisenberg was in charge of German research on the atomic bomb. After the war, he was named director of the Max Planck Institute for Physics in Göttingen. He was also an accomplished mountain climber. Heisenberg was awarded the Nobel Prize in Physics in 1932.

1.26 QUANTUM MECHANICAL EXPLANATION

According to quantum mechanical picture, a moving particle is to be treated as a wave packet (group of waves). The particle may be found anywhere within the wave packet. This shows that the position of the particle is uncertain within the limits of the wave packet. Moreover, the wave packet has a velocity spread. So, there is uncertainty about both the velocity and momentum of the particle.

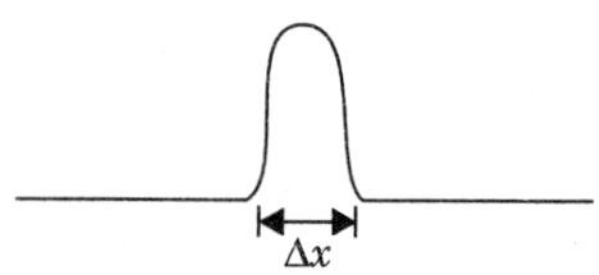

Fig. 1.10

If the wave packet is narrow as shown in Fig. 1.10, the position of the particle can be determined fairly accurately. However, the wavelength and hence the momentum of the particle cannot be measured accurately. This is due to lack of enough waves for exact measurement.

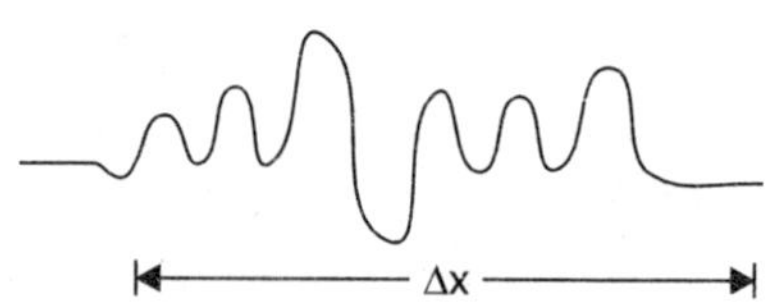

Fig. 1.11

If the wave packet is wide, then the wavelength can be determined fairly accurately but the determination of the position of the particle becomes uncertain.

We conclude from the above discussion that it is impossible to determine the exact position and the exact momentum of a particle simultaneously.

1.27 DERIVATION OF ENERGY-TIME UNCERTAINTY RELATION FROM POSITION-MOMENTUM UNCERTAINTY RELATION

The uncertainty relation in terms of position and momentum is :

$$\Delta x \, \Delta p \geq \frac{\hbar}{2}$$

The kinetic energy of a free particle of mass m moving with velocity v is given by

$$E = \frac{1}{2} mv^2 = \frac{p^2}{2m}$$

where $p(= mv)$ is the momentum of the particle.

If ΔE and Δp are the uncertainties in energy and momentum respectively, then

$$\Delta E = \frac{2p\Delta p}{2m} \quad \text{or} \quad \Delta E = \frac{p \, \Delta p}{m}$$

or $$\Delta E = \frac{mv \, \Delta p}{m} \quad \text{or} \quad \Delta E = v \, \Delta p$$

or $$\Delta E = \frac{\Delta x}{\Delta t} \Delta p \quad \text{or} \quad \Delta E \, \Delta t = \Delta x \, \Delta p$$

But $$\Delta x \, \Delta p \geq \frac{\hbar}{2} \quad \therefore \quad \Delta E \, \Delta t \geq \frac{\hbar}{2}.$$

1.28 DERIVATION OF ANGULAR MOMENTUM AND ANGULAR POSITION UNCERTAINTY RELATION FROM POSITION-MOMENTUM UNCERTAINTY RELATION

Momentum, $p = mv = mr\omega$

Angular momentum, $J = mvr = mr^2 \omega$

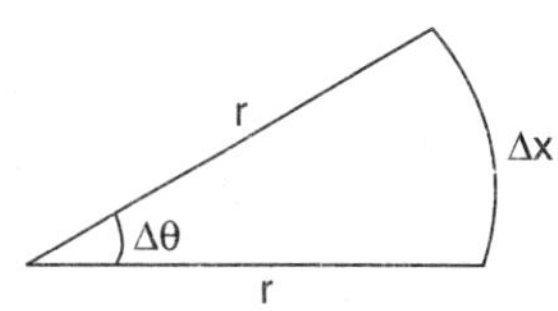

Fig. 1.12

$$p = \frac{(mr\omega)r}{r} \quad \text{or} \quad p = \frac{mr^2\omega}{r}$$

or $$p = \frac{J}{r}$$

$$\Delta p = \frac{\Delta J}{r} \qquad ...(i)$$

Again $x = r\theta, \quad \Delta x = r\Delta\theta$ ...(*ii*)

Multiplying (*ii*) and (*i*), we get

$$\Delta x \, \Delta p = r \, \Delta\theta \times \frac{\Delta J}{r} \quad \text{or} \quad \Delta x \, \Delta p = \Delta J \, \Delta\theta$$

But $$\Delta x \, \Delta p \geq \frac{\hbar}{2}$$

$$\therefore \quad \Delta J \, \Delta\theta \geq \frac{\hbar}{2}$$

1.29 UNCERTAINTY PRINCIPLE AND FAILURE OF BOHR MODEL

One can now understand the reasons for the failure of the Bohr model. In Bohr model, an electron is regarded as a charged particle moving in well-defined circular orbits about the nucleus.

The wave character of the electron is not considered in Bohr model. Further, an orbit is a clearly-defined path. This path can be completely defined only if both the exact position and the exact velocity of the electron at the same time are known. This is not possible according to the Heisenberg uncertainty principle. *Bohr model of the hydrogen atom, therefore, not only ignores dual behaviour of matter but also contradicts Heisenberg uncertainty principle*. In view of these inherent weaknesses in the Bohr model, there was no point in extending Bohr model to other atoms. What was needed was an insight into the structure of the atom, which took into account wave-particle duality of matter and was consistent with Heisenberg uncertainty principle. This came with the advent of quantum mechanics.

1.30 UNCERTAINTY PRINCIPLE IS SIGNIFICANT FOR MICROSCOPIC OBJECTS

The effect of Heisenberg Uncertainty Principle is significant only for motion of microscopic objects and is negligible for that of macroscopic objects. This can be seen from the following examples.

If uncertainty principle is applied to an object of mass, say about a milligram (10^{-6} kg), then

$$\Delta v.\Delta x = \frac{h}{4\pi.m} = \frac{6.626 \times 10^{-34}\ \text{Js}}{4 \times 3.1416 \times 10^{-6}} \approx 10^{-28}\ \text{m}^2\,\text{s}^{-1}$$

The value of $\Delta v.\Delta x$ obtained is extremely small and therefore one may say that in dealing with milligram-sized or heavier objects, uncertainties of such small dimensions are hardly of any real consequence.

On the other hand, in the case of a microscopic object like an electron, $\Delta v.\Delta x$ obtained is much larger and such uncertainties are of real consequence. For example, for an electron whose mass is 9.11×10^{-31} kg, according to Heisenberg uncertainty principle

$$\Delta v.\Delta x = \frac{h}{4\pi.m} = \frac{6.626 \times 10^{-34}\ \text{Js}}{4 \times 3.1416 \times 9.11 \times 10^{-31}\ \text{kg}}$$
$$\approx 10^{-4}\ \text{m}^2\text{s}^{-1}.$$

It, therefore, means that if one tries to find the exact location of the electron, say to an uncertainty of only 10^{-8} m, then the uncertainty Δv in velocity would be $\frac{10^{-4}\ \text{m}^2\text{s}^{-1}}{10^{-8}\ \text{m}} \approx 10^{+4}$ m s^{-1}, which is so large that the classical picture of electrons moving in Bohr's orbits (fixed) cannot hold good. It, therefore, means that the precise statements of the position and momentum of electrons have to be replaced by the statements of probability that the electron has a given position and momentum. This is what happens in the quantum mechanical model of atom.

1.31 UNCERTAINTY PRINCIPLE AND TRAJECTORY OF ELECTRON

One of the important implications of the Heisenberg Uncertainty Principle is that it rules out existence of define paths or trajectories. The trajectory of an object is determined by its location and velocity at various moments. If we know where a body is at a particular instant and if we also know its velocity and the forces acting on it at that instant, we can tell where the body would be sometime later. We, therefore, conclude that the position of an object and its velocity fix its trajectory. Since for a sub-atomic object such as an electron, it is not possible to simultaneously determine the position and velocity at any given instant to an arbitrary degree of precision therefore, it is not possible to talk of the trajectory of an electron.

1.32 UNCERTAINTY PRINCIPLE—A FUNDAMENTAL LIMITATION OF NATURE

It should be borne in mind that the uncertainty is not due to lack of sufficiently refined techniques but it is due to the fact that we cannot observe microscopic things without disturbing them. No instrument can observe the position of an electron without affecting its motion. In other words, uncertainty principle is the fundamental limitation of nature. Thus, we cannot design an experiment to obtain an accurate value of both the position and momentum for microscopic objects. *However, in our daily life, this principle has no significance. This is because we come across only large objects, i.e., the objects which we can observe with naked eye without altering their motion. The position and velocity of these objects can be determined accurately because in these cases, during the interaction between the object and the measuring device, the changes in* position and velocity are negligible.

1.33 PHYSICAL CONCEPT OF UNCERTAINTY PRINCIPLE

To determine the position of an object, we must be able to see the object. This can be done with the help of light of suitable wavelength. When a beam of light falls on an object, the photons of this incident light are scattered at the object and the reflected light enters our eye. Now, if the object is large, its position and velocity will not change by the impact of the striking photons. Thus, it will be possible to determine both the position and velocity of the object simultaneously.

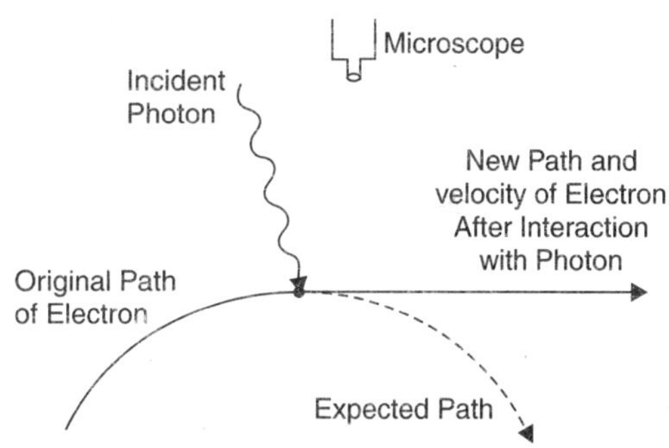

Fig. 1.13. The change in the path of electron by the impact of striking photon.

However, in case of microscopic objects, such as electrons, the interaction of the striking photon of light with the particle will cause appreciable displacement of the particle from its normal path. As a result of this, the particle undergoes a considerable change in its path and velocity (or momentum) due to the impact of a single photon used to observe it. Thus, the very act of measuring the position of the microscopic particle causes a change in its momentum. Fig. 1.13 shows as to how the position of an electron changes by the impact of photon. The collision of photon with the electron sends it in an unpredictable direction.

Thus, it is not possible to determine simultaneously the exact position and momentum of the electron or of any other microscopic object.

Let us now consider a thought experiment for determining the position of an electron by using photons of light. According to principles of optics, if we use light of wavelength λ, then the position of electron cannot be located more accurately then $\pm \lambda$. The shorter the wavelength, the greater is the accuracy. Therefore, to observe the position of the electron accurately, light of appropriately small wavelength should be used. But the photons of radiations of smaller wavelength will have higher momentum ($p = h/\lambda$). When even a single photon of this light used to observe the position of electron strikes against it, a large amount of momentum will be transferred to the electron at the time of collision. This will change the velocity of the electron and consequently will result into greater uncertainty in velocity or momentum.

On the other hand, in order to minimise the change in momentum, we have to use light having photons with small values of p. This will require radiations of longer wavelengths ($\lambda = h/p$). Now, if we use photons of light having larger wavelengths (low momentum), the velocity or momentum will not change appreciably but we will not be able to measure position accurately with larger wavelength. Therefore, uncertainty in position will increase. Thus, we cannot simultaneously measure the position and momentum of a small moving object like electron accurately.

1.34 EXAMPLES OF POSITION-MOMENTUM UNCERTAINTY

(*i*) **Determination of the position of a particle by Heisenberg's Gamma Ray Microscope.**

Let us employ a microscope of high resolving power to observe the position of an electron. If Δx be the smallest distance between the two points that can be just resolved by the microscope, then

$$\Delta x = \frac{\lambda}{2 \sin \theta}$$

where λ is the wavelength of the light used and θ is the semi-vertical angle of the cone of light rays entering the objective. Clearly, Δx represents the uncertainty in determining the position of the particle.

Since $\Delta x \propto \lambda$ therefore uncertainty can be minimised by using radiations of shortest wavelength such as γ-rays. But if we use γ-rays, Compton effect is involved.

The electron is seen by the γ-radiation scattered by it into the microscope. During this scattering, the electron will recoil and there will be a change in its momentum. In order that this change may be as small as possible, consider a *single* γ-ray photon incident on it along the x-axis. Then a scattered photon of wavelength λ will enter the objective anywhere between OA and OB. The momentum of the scattered photon is $\frac{h}{\lambda}$. If it enters the objective along OA, its component along the x-axis is $\frac{h}{\lambda}$ sin θ, So, momentum imparted to the electron along x-axis

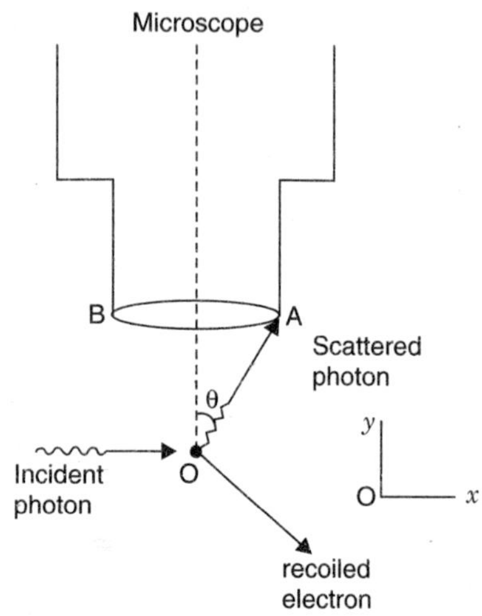

Fig. 1.14

$$= \frac{h}{\lambda'} - \frac{h}{\lambda} \sin \theta \qquad ...(i)$$

where λ' is the wavelength of the γ-radiation before scattering.

If the scattered photon enters the microscope along OB, then its component along the x-axis will be $-\frac{h}{\lambda}$ sin θ.

So, the momentum imparted to the electron in the x-direction

$$= \frac{h}{\lambda'} - \left(-\frac{h}{\lambda} \sin \theta\right) = \frac{h}{\lambda'} + \frac{h}{\lambda} \sin \theta \qquad ...(ii)$$

The momentum given to the electron can, therefore, have any value between those given by equations (*i*) and (*ii*). The uncertainty or the error involved in the measurement of the momentum of the electron, therefore, is given by

$$\Delta p_x = \left(\frac{h}{\lambda'} + \frac{h}{\lambda}\sin\theta\right) - \left(\frac{h}{\lambda'} - \frac{h}{\lambda}\sin\theta\right) = \frac{2h}{\lambda}\sin\theta$$

$$\therefore \qquad \Delta x \,.\, \Delta p_x = \frac{\lambda}{2\sin\theta}\frac{2h}{\lambda}\sin\theta = h$$

However, a more rigorous treatment gives $\Delta x \,.\, \Delta p_x = \frac{\hbar}{2}$.

(*ii*) **Diffraction of electrons by a slit.**

Consider a narrow slit AB *of width* Δx. Let a beam of electrons fall on this slit. After passing through the slit, the electron beam spreads out *i.e.,* it suffers diffraction. The first minimum of the diffraction pattern is given by the equation $d \sin\theta = n\lambda$ where $d = \Delta x$, $n = 1$, and θ is the angle through which the electron beam is diffracted.

$$\therefore \qquad \Delta x \sin\theta = \lambda \qquad ...(i)$$

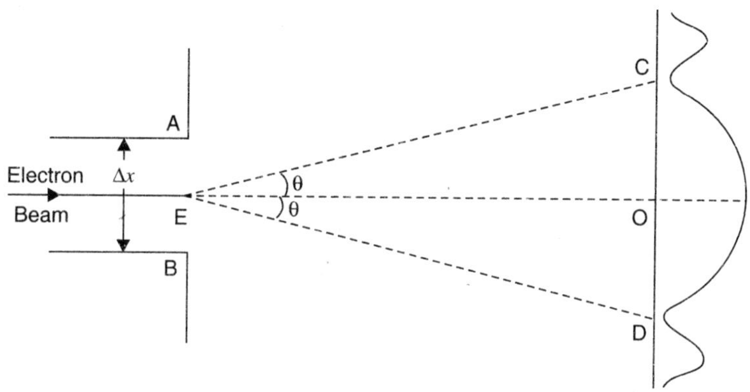

Fig. 1.15

It is not possible to locate the exact position of the slit where the electron beam enters the slit. Obviously, the uncertainty in the measurement of the position of the electron beam in the slit in a direction perpendicular to the direction of incident beam is equal to the width of the slit.

$$\Delta x = \frac{\lambda}{\sin\theta} \qquad ...(ii)$$

Let $\vec{p}$ be the momentum of the electron. If the electron moves in the direction EC after diffraction, then momentum of the electron in a direction perpendicular to the initial direction of the electron *i.e.,* along OC $= p \sin\theta$. If the electron after diffraction goes in the direction ED, then momentum of the electron in the direction OD $= -p\sin\theta$. As the electron may be any where within the diffraction pattern from angle $-\theta$ to $+\theta$, the perpendicular component of the momentum of the electron may be any where between $-p\sin\theta$ and $+p\sin\theta$.

$\therefore$ Uncertainty in the measurement of momentum of the electron

$$\Delta p = p\sin\theta - (-p\sin\theta) = 2p\sin\theta$$

$$= 2p\sin\theta$$

But $\qquad \lambda = \frac{h}{p}$ or $p = \frac{h}{\lambda}$ $\quad\therefore\quad \Delta p = 2\frac{h}{\lambda}\sin\theta \qquad ...(iii)$

From (*ii*) and (*iii*), we have

$$\Delta x \,.\, \Delta p = \frac{\lambda}{\sin\theta} \cdot \frac{2h}{\lambda} \sin\theta = 2h$$

This is clearly in agreement with the uncertainty principle.

1.35 APPLICATIONS OF UNCERTAINTY PRINCIPLE

(*i*) **Why electron cannot exist in the nucleus ?**

On the basis of Heisenberg's uncertainty principle, it can be shown as to why electron cannot exist within the atomic nucleus. The radius of the atomic nucleus is of the order of 10^{-15} m. Now, if the electron were to exist within the nucleus, then the maximum uncertainty in its position would have been 10^{-15} m.

or $$\Delta x = 10^{-15} \text{ m}$$

Now according to uncertainty principle,

$$\Delta x \times \Delta p \geq \frac{h}{4\pi} \quad \text{or} \quad \Delta x \times m\Delta v \geq \frac{h}{4\pi}$$

or $$\Delta v \geq \frac{h}{4\pi\, m\Delta x}$$

Mass of electron, $m = 9.1 \times 10^{-31}$ kg,

$\Delta x = 1 \times 10^{-15}$ m

$$\therefore \quad \Delta v = \frac{6.6 \times 10^{-34}}{4 \times 3.14 \times 9.1 \times 10^{-31} \times 1 \times 10^{-15}} \text{ m s}^{-1}$$

$$= 5.77 \times 10^{10} \text{ m s}^{-1}.$$

The value of uncertainty in velocity, Δv, is much higher than the velocity of light (3.0×10^8 m s^{-1}). Hence an electron cannot be found within the atomic nucleus.

(*ii*) **Existence of protons, neutrons and α-particles within the nucleus**

If a particular type of particle is to exist within the nucleus, then the uncertainty in its *position* in the nucleus should not exceed the radius of the nucleus. The radius of the typical atomic nucleus is of the order of 10^{-14} m. So, if a proton, or a neutron or an α-particle is to exist within the nucleus, the uncertainty in their position is given by

$$\Delta x \geq 10^{-14} \text{ m}$$

According to Heisenberg's uncertainty principle

$$\Delta x \,.\, \Delta p_x \approx \frac{\hbar}{2}$$

where Δp_x is the uncertainty in the x-component of momentum or the momentum itself and $\hbar = h/2\pi$.

$\therefore$ Uncertainty in the momentum of the proton, or neutron or α-particle, if it is to exist in the nucleus, is given by

$$\Delta p_x \geq \frac{h}{2\Delta x} = \frac{6.6 \times 10^{-34}}{2 \times 2\pi \times 10^{-14}} = 5.3 \times 10^{-21} \text{ kg m s}^{-1}$$

This means that the total momentum of the proton or neutron or α-particle in the nucleus must be at least of the order of 5.3×10^{-21} kg m s^{-1}.

The rest mass of the proton or the neutron is of the order of $m_0 = 1.67 \times 10^{-27}$ kg. Hence the value of momentum 5.3×10^{-21} kg m s^{-1} as calculated above is *non-relativistic* as the value of v comes out to be

$$\frac{5.3 \times 10^{-21}}{1.67 \times 10^{-27}} = 3.1 \times 10^{6} \text{ m s}^{-1}$$

The corresponding value of kinetic energy is given by

$$E_k = \frac{p^2}{2m_0} = \frac{(5.3 \times 10^{-21})^2}{2 \times 1.67 \times 10^{-27}} = 8.4 \times 10^{-15} \text{ J}$$

$$= \frac{8.4 \times 10^{-15}}{1.6 \times 10^{-19}} = 52 \text{ keV}$$

The rest mass of the α-particle is nearly four times the proton mass. Thus, the α-particle should have a minimum kinetic energy of 13 keV. Since the energy carried by the protons or neutrons emitted by the nuclei are greater than 52 keV and for α-particles more than 13 keV, these particles can exist in the nuclei.

(*iii*) **Zero point energy**

Consider a particle confined to move under the influence of a potential of linear dimensions l. Since the particle can exist *any where* in this region, the uncertainty in its position $\Delta x = l/2$.

According to uncertainty principle, if Δp_x is the uncertainty in the x-component of its momentum, then

$$\Delta p_x \,.\, \Delta x = \frac{\hbar}{2} \qquad \text{or} \qquad \Delta p_x = \frac{\hbar}{2\Delta x} = \frac{\hbar}{l}$$

Suppose the minimum value of the momentum of the particle is equal to its uncertainty Δp_x.

Then $$p_x = \Delta p_x = \frac{\hbar}{l}$$

If m_0 is the rest mass of the particle, then the lowest value of kinetic energy in a potential of finite dimension is given by

$$E = \frac{p^2}{2m_0} = \frac{p_x^2}{2m_0} = \frac{\hbar^2}{2m_0 l^2}$$

This is the minimum energy possessed by a particle confined to a box of finite length l. *Thus we conclude that a particle confined to a region of space cannot have zero kinetic energy.*

(*iv*) **Energy of a particle in one-dimensional box**

Let us consider a particle of mass m in one-dimensional box of length l. If $(\Delta x)_{\text{max.}}$ represents the maximum uncertainty in the position of the particle, then

$$(\Delta x)_{\text{max.}} = l$$

We know that $$\Delta x \, \Delta p \approx \hbar$$

$$\Rightarrow \qquad \Delta p = \frac{\hbar}{\Delta x}$$

The uncertainty in momentum must be less than momentum itself.

So, the minimum momentum of the particle is $\frac{\hbar}{l}$.

Kinetic energy of the particle, $T = \frac{p^2}{2m} = \frac{\hbar^2}{2ml^2}$

This is the expression for the minimum kinetic energy of a particle in a box.

(*v*) **Radius of Bohr's first orbit**

We can determine the radius of the first Bohr orbit by making use of uncertainty principle.

If Δx and Δp are the uncertainties in the simultaneous measurements of position and momentum of the electron in the first orbit, then

$$\Delta x \, \Delta p \geq \hbar \quad \text{or} \quad \Delta p \geq \frac{\hbar}{\Delta x} \qquad ...(i)$$

Total energy of electron, E = kinetic energy T + potential energy V

Uncertainty in total energy is given by

$$\Delta E = \Delta T + \Delta V \qquad ...(ii)$$

Kinetic energy of electron moving with velocity v, $T = \frac{1}{2} mv^2$

$$\Delta T = \frac{1}{2} m (\Delta v)^2 \quad \text{or} \quad \Delta T = \frac{(m\Delta v)^2}{2m}$$

or

$$\Delta T = \frac{(\Delta p)^2}{2m} \qquad ...(iii)$$

Using equation (*i*), $\Delta T = \frac{\hbar^2}{2m(\Delta x)^2}$ ...(*iv*)

Potential energy of electron, $V = -\frac{Ze^2}{x}$

$$\Delta V = -\frac{Ze^2}{\Delta x} \qquad ...(v)$$

Here ΔV is the uncertainty in the potential energy of the electron.

Combining equations (*ii*), (*iv*) and (*v*), we get

$$\Delta E = \frac{\hbar^2}{2m(\Delta x)^2} - \frac{Ze^2}{\Delta x} \qquad ...(vi)$$

For minimum value of uncertainty in energy,

$$\frac{d(\Delta E)}{d(\Delta x)} = 0 \quad \text{and} \quad \frac{d^2(\Delta E)}{d(\Delta x)^2} \text{ is positive.}$$

Differentiating equation (*vi*) w.r.t. x, we get

$$\frac{d(\Delta E)}{d(\Delta x)} = -\frac{\hbar^2}{m(\Delta x)^3} + \frac{Ze^2}{(\Delta x)^2} = 0$$

or

$$\frac{\hbar^2}{m(\Delta x)^3} = \frac{Ze^2}{(\Delta x)^2} \quad \text{or} \quad \Delta x = \frac{\hbar^2}{mZe^2} \qquad ...(vii)$$

Differentiating again, we get

$$\frac{d^2(\Delta E)}{d(\Delta x)^2} = \frac{3\hbar^2}{m\left(\frac{\hbar^2}{mZe^2}\right)(\Delta x)^3} - \frac{2Ze^2}{(\Delta x)^3}$$

$$= \frac{3Ze^2}{(\Delta x)^3} - \frac{2Ze^2}{(\Delta x)^3} = \frac{Ze^2}{(\Delta x)^3}, \text{ which is positive.}$$

So, it follows from equation (*vii*) that

$$r = \frac{\hbar^2}{mZe^2} = \frac{h^2}{4\pi^2 mZe^2}$$

which is the expression for the radius of Bohr's first orbit.

(*vi*) **Finite width of spectral lines**

$$\Delta E \; \Delta t \approx \hbar$$

The life time of an electron in an excited orbit is 10^{-8} s nearly. So, Δt has a finite value.

Thus, ΔE must also have some finite value.

From here, we can conclude that the excited energy levels have a finite energy spread. So, an electron jumping from higher level to lower level cannot emit truly monochromatic radiation. Thus, we can conclude that the spectral lines must have a natural finite width. The spectral lines can never be infinitely sharp.

KNOWLEDGE PLUS

The uncertainty principle gives the minimum absolute uncertainty in the measurement of conjugate quantities like (x, p_x), (y, p_y), (z, p_z), (t, E) and (θ, L), The actual uncertainties are much larger and include contributions on account of errors due to measuring instruments, conditions of performance of experiment and certain other causes. The uncertainty principle is applicable to all objects, but its consequences are significant only for such minute particles of microscopic size as electrons, protons, neutrons, etc. because in the case of *macroscopic* (large size) objects, the uncertainties in the measurement of these quantities are so large that their products exceed the value of Planck's constant h by many orders of magnitude. Thus, due to the smallness of Planck's constant, the uncertainty principle becomes meaningless for measurements in connection with macroscopic bodies.

NUMERICAL EXAMPLES BASED ON UNCERTAINTY PRINCIPLE

■ **Example 1.39.** *Calculate the uncertainty in the momentum of an electron if it is confined to a linear region of length 1×10^{-10} metre. Given : $h = 6.626 \times 10^{-34}$ Js and $\pi = 3.143$.*

Solution. According to uncertainty principle,

$$\Delta x . \Delta p = \frac{h}{4\pi} \text{ or } \Delta p = \frac{h}{4\pi \, \Delta x}$$

$$\Delta p = \frac{6.626 \times 10^{-34} \text{ kg m}^2 \text{ s}^{-1}}{4 \times 3.143 \times 10^{-10} \text{ m}}$$

$$= \mathbf{5.27 \times 10^{-25} \text{ kg m s}^{-1}}.$$

■ **Example 1.40.** *An electron has a speed of 40 m s^{-1} accurate upto 99.99%. What is the uncertainty in locating its position ? (Mass of electron = 9.1×10^{-31} kg).*

Solution. According to uncertainty principle,

$$\Delta x . m\Delta v = \frac{h}{4\pi} \text{ or } \Delta x = \frac{h}{4\pi \, m\Delta v}$$

$$\Delta v = 40 \text{ m s}^{-1} \text{ (accurate upto 99.99\%)}$$

$$= \frac{100 - 99.99}{100} \times 40 \text{ m s}^{-1}$$

$$= \frac{0.01}{100} \times 40 = 4 \times 10^{-3} \text{ m s}^{-1}$$

$$m = 9.1 \times 10^{-31} \text{ kg},$$

$$h = 6.626 \times 10^{-34} \text{ kg m}^2 \text{ s}^{-1}, \pi = 3.143$$

$$\therefore \quad \Delta x = \frac{6.626 \times 10^{-34} \text{ kg m}^2 \text{ s}^{-1}}{4 \times 3.143 \times 9.1 \times 10^{-31} \text{ kg} \times 4 \times 10^{-3} \text{ m s}^{-1}}$$

$$= \mathbf{1.45 \times 10^{-2} \ m.}$$

■ **Example 1.41.** *The position and momentum of 0.5 keV electron are simultaneously determined. If its position is located within 0.2 nm, what is the percentage, uncertainty in its momentum ?*

Solution. Energy of the electron,

$$E = 0.5 \times 10^3 \text{ eV}$$
$$= 0.5 \times 10^{+3} \times 1.6 \times 10^{-19} \text{ J}$$
$$= 8 \times 10^{-17} \text{ J}$$

As it is a non-relativistic case, momentum of the electron

$$p = \sqrt{2mE}$$
$$= (2 \times 9 \times 10^{-31} \times 8 \times 10^{-17})^{1/2}$$
$$= 1.2 \times 10^{-23} \text{ kg m s}^{-1}$$

Uncertainty in position $\Delta x = 0.2 \times 10^{-9}$ m $= 2 \times 10^{-10}$ m

Uncertainty in momentum Δp is given by

$$\Delta p \,.\, \Delta x = \frac{\hbar}{2}$$

$$\therefore \quad \Delta p = \frac{\hbar}{2\Delta x} = \frac{h}{2 \times 2\pi \times \Delta x}$$

$$= \frac{7 \times 6.6 \times 10^{-34}}{2 \times 2 \times 22 \times 2 \times 10^{-10}}$$

$$= 2.6 \times 10^{-25} \text{ kg m s}^{-1}$$

$\therefore$ percentage uncertainty in momentum

$$= \frac{\Delta p}{p} \times 100$$

$$= \frac{2.6 \times 10^{-25}}{1.2 \times 10^{-23}} \times 100 = \mathbf{2.2\%.}$$

■ **Example 1.42.** *The uncertainty in measuring the speed of an accelerated electron is 1.2×10^5 m s^{-1}. Calculate the uncertainty in finding its location while it is still in motion.*

(Given : $m_e = 9.31 \times 10^{-31}$ kg).

Solution. According to uncertainty principle,

$$\Delta x \,.\, m\Delta v = \frac{h}{4\pi} \quad \text{or} \quad \Delta x = \frac{h}{4\pi \, m\Delta v}$$

$h = 6.626 \times 10^{-34}$ kg m^2 s^{-1},

$m_e = 9.31 \times 10^{-31}$ kg

$\Delta v = 1.2 \times 10^5$ m s^{-1}

$$\Delta x = \frac{6.626 \times 10^{-34} \text{ kg m}^2 \text{ s}^{-1}}{4 \times 3.143 \times 9.31 \times 10^{-31} \text{ kg} \times 1.2 \times 10^{-5} \text{ m s}^{-1}}$$

$$= \mathbf{4.72 \times 10^{-10} \ m \ s^{-1}.}$$

■ **Example 1.43.** *An electron is confined to a box of length 10^{-9} m. Calculate the minimum uncertainty in the measurement of its velocity.*

Solution. According to Heisenberg's position-momentum uncertainty

$$\Delta x \,.\, \Delta p_x \approx h$$

When Δx is maximum, Δp_x is minimum

$$\therefore \quad (\Delta x)_{max.} \,.\, (\Delta p_x)_{min.} \approx h$$

Now $(\Delta x)_{max.} = 10^{-9}$ m

$$\therefore \quad (\Delta p_x)_{min.} = \frac{h}{(\Delta x)_{max.}} = \frac{6.62 \times 10^{-34}}{10^{-9}}$$

$$= 6.62 \times 10^{-25} \text{ kg m s}^{-1}$$

Again $(\Delta p_x)_{min.} = m(\Delta v_x)_{min.}$

$$(\Delta v_x)_{min.} = \frac{(\Delta p_x)_{min.}}{m} = \frac{6.62 \times 10^{-25}}{9.1 \times 10^{-31}} \text{ m s}^{-1}$$

$$= \mathbf{7.27 \times 10^5 \ m \ s^{-1}.}$$

■ **Example 1.44.** *Using the mathematical expression for the uncertainty principle, show that the simultaneous measurement of position and momentum of a particle with absolute accuracy is not possible.*

Solution. According to uncertainty principle,

$$\Delta x \,.\, \Delta p = \frac{h}{4\pi}$$

Case I. *Measuring the exact position of the particle*

If the position of the particle is measured with absolute accuracy, then

$$\Delta x = 0$$

$$\Delta p = \frac{h}{4\pi \, \Delta x} = \frac{h}{4\pi \times \text{Zero}} = \text{Infinity } (\infty)$$

This means that uncertainty in measuring the momentum (Δp) will be infinity or the velocity of the particle cannot be determined.

Case II. *Measuring the exact momentum of the particle.*

If the momentum of the particle is measured with absolute accuracy, then $\Delta p = 0$

$$\Delta x = \frac{h}{4\pi\,\Delta p} = \frac{h}{4\pi \times \text{Zero}} = \text{Infinity } (\infty)$$

This clearly shows that the position of the particle cannot be determined.

Thus, both the momentum and the position of a particle cannot be determined simultaneously with absolute accuracy.

■ **Example 1.45.** *Life time of a nucleus in the excited state is 10^{-12} s. Calculate the probable uncertainty in energy and frequency of a γ-ray photon emitted by it.*

Solution. According to Heisenberg's time energy uncertainty principle

$$\Delta E \,.\, \Delta t = \frac{h}{2\pi} = \hbar$$

As the life time of a nucleus in the excited state is 10^{-12} s, we have $\Delta t = 10^{-12}$ s

$$\therefore \quad \Delta E = \frac{h}{2\pi\,\Delta t} \geq \frac{6.62 \times 10^{-34}}{2\pi \times 10^{-12}} \text{ J}$$

$$= \mathbf{1.054 \times 10^{-22}\ J}$$

The frequency of γ-ray photon is uncertain by an amount

$$\Delta \nu = \frac{\Delta E}{h} = \frac{1.054 \times 10^{-22}}{6.62 \times 10^{-34}} \text{ Hz}$$

$$= \mathbf{1.59 \times 10^{11}\ Hz.}$$

■ **Example 1.46.** *A base ball of mass 200 g is moving with a velocity of 3×10^2 cm s^{-1}. If we can locate the base ball within an error equal to the magnitude of the wavelength of light used (5000 Å), how will the uncertainty in momentum be compared with the total momentum of the base ball ?*

Solution. Step I. *Calculation of uncertainty in momentum of the base ball.*

According to uncertainty principle,

$$\Delta x \,.\, \Delta p = \frac{h}{4\pi} \text{ or } \Delta p = \frac{h}{4\pi \times \Delta x}$$

$\Delta x = 5000$ Å $= 5000 \times 10^{-10}$ m $= 5 \times 10^{-7}$ m

$\pi = 3.143$, $h = 6.626 \times 10^{-34}$ kg m^2 s^{-1}.

$$\Delta p = \frac{6.626 \times 10^{-34} \text{ kg m}^2 \text{ s}^{-1}}{4 \times 3.143 \times 5 \times 10^{-7} \text{ m}}$$

$$= 1.05 \times 10^{-28} \text{ kg m s}^{-1}.$$

Step II. *Calculation of $\Delta p / p$*

Momentum $(p) = m \times v$

$m = 200$ g $= 200 \times 10^{-3}$ kg

$v = 3 \times 10^2$ cm s^{-1}

$= 3 \times 10^2 \times 10^{-2}$ m s^{-1} $= 3$ m s^{-1}

$p = (200 \times 10^{-3}$ kg$) \times (3$ m s$^{-1})$

$= 600 \times 10^{-3}$ kg m s^{-1}

$$\therefore \quad \frac{\Delta p}{p} = \frac{1.05 \times 10^{-28} \text{ kg m s}^{-1}}{600 \times 10^{-3} \text{ kg m s}^{-1}}$$

$$= \mathbf{1.75 \times 10^{-28}.}$$

■ **Example 1.47.** *The electron in hydrogen atom may be thought of as confined to a nucleus of radius = 5×10^{-11} metre. Calculate the minimum uncertainty in the momentum of the electron. Also calculate the minimum kinetic energy of the electron.*

Given : $m_e = 9 \times 10^{-31}$ kg ; 1 eV = 1.6×10^{-19} J, h = 6.62×10^{-34} Js.

Solution. According to the Heisenberg's uncertainty principle

$$\Delta x \,.\, \Delta p \geq \frac{\hbar}{2} \geq \frac{h}{4\pi}$$

where Δx is the uncertainty in position and Δp_x the minimum uncertainty in the momentum of a particle. As the electron is confined to a radius of 5×10^{-11} metre, the uncertainty in its position, $\Delta x = 5 \times 10^{-11}$ metre.

$$\therefore \quad \Delta p_x \geq \frac{h}{4\pi\,\Delta x} \geq \frac{6.62 \times 10^{-34}}{4\pi \times 5 \times 10^{-11}}$$

$$= 1.054 \times 10^{-24} \text{ kg m s}^{-1}$$

$$\text{Now kinetic energy E} = \frac{p^2}{2m}$$

$$\therefore \quad \Delta E = \frac{\Delta p^2}{2m} \geq \frac{(1.054 \times 10^{-24})^2}{2 \times 9 \times 10^{-31}} \text{ J}$$

$$\geq 6.142 \times 10^{-19} \text{ J}$$

$$\geq \frac{6.142 \times 10^{-19}}{1.6 \times 10^{-19}} \text{ eV} = \mathbf{3.84\ eV.}$$

■ **Example 1.48.** *Calculate the uncertainty in the position of a dust particle with mass equal to 1 mg if uncertainty in its velocity is 5.5×10^{-20} m s^{-1}.*

Solution. According to uncertainty principle,

$$\Delta x \,.\, m\Delta v = \frac{h}{4\pi} \text{ or } \Delta x = \frac{h}{4\pi m\Delta v}$$

$\Delta v = 5.5 \times 10^{-20}$ m s^{-1},

$m = 1$ mg $= 10^{-6}$ kg, $\pi = 3.143$,

$h = 6.626 \times 10^{-34}$ kg m^2 s^{-1}.

$$\therefore \quad \Delta x = \frac{6.626 \times 10^{-34} \text{ kg m}^2 \text{ s}^{-1}}{4 \times 3.143 \times 10^{-6} \text{ kg} \times 5.5 \times 10^{-20} \text{ m s}^{-1}}$$

$= \mathbf{9.58 \times 10^{-10}}$ **m.**

■ **Example 1.49.** *A proton is accelerated to one-tenth of velocity of light. If the velocity can be measured with a precision of ± 0.5%, what must be the uncertainty in its position ? ($m_p = 1.675 \times 10^{-27}$ kg).*

Solution. According to uncertainty principle,

$$\Delta x \,.\, m\Delta v = \frac{h}{4\pi} \text{ or } \Delta x = \frac{h}{4\pi\, m\, \Delta v}$$

Velocity of proton, $v = \frac{1}{10} \times 3 \times 10^8$ m s^{-1}

$= 3 \times 10^7$ m s^{-1},

$\Delta v = 0.5\%$ of 3×10^7

$= 0.5 \times 10^{-2} \times 3 \times 10^7 = 1.5 \times 10^5$ m s^{-1}.

$m = 1.675 \times 10^{-27}$ kg,

$h = 6.626 \times 10^{-34}$ kg m^2 s^{-1}

$$\Delta x = \frac{6.626 \times 10^{-34} \text{ kg m}^2 \text{ s}^{-1}}{4 \times 3.143 \times 1.675 \times 10^{-27} \text{ kg} \times 1.5 \times 10^5 \text{ ms}^{-1}}$$

$= \mathbf{2.1 \times 10^{-13}}$ **m.**

■ **Example 1.50.** *The uncertainty in position and velocity of a particle are 10^{-10} m and 5.27×10^{-24} m s^{-1} respectively. Calculate the mass of the particle. ($h = 6.625 \times 10^{-34}$ Js).*

Solution. According to uncertainty principle,

$$\Delta x \,.\, m\Delta v = \frac{h}{4\pi} \text{ or } m = \frac{h}{4\pi \times \Delta x \times \Delta v}$$

$h = 6.625 \times 10^{-34}$ Js

$= 6.625 \times 10^{-34}$ kg m^2 s^{-1},

$\Delta x = 10^{-10}$ m ; $\Delta v = 5.27 \times 10^{-24}$ m s^{-1}.

$$m = \frac{6.625 \times 10^{-34} \text{ kg m}^2 \text{ s}^{-1}}{4 \times 3.143 \times 10^{-10} \text{ m} \times 5.27 \times 10^{-24} \text{ m s}^{-1}}$$

$= \mathbf{0.099}$ **kg.**

■ **Example 1.51.** *Calculate the uncertainty in velocity of a cricket ball of mass 150 g if uncertainty in position is of the order of 1 Å.*

Solution. According to uncertainty principle,

$$\Delta x \,.\, \Delta p = \frac{h}{4\pi}$$

or $$\Delta x \,.\, m\Delta v = \frac{h}{4\pi}$$

or $$\Delta v = \frac{h}{4\pi\, m\, \Delta v}$$

$h = 6.626 \times 10^{-34}$ kg m^2 s^{-1},

$m = 150$ g $= 0.15$ kg,

$\Delta x = 1$ Å $= 1 \times 10^{-10}$ m

$$\Delta v = \frac{6.626 \times 10^{-34} \text{ kg m}^2 \text{ s}^{-1}}{4 \times 3.143 \times 0.15 \text{ kg} \times 10^{-10} \text{ m}}$$

$= \mathbf{3.51 \times 10^{-24}}$ **m s^{-1}.**

■ **Example 1.52.** *Calculate the product of uncertainties in displacement and velocity for an electron of mass 9.1×10^{-31} kg according to Heisenberg's uncertainty principle.*

($h = 6.6 \times 10^{-34}$ kg m^2 s^{-1}).

Solution. According to Heisenberg's uncertainty principle,

$$\Delta x \times \Delta p \geq \frac{h}{4\pi} \text{ or } \Delta x \times m\Delta v \geq \frac{h}{4\pi}$$

($\because$ m is constant)

or $$\Delta x \times \Delta v \geq \frac{h}{4m\pi}$$

$m = 9.1 \times 10^{-31}$ kg,

$h = 6.6 \times 10^{-34}$ kg m^2 s^{-1}, $\pi = 3.14$

$$\Delta x \times \Delta v \geq \frac{6.6 \times 10^{-34} \text{ kg m}^2 \text{ s}^{-1}}{4 \times 9.1 \times 10^{-31} \text{ kg} \times 3.14}$$

$$\Delta x \times \Delta v \geq \frac{6.6}{4 \times 9.1 \times 3.14} \times 10^{-3} \text{ m}^2 \text{s}^{-1}$$

or $\Delta x \times \Delta v = 5.77 \times 10^{-5}$ m^2 s^{-1}

Thus, product of uncertainties in position and velocity of an electron is equal to or greater than **5.77×10^{-5} m^2 s^{-1}.**

■ **Example 1.53.** *An electron has a speed of 500 m s^{-1} with uncertainty of 0.02%. What is the uncertainty in locating its position ?*

Solution. Velocity of electron = 500 m s^{-1}

Uncertainty in velocity, $\Delta v = 500 \times 0.02\%$

$$= \frac{500 \times 0.02}{100} = 0.1 \text{ m s}^{-1}$$

Mass of electron, $m = 9.1 \times 10^{-31}$ kg

$h = 6.6 \times 10^{-34}$ kg m^2 s^{-1}

According to uncertainty principle,

$$\Delta x \times \Delta p \geq \frac{h}{4\pi}$$

or
$$\Delta x \times m\,\Delta v \geq \frac{h}{4\pi}$$

or
$$\Delta x \geq \frac{h}{4\pi m\,\Delta v}$$

$$\geq \frac{6.6 \times 10^{-34} \text{ kg m}^2 \text{ s}^{-1}}{4 \times \frac{22}{7} \times 9.1 \times 10^{-31} \times 0.1}$$

$$\geq \mathbf{5.77 \times 10^{-4} \text{ m}}$$

$\therefore$ Uncertainty in position = **5.77×10^{-4} m.**

■ **Example 1.54.** *Calculate the uncertainty in the position of an electron if the uncertainty in its velocity is 5.7×10^5 m s^{-1}. ($h = 6.6 \times 10^{-34}$ J s and mass of electron = 9.1×10^{-31} kg)*

Solution. According to Heisenberg's uncertainty principle.

$$\Delta x \times \Delta p \geq \frac{h}{4\pi} \quad \text{or} \quad \Delta x \times m\Delta v \geq \frac{h}{4\pi}$$

$$\therefore \quad \Delta x \geq \frac{h}{4\pi\, m\Delta v}$$

$\Delta v = 5.7 \times 10^5$ m s^{-1}, $m = 9.1 \times 10^{-31}$ kg,

$h = 6.6 \times 10^{-34}$ J s

$$\therefore \quad \Delta x = \frac{6.6 \times 10^{-34} \text{ kg m}^2 \text{ s}^{-1}}{4 \times 3.142 \times 9.1 \times 10^{-31} \text{ kg} \times 5.7 \times 10^5 \text{ m s}^{-1}}$$

$$= \mathbf{1.012 \times 10^{-10} \text{ m.}}$$

■ **Example 1.55.** *A microscope using suitable photons is employed to locate an electron in an atom within a distance of 0.1 Å. What is the uncertainty involved in the measurement of its velocity ? Mass of electron = 9.11×10^{-31} kg and h = 6.626×10^{-34} Js.*

Solution. $\Delta x \times \Delta p = \frac{h}{4\pi}$ or $\Delta x \times m\Delta v = \frac{h}{4\pi}$

$$\Delta v = \frac{h}{4\pi \times \Delta x \times m}$$

$$\Delta v = \frac{6.626 \times 10^{-34} \text{ J s}}{4 \times 3.14 \times 0.1 \times 10^{-10} \text{ m} \times 9.11 \times 10^{-31} \text{ kg}}$$

$= 0.579 \times 10^7$ m s^{-1} (1 J = 1 kg m^2 s^{-2})

$= \mathbf{5.79 \times 10^6}$ **m s^{-1}.**

■ **Example 1.56.** *A cricket ball weighing 100 g is to be located within 1 nm. What is the uncertainty in its velocity ? Comment on your result.*

Solution. Mass of cricket ball

= 100 g = 100×10^{-3} kg

= 0.1 kg

Uncertainty in position,

$\Delta x = 1$ nm $= 1 \times 10^{-9}$ m

($\because$ 1 nm = 10^{-9} m)

According to uncertainty principle,

$$\Delta x \times \Delta p = \frac{h}{4\pi} \quad \text{or} \quad \Delta x \times m\Delta v = \frac{h}{4\pi}$$

or
$$\Delta v = \frac{h}{4\pi\, \Delta x m}$$

or
$$\Delta v = \frac{6.626 \times 10^{-34} \text{ kg m}^2 \text{ s}^{-1}}{4 \times \frac{22}{7} \times (10^{-9} \text{ m}) \times (0.1 \text{ kg})}$$

$$= \frac{6.626}{4 \times \frac{22}{7}} \times 10^{-24} \text{ m s}^{-1}$$

$= \mathbf{5.27 \times 10^{-25}}$ **m s^{-1}.**

Comment. When we compare this value of uncertainty in velocity with the usual velocity of a cricket ball, we find that this uncertainty in velocity is completely negligible. This means that we can disregard the uncertainty principle for the motion of *macroscopic bodies.* Thus, we can say that for such cases, the position and velocity can be simultaneously determined to a reasonable degree of precision.

■ **Example 1.57.** *The mass of an electron is 9.11 × 10^{-31} kg. Calculate the uncertainty in its velocity, if* *the uncertainty in its position is of the order of ± 10 pm. ($h = 6.6 \times 10^{-34}$ kg m^2 s^{-1}).*

Solution. Uncertainty in position,

$$\Delta x = 10 \text{ pm} = 10 \times 10^{-12} \text{ m} = 10^{-11} \text{ m} \quad (\because \; 1 \text{ pm} = 10^{-12} \text{ m})$$

$$m = 9.11 \times 10^{-31} \text{ kg},$$

$$h = 6.6 \times 10^{-34} \text{ kg m}^2 \text{ s}^{-1}$$

According to Heisenberg's uncertainty principle,

$$\Delta x \times \Delta p = \frac{h}{4\pi} \text{ or } \Delta x \times m\Delta v = \frac{h}{4\pi} \quad (\because \; m \text{ is constant})$$

or $$\Delta v = \frac{h}{4\pi\, m\Delta x}$$

or $$\Delta v = \frac{6.6 \times 10^{-34} \text{ kg m}^2 \text{ s}^{-1}}{4 \times \frac{22}{7} \times 9.11 \times 10^{-31} \text{ kg} \times 10^{-11} \text{ m}}$$

$$= \frac{6.6}{4 \times \frac{22}{7} \times 9.11} \times 10^8 \text{ m s}^{-1}$$

$$= \mathbf{5.76 \times 10^6 \text{ m s}^{-1}}.$$

■ **Example 1.58.** *Calculate the uncertainty in the position of an electron if uncertainty in its velocity is (i) 0.001% (ii) zero.*

(The mass of electron = 9.1×10^{-31} kg, velocity of electron = 300 m s^{-1}).

Solution. Mass of electron = 9.1×10^{-31} kg,

$h = 6.6 \times 10^{-34}$ J s

(*i*) When Δv = 0.001% of velocity of electron

$$= \frac{0.001 \times 300}{100} \text{ m s}^{-1} = 3.0 \times 10^{-3} \text{ m s}^{-1}$$

Now, $$\Delta x \times m\, \Delta v = \frac{h}{4\pi}$$

$$\therefore \quad \Delta x = \frac{h}{4\pi\, m\, \Delta v}$$

$$= \frac{6.6 \times 10^{-34} \text{ kg m}^2 \text{ s}^{-1}}{4 \times 3.14 \times 9.1 \times 10^{-31} \text{ kg} \times 3.0 \times 10^{-3}}$$

$$= \mathbf{1.92 \times 10^{-2} \text{ m}}.$$

(*ii*) When $\Delta v = 0$

$$\Delta x = \frac{h}{4\pi\, m\, \Delta v}$$

If Δv is zero, then denominator in the above expression becomes zero and, therefore, uncertainty in position is **infinity.**

EXERCISES

36. Calculate the uncertainty in velocity of a cricket ball of mass 150 g if the uncertainty in position, (Δx) is of the order 1 Å ($h = 6.63 \times 10^{-34}$ J s). [**Ans.** 3.5×10^{-24} m s^{-1}]

37. Calculate the uncertainty in the velocity of a bullet weighing 10 g whose position is known with ± 0.01 mm. [**Ans.** 5.27×10^{-28} m s^{-1}]

38. What is the minimum uncertainty in the position of a bullet of mass 5 g that is known to have speed somewhere between 550.00001 m s^{-1} and 550.00000 m s^{-1} ? [**Ans.** 1.05×10^{-27} m]

39. A proton is accelerated to one-tenth of the velocity of light. If its velocity can be measured with a precision of ± 1%, what must be its uncertainty in position ? ($h = 6.6 \times 10^{-34}$ J s, mass of proton = 1.66×10^{-27} kg). [**Ans.** 1.06×10^{-13} m]

40. Calculate the uncertainty in position of an electron if the uncertainty in velocity is 5.7×10^5 m s^{-1} ($h = 6.6 \times 10^{-34}$ kg m^2 s^{-1}, mass of electron = 9.1×10^{-31} kg). [**Ans.** 1×10^{-10} m]

41. Calculate the minimum uncertainty in velocity of a particle of mass 1.1×10^{-27} kg if uncertainty in its position is 3×10^{-10} cm. ($h = 6.6 \times 10^{-34}$ kg m^2 s^{-1}) [**Ans.** 1.59×10^4 m s^{-1}]

42. The uncertainties in the position and velocity of a particle are : 10^{-10} m and 5.27×10^{-24} m s^{-1}. Calculate the mass of particle ($h = 6.626 \times 10^{-34}$ Js). [**Ans.** 0.10 kg]

43. A cricket ball weighing 100 g is located within 1 nm. What is the uncertainty in the velocity ? [**Ans.** 5.27×10^{-25} m s^{-1}]

44. What will be uncertainty in the velocity of an electron if it is located within 5×10^{-5} Å ? [**Ans.** 1.16×10^{-2} m]

45. The mass of electron is 9.11×10^{-31} kg. Calculate the uncertainty in its velocity if the uncertainty in position is of the order of $\pm$ 10 pm. ($h = 6.6 \times 10^{-34}$ kg m^2 s^{-1}). [**Ans.** 5.76×10^6 m s^{-1}]

46. The speed of an electron moving at 600 m s^{-1} is measured to the accuracy of 0.005%. What will be the minimum error in determining its position ? [**Ans.** 1.93×10^{-3} m]

47. Calculate the product of uncertainty in displacement and velocity for an electron of mass 9.1×10^{-31} kg according to Heisenberg's uncertainty principle. [**Ans.** 5.79×10^{-5} m^2 s^{-1}]

48. An electron has a speed of 500 m s^{-1} with uncertainty of 0.02%. What is the uncertainty in locating its position ? [**Ans.** 5.79×10^{-4} m]

49. Calculate the uncertainty in the velocity of a particle of mass 1.1×10^{-27} kg if the uncertainty in its position is 3×10^{-10} cm. [**Ans.** 1.6×10^4 m s^{-1}]

1.36 TOWARDS QUANTUM MECHANICAL MODEL OF THE ATOM

In view of the shortcomings of the Bohr model, attempts were made to develop a more suitable and general model for atoms. Two important developments which contributed significantly in the formulation of such a model were :

(*i*) Dual behaviour of matter, and

(*ii*) Heisenberg uncertainty principle.

1.37 QUANTUM MECHANICAL MODEL OF ATOM

Classical mechanics, based on Newton's laws of motion, successfully describes the motion of all macroscopic objects such as a falling stone, orbiting planets etc., which have essentially a particle-like behaviour. However, it fails when applied to microscopic objects like electrons, atoms, molecules etc. This is mainly because of the fact that classical mechanics ignores the concept of dual behaviour of matter and the uncertainty principle. The branch of science that takes into account this dual behaviour of matter is called **quantum mechanics.**

Quantum mechanics is a theoretical science that deals with the study of the motions of the microscopic objects that have both observable wave-like and particle-like properties .It specifies the laws of motion that these objects obey. When quantum mechanics is applied to macroscopic objects (for which wave-like properties are insignificant), the results are the same as those from the classical mechanics.

Quantum mechanics was developed independently in 1926 by Werner Heisenberg and Erwin Schrödinger. Here, however, we shall be discussing the quantum mechanics of Schrödinger which is based on the ideas of wave motion. The fundamental equation of quantum mechanics is the **Schrödinger equation.** This was proposed by Schrödinger and it won him the Nobel Prize in Physics in 1933.

In 1927, Erwin Schrödinger described the behaviour of electrons around the nucleus by a mathematical equation known as **Schrödinger wave equation.** The Schrödinger wave equation is :

$$\frac{\partial^2\psi}{\partial x^2} + \frac{\partial^2\psi}{\partial y^2} + \frac{\partial^2\psi}{\partial z^2} + \frac{8\pi^2 m}{h^2}(E - V)\psi = 0$$

where x, y and z are three space co-ordinates, m is the mass of the electron, h is Planck's constant, E is the total energy and V is the potential energy of the electron, ψ (Greek letter *psi*) is the amplitude of the wave, called wave function and $\frac{\partial^2\psi}{\partial x^2}$ refers to the second derivative of ψ with respect to x only and so on.

When the Schrödinger wave equation is solved, the solution gives information about the various energy states and other measurable properties of the system. For example, the solution of Schrödinger wave equation for an electron in a hydrogen atom provides information about the various energy states and other properties such as radiation frequencies emitted or absorbed for the atom. The solutions are called **wave functions** and are denoted by the symbol ψ.

1.38 PHYSICAL SIGNIFICANCE OF WAVE FUNCTION

In the physical sense, ψ gives the *amplitude of the wave associated* with the electron. We know that in the case of light waves, the square of the amplitude of the wave at a point is proportional to the intensity of light. Extending the same concept to electron wave motion, the square of the wave function, ψ^2 may be taken as *intensity of electron at any point.* In other words, ψ^2 determines the probability of finding the moving electron in a given region *i.e.,* it gives the probability density. Thus ψ^2 has been called the **probability density and ψ the probability amplitude.** Thus, the solutions of Schrödinger wave equation replace the discrete energy levels or orbits proposed by Bohr and led to the concept of *most probable regions in space* in terms of ψ^2. A large value of ψ^2 means a high probability of finding the electron at that place and a small value of ψ^2 means low probability. If ψ^2 is almost zero at a particular point, it means that the probability of finding the electron at that point is negligible. Therefore, *the wave mechanics approach gives meaningful wave functions which describe the position and energy levels of electrons in an atom.*

Erwin Schrödinger (1887–1961)

Erwin Schrödinger, an Austrian physicist received his Ph.D. in theoretical physics from the University of Vienna in 1910. In 1927, Schrödinger succeeded Max Planck at the University of Berlin at Planck's request. In 1933, Schrödinger left Berlin because of his opposition to Hitler and Nazi policies and returned to Austria in 1936. After the invasion of Austria by Germany, Schrödinger was forcibly removed from his professorship. He then moved to Dublin, Ireland where he remained for seventeen years. Schrödinger shared the Nobel Prize for Physics with P.A.M. Dirac in 1933.

1.39 TIME-INDEPENDENT SCHRÖDINGER EQUATION

In wave mechanics, a material particle is equivalent to a wave packet. For locating the position of the particle within the wave packet, Schrödinger gave an equation.

There are many situations in which the potential energy, force etc. vary with the position of the particle but do not depend upon time. The differential equation dealing with such a situation is called steady-state Schrödinger equation or time-independent Schrödinger equation.

According to classical wave optics, the differential equation of a wave motion is given by

$$\frac{\partial^2 \phi}{\partial x^2} + \frac{\partial^2 \phi}{\partial y^2} + \frac{\partial^2 \phi}{\partial z^2} = \frac{1}{c^2}\frac{\partial^2 \phi}{\partial t^2}$$

where c is the velocity of light and ϕ is a function of space co-ordinates x, y, z and time t.

According to de Broglie's concept of matter wave, a moving particle is associated with a wave.

Schrödinger introduced a mathematical function ψ associated with a moving particle. This function, called *wave function,* is a complex function of the space co-ordinates of the particle and time.

In analogy with optics, the differential equation of wave motion of a particle is written as :

$$\frac{\partial^2 \Psi}{\partial x^2} + \frac{\partial^2 \Psi}{\partial y^2} + \frac{\partial^2 \Psi}{\partial z^2} = \frac{1}{v^2} \frac{\partial^2 \Psi}{\partial t^2} \qquad \ldots(1)$$

where v is the velocity of the particle.

The solution of equation (1) is given by

$$\Psi(\vec{r}, t) = \Psi_0(\vec{r})\, e^{-i\omega t} \qquad \ldots(2)$$

where Ψ_0 is the amplitude at the point under consideration. It is a function of position co-ordinates and not time.

Differentiating equation (2) with respect to t, we get

$$\frac{\partial \Psi}{\partial t} = -i\omega\, \Psi_0(\vec{r})\, e^{-i\omega t}$$

Differentiating again,

$$\frac{\partial^2 \Psi}{\partial t^2} = (-i\omega)(-i\omega)\, \Psi_0(\vec{r})\, e^{-i\omega t}$$

or

$$\frac{\partial^2 \Psi}{\partial t^2} = -\omega^2\, \Psi_0(\vec{r})\, e^{-i\omega t} \quad \text{or} \quad \frac{\partial^2 \Psi}{\partial t^2} = -\omega^2\, \Psi(\vec{r}, t)$$

Putting this value in equation (1), we get

$$\frac{\partial^2 \Psi}{\partial x^2} + \frac{\partial^2 \Psi}{\partial y^2} + \frac{\partial^2 \Psi}{\partial z^2} = -\frac{\omega^2}{v^2} \Psi \qquad \ldots(3)$$

But

$$\omega = 2\pi\nu = \frac{2\pi v}{\lambda} \quad \text{or} \quad \frac{\omega}{v} = \frac{2\pi}{\lambda} \qquad \ldots(4)$$

Also,

$$\nabla^2 \Psi = \frac{\partial^2 \Psi}{\partial x^2} + \frac{\partial^2 \Psi}{\partial y^2} + \frac{\partial^2 \Psi}{\partial z^2} \qquad \ldots(5)$$

Combining (4) and (5) with (3), we get

$$\nabla^2 \Psi = -\frac{4\pi^2}{\lambda^2} \Psi$$

or

$$\nabla^2 \Psi + \frac{4\pi^2}{\lambda^2} \Psi = 0 \qquad \ldots(6)$$

Let us now introduce the wave-mechanical concept.

Using de Broglie wave equation $\lambda = \frac{h}{mv}$, we get

$$\nabla^2 \Psi + \frac{4\pi^2}{h^2/m^2 v^2} \Psi = 0$$

or

$$\nabla^2 \Psi + \frac{4\pi^2 m^2 v^2}{h^2} \Psi = 0 \qquad \ldots(7)$$

Let E and V be the total energy and potential energy respectively.

Then, kinetic energy, $\frac{1}{2} mv^2 = \mathrm{E} - \mathrm{V}$

or $$mv^2 = 2(E - V)$$

or $$m^2v^2 = 2m\,(E - V)$$

From equation (7), $\nabla^2\Psi + \dfrac{8\pi^2 m}{h^2}(E - V)\,\Psi = 0$

or $$\nabla^2\Psi + \frac{2m}{\left(\frac{h}{2\pi}\right)^2}(E - V)\,\Psi = 0$$

But $$\hbar = \frac{h}{2\pi}$$

$$\therefore \quad \nabla^2\Psi + \frac{2m}{\hbar^2}(E - V)\,\Psi = 0 \qquad \ldots(8)$$

This equation is called Schrödinger's time-independent wave equation or steady-state wave equation. This equation is used in explaining stationary phenomena.

Special case of free particle

For the case of a free particle, V = 0

Putting V = 0 in equation (8), we get

$$\nabla^2\Psi + \frac{2m}{\hbar^2}E\,\Psi = 0$$

1.40 TIME-DEPENDENT SCHRÖDINGER EQUATION

A plane wave travelling in the x-direction is represented by :

$$\xi = A\,e^{i(\omega t - kx)} \qquad \ldots(1)$$

where $\omega\,(= 2\pi\nu)$ is the angular frequency and $k\left(= \dfrac{2\pi}{\lambda}\right)$ is the propagation constant.

Now, $$E = h\nu \quad \text{or} \quad \nu = \frac{E}{h}\ ;\ \omega t = 2\pi\nu t = 2\pi\frac{E}{h}t = \frac{Et}{\frac{h}{2\pi}} = \frac{Et}{\hbar}$$

Again, $$p_x = \frac{h}{\lambda} \quad \text{or} \quad \frac{1}{\lambda} = \frac{p_x}{h}\ ;\ kx = \frac{2\pi}{\lambda}x = 2\pi\frac{p_x x}{h} = \frac{p_x}{\frac{h}{2\pi}}x = \frac{p_x}{\hbar}x$$

From equation (1) $\quad \xi = Ae^{i\left(\frac{E}{\hbar}t - \frac{p_x}{\hbar}x\right)} \quad$ or $\quad \xi = Ae^{\frac{i}{\hbar}(Et - p_x x)}$.

If we consider analogy between electron beams and photons, we find that the electrons in a beam travelling in the x-direction are controlled by pilot waves with a wave function ψ. This wave function plays the same role for electrons and other microscopic particles as ξ for sound waves or E for electromagnetic waves.

$\Psi(x, t)$, amplitude of wave at a point x at time t, is given by

$$\Psi(x, t) = Ae^{\frac{i}{\hbar}(Et - xp_x)}$$

In wave mechanics, we use the exponential form with a negative sign for a wave moving along the positive direction of x-axis.

$$\therefore \quad \Psi(x, t) = Ae^{-\frac{i}{\hbar}(Et - xp_x)} = Ae^{\frac{i}{\hbar}(xp_x - Et)} \qquad \ldots(2)$$

Differentiating equation (2) with respect to time t, we get

$$\frac{\partial \Psi}{\partial t} = -\frac{i}{\hbar} \mathrm{E}\, \mathrm{A} e^{\frac{i}{\hbar}(xp_x - \mathrm{E}t)}$$

$$\frac{\partial \Psi}{\partial t} = -\frac{i}{\hbar} \mathrm{E}\, \Psi \qquad \text{or} \qquad \mathrm{E}\Psi = -\frac{\hbar}{i}\frac{\partial \Psi}{\partial t} \qquad \ldots(3)$$

Differentiating equation (2) w.r.t. 'x', we get

$$\frac{\partial \Psi}{\partial x} = \frac{i}{\hbar} p_x \mathrm{A}\, e^{\frac{i}{\hbar}(xp_x - \mathrm{E}t)}$$

Differentiating again, $\dfrac{\partial^2 \Psi}{\partial x^2} = \left(\dfrac{i}{\hbar} p_x\right) \mathrm{A} \left(\dfrac{i}{\hbar} p_x\right) e^{\frac{i}{\hbar}(xp_x - \mathrm{E}t)}$

or
$$\frac{\partial^2 \Psi}{\partial x^2} = -\frac{p_x^{\,2}}{\hbar^2} \mathrm{A}\, e^{\frac{i}{\hbar}(xp_x - \mathrm{E}t)} \qquad \text{or} \qquad \frac{\partial^2 \Psi}{\partial x^2} = -\frac{p_x^{\,2}}{\hbar^2} \Psi$$

or
$$p_x^{\,2}\, \Psi = -\hbar^2 \frac{\partial^2 \Psi}{\partial x^2} \qquad \ldots(4)$$

Let us consider a particle of mass m moving with non-relativistic velocity v piloted by the wave function Ψ. The total energy E of the particle is the sum of kinetic energy $\frac{1}{2}mv^2$ and potential energy V of the particle

$$\mathrm{E} = \frac{1}{2}mv^2 + \mathrm{V}$$

But
$$\frac{1}{2}mv^2 = \frac{1}{2m}m^2v^2 = \frac{p_x^{\,2}}{2m}$$

where p_x is the momentum of the particle.

$$\therefore \qquad \mathrm{E} = \frac{p_x^{\,2}}{2m} + \mathrm{V}$$

Multiplying both sides by Ψ, we get

$$\mathrm{E}\,\Psi = \frac{p_x^{\,2}}{2m}\Psi + \mathrm{V}\,\Psi$$

Putting the value of E Ψ from equation(3), we get

$$-\frac{\hbar}{i}\frac{\partial \Psi}{\partial t} = \frac{p_x^{\,2}}{2m}\Psi + \mathrm{V}\,\Psi$$

Putting the value of $p_x^{\,2}\,\Psi$ from equation (4), we get

$$-\frac{\hbar}{i}\frac{\partial \Psi}{\partial t} = -\frac{\hbar^2}{2m}\frac{\partial^2 \Psi}{\partial x^2} + \mathrm{V}\Psi$$

or
$$-\frac{\hbar^2}{2m}\frac{\partial^2 \Psi}{\partial x^2} + \mathrm{V}\,\Psi = -\frac{\hbar}{i}\frac{\partial \Psi}{\partial t} \qquad \text{or} \qquad -\frac{\hbar^2}{2m}\frac{\partial^2 \Psi}{\partial x^2} + \mathrm{V}\,\Psi = i\,\hbar\frac{\partial \Psi}{\partial t} \qquad \ldots(5)$$

Since Ψ is a function of x and t therefore the above equation may be written as :

$$-\frac{\hbar^2}{2m}\frac{\partial^2}{\partial x^2}\Psi(x,t) + \mathrm{V}\,\Psi(x,t) = i\,\hbar\frac{\partial}{\partial t}\Psi\,(x,t)$$

This equation is Schrödinger's one-dimensional wave equation for waves associated with a particle moving in a potential field V. This is a time-dependent wave equation.

Generalising this equation for the motion of a particle in three-dimensions,

$$-\frac{\hbar^2}{2m}\left[\frac{\partial^2\Psi}{\partial x^2}+\frac{\partial^2\Psi}{\partial y^2}+\frac{\partial^2\Psi}{\partial z^2}\right]+\mathrm{V}\,\Psi=i\,\hbar\,\frac{\partial\Psi}{\partial t}$$

or

$$-\frac{\hbar^2}{2m}\nabla^2\,\Psi(\vec{r},t)+\mathrm{V}\,\Psi(\vec{r},t)=i\,\hbar\,\frac{\partial\Psi(\vec{r},t)}{\partial t} \quad \text{...(6)}$$

This equation is Schrödinger's time-dependent three-dimensional quantum-mechanical wave equation. This equation is for waves associated with a particle moving in a potential field V. This equation is used only for non-relativistic problems. When particle moves with a speed comparable to that of light, a more elaborate formulation is required.

Case of free particle

When no force acts on a particle, the particle is said to be free.

The potential energy V of a free particle is zero. The total energy E of the free particle is only kinetic.

Putting V = 0 in equation (5), we have

$$-\frac{\hbar^2}{2m}\frac{\partial^2\Psi}{\partial x^2}=i\,\hbar\,\frac{\partial\Psi}{\partial t}$$

This equation is Schrödinger's one-dimensional time-dependent wave equation for a free particle.

For the three-dimensional case,

$$-\frac{\hbar^2}{2m}\left[\frac{\partial^2\Psi}{\partial x^2}+\frac{\partial^2\Psi}{\partial y^2}+\frac{\partial^2\Psi}{\partial z^2}\right]=i\,\hbar\,\frac{\partial\Psi}{\partial t} \quad \text{or} \quad -\frac{\hbar^2}{2m}\nabla^2\,\Psi(\vec{r},t)=i\,\hbar\,\frac{\partial\Psi(\vec{r},t)}{\partial t}$$

or

$$-\frac{\hbar^2}{2m}\nabla^2\,\Psi=i\,\hbar\,\frac{\partial\Psi}{\partial t}$$

This equation is Schrödinger's time-dependent three-dimensional wave equation for a moving free particle.

1.41 DERIVATION OF TIME-INDEPENDENT SCHRÖDINGER WAVE EQUATION FROM TIME-DEPENDENT SCHRÖDINGER EQUATION

The three-dimensional time-dependent Schrödinger's wave equation is given by

$$-\frac{\hbar^2}{2m}\nabla^2\,\Psi(\vec{r},t)+\mathrm{V}\,\Psi(\vec{r},t)=i\,\hbar\,\frac{\partial\Psi(\vec{r},t)}{\partial t}$$

The one-dimensional Schrödinger wave equation is

$$-\frac{\hbar^2}{2m}\frac{\partial^2\Psi(x,t)}{\partial x^2}+\mathrm{V}\,\Psi(x,t)=i\,\hbar\,\frac{\partial\Psi(x,t)}{\partial t}$$

In a simple form, $-\dfrac{\hbar^2}{2m}\dfrac{\partial^2\Psi}{\partial x^2}+\mathrm{V}\,\Psi=i\,\hbar\,\dfrac{\partial\Psi}{\partial t}$

If the potential energy V does not depend explicitly on time and is a function of x only and the total energy E is constant, then the wave function $\Psi(x,t)$ can be expressed as the product of two separate functions $\psi(x)$ and $f(t)$.

$$\Psi(x, t) = \psi(x) f(t)$$

This could be briefly expressed as :

$$\Psi(x, t) = \psi f$$

Now $\qquad \dfrac{\partial \Psi}{\partial x} = f \dfrac{\partial \psi}{\partial x}$ and $\dfrac{\partial^2 \Psi}{\partial x^2} = f \dfrac{\partial^2 \psi}{\partial x^2}$

Also, $\qquad \dfrac{\partial \Psi}{\partial t} = \psi \dfrac{\partial f}{\partial t}$

As established earlier, $\quad \Psi = A\, e^{\frac{i}{\hbar}(xp_x - Et)}$

$$\frac{\partial \Psi}{\partial t} = -\frac{i}{\hbar} EA\, e^{\frac{i}{\hbar}(xp_x - Et)} \quad \text{or} \quad \frac{\partial \Psi}{\partial t} = -\frac{i}{\hbar} E\Psi$$

or
$$\frac{\partial \Psi}{\partial t} = -\frac{i}{\hbar} E\psi f$$

Substituting values in $-\dfrac{\hbar^2}{2m}\dfrac{\partial^2 \Psi}{\partial x^2} + V\,\Psi = i\,\hbar \dfrac{\partial \Psi}{\partial t}$, we get

$$-\frac{\hbar^2}{2m} f \frac{\partial^2 \psi}{\partial x^2} + V f \psi = i\,\hbar\left(-\frac{i}{\hbar} E\,\psi f\right)$$

or
$$-\frac{\hbar^2}{2m}\frac{\partial^2 \psi}{\partial x^2} + V\,\psi = E\,\psi \quad \text{or} \quad \frac{\hbar^2}{2m}\frac{\partial^2 \psi}{\partial x^2} - V\,\psi = -E\,\psi$$

or
$$\frac{\hbar^2}{2m}\frac{\partial^2 \psi}{\partial x^2} + (E - V)\,\psi = 0 \quad \text{or} \quad \frac{\partial^2 \psi}{\partial x^2} + \frac{2m}{\hbar^2}(E - V)\,\psi = 0$$

Since ψ is a function of only x therefore this equation may be stated explicitly as :

$$\frac{\partial^2 \psi(x)}{\partial x^2} + \frac{2m}{\hbar^2}(E - V)\,\psi(x) = 0$$

This is Schrödinger's one-dimensional time-independent equation.

The time-independent or stationary state form of Schrödinger's quantum-mechanical wave equation for the motion of a particle in three dimensions is given by

$$\left(\frac{\partial^2 \psi}{\partial x^2} + \frac{\partial^2 \psi}{\partial y^2} + \frac{\partial^2 \psi}{\partial z^2}\right) + \frac{2m}{\hbar^2}(E - V)\,\psi = 0$$

or
$$\nabla^2 \psi + \frac{2m}{\hbar^2}(E - V)\,\psi = 0.$$

In three-dimensions, ψ is a function of x, y, z or $\vec{r}$. So, the above equation can be expressed more explicitly as follows.

$$\nabla^2 \psi(\vec{r}) + \frac{2m}{\hbar^2}(E - V)\,\psi(\vec{r}) = 0$$

1.42 TIME-DEPENDENT PART OF WAVE FUNCTION Ψ

We know that $\qquad \dfrac{\partial \Psi}{\partial t} = -\dfrac{i}{\hbar} E\,\psi f$

$$\frac{\partial}{\partial t}(\psi f) = -\frac{i}{\hbar}\text{E}\psi f \quad \text{or} \quad \psi\frac{\partial f}{\partial t} = -\frac{i}{\hbar}\text{E}\psi f$$

or
$$\frac{\partial f}{\partial t} = -\frac{i}{\hbar}\text{E}f \quad \text{or} \quad -\frac{\hbar}{i}\frac{\partial f}{\partial t} = \text{E}f$$

or
$$i\hbar\frac{\partial f}{\partial t} - \text{E}f = 0$$

Since f is a function of only t therefore partial derivatives can be replaced by total derivatives.

$$\therefore \quad i\hbar\frac{\partial f}{\partial t} - \text{E}f = 0$$

More explicitly, $i\hbar\dfrac{\partial f(t)}{\partial t} - \text{E}f(t) = 0$

This is the equation for the time-dependent part of the wave function ψ.

1.43 EIGEN VALUES AND EIGEN FUNCTIONS

The values of energy E_n for which Schrödinger steady-state equation can be solved are called eigen values. The corresponding wave functions ψ_n are called eigen-functions. These terms come from the German *Eigenwert* meaning "proper or characteristic value" and *Eigen function*, "proper or characteristic function".

The discrete energy levels of the hydrogen atom given by

$$\text{E}_n = -\frac{me^4}{32\pi^2 \epsilon_0^2 \hbar^2}\left(\frac{1}{n^2}\right), \; n = 1, 2, 3, \ldots\ldots$$

are an example of a set of eigen values. It is interesting to mention here that these particular values of E are the only ones that yield acceptable wave functions for the electron in hydrogen atom.

An important example of a dynamical variable other than total energy that is found to be quantised in stable systems is angular momentum L. In the case of the hydrogen atom, we shall find that the eigen values of the magnitude of the total angular momentum are specified by

$$\text{L} = \sqrt{l(l+1)}\hbar, \; l = 0, 1, 2, \ldots\ldots (n-1)$$

Let us now discuss the non-relativistic case of a particle of mass m moving with velocity v. In this case, the total energy is given by

$$\text{E} = \frac{1}{2}mv^2 + \text{V}$$

where $\frac{1}{2}mv^2$ is the kinetic energy and V is the potential energy of the particle.

But
$$\frac{mv^2}{2} = \frac{m^2v^2}{2m} = \frac{p_x^{\,2}}{2m}$$

$$\therefore \quad \text{E} = \frac{p_x^{\,2}}{2m} + \text{V} \quad \text{or} \quad \text{E} - \text{V} = \frac{p_x^{\,2}}{2m}$$

Replacing p_x by p, we get

$$\text{E} - \text{V} = \frac{p^2}{2m}$$

Now, the equation $\frac{\partial^2 \psi}{\partial x^2} + \frac{2m}{\hbar^2}(E - V)\,\psi = 0$

becomes $$\frac{\partial^2 \psi}{\partial x^2} + \frac{2m}{\hbar^2}\frac{p^2}{2m}\psi = 0 \quad \text{or} \quad \frac{\partial^2 \psi}{\partial x^2} + \frac{p^2}{\hbar^2}\psi = 0$$

We shall get several solutions of the above equation for different values of total energy and momentum. The wave function ψ_1 is obtained when momentum is p_1 and total energy is E_1. The wave function ψ_2 is obtained when momentum is p_2 and total energy is E_2. Similarly, wave function ψ_n is obtained when momentum is p_n and total energy is E_n. The wave functions $\psi_1, \psi_2, \ldots\ldots,$ ψ_n are called eigen functions for eigen values $p_1, p_2, \ldots\ldots, p_n$ of momentum and for eigen values E_1, $E_2, \ldots\ldots, E_n$ of total energy.

1.44 WHY SCHRÖDINGER'S WAVE EQUATION IS NOT VALID FOR RELATIVISTIC PARTICLES ?

In deriving Schrödinger's wave equation, we used non-relativistic (classical) expression for total energy $E = \frac{p^2}{2m} + V$. Also, we used non-relativistic formula (mv) for the momentum of the particle. The formula $\left(\frac{1}{2}mv^2\right)$ used for kinetic energy is also a non-relativistic formula.

1.45 FREE PARTICLE IN A ONE-DIMENSIONAL BOX (AN APPLICATION OF QUANTUM MECHANICS)

Consider a particle of rest mass m confined to move in a rectangular box under the following conditions :

(i) The walls of the box are rigid, hard and elastic. When the particle strikes the wall, it rebounds with the same kinetic energy. So, its total energy E remains constant.

(ii) The walls of the box are non-penetrable. This implies that the potential energy is so large outside the box that the total energy of the particle can never exceed it. The particle, therefore, always remains within the box. There is just no chance for the particle to be found outside the box.

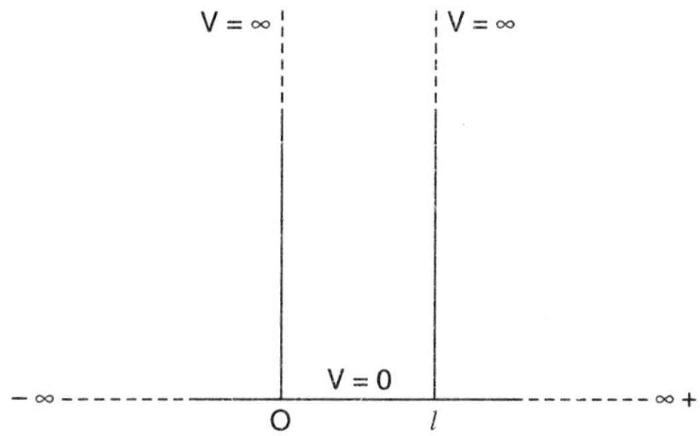

Fig. 1.16

(iii) The box is one-dimensional. The particle moves only along a straight line, say x-axis. The length of the box is l.

(iv) The particle is free. The potential energy V of the particle inside the box is zero. Since the potential energy is independent of time therefore Schrödinger's time independent wave equation is used to study the state of system. Fig. 1.16 shows the potential energy at various points. Because of its appearance, it is popularly known as infinite rectangular or square potential well.

Schrödinger's Wave Equation. Schrödinger's one-dimensional, time independent, quantum mechanical wave equation is :

$$\frac{\partial^2\psi(x)}{\partial x^2}+\frac{2m}{\hbar^2}(E-V)\,\psi(x)=0$$

For the case of a free particle, V = 0. So, the Eigen value equation for a free particle in a one-dimensional box is given by

$$\frac{\partial^2\psi(x)}{\partial x^2}+\frac{2m}{\hbar^2}E\,\psi(x)=0. \qquad ...(i)$$

This is valid for non-relativistic conditions.

Solution. Since the free particle remains within the box of length l therefore $\psi(x)$ has a finite value *i.e.*, $\psi(x) \neq 0$ for values of x lying between 0 and l.

Mathematically, $\psi(x) \neq 0$ for $0 < x < l$

This could also be written as under :

$$\psi(x) = 0 \text{ for } x < 0 \text{ and for } x > l.$$

Since the potential energy of the free particle is zero therefore the total energy E is equal to the kinetic energy and is always positive.

Now
$$E=\frac{1}{2}mv^2=\frac{m^2v^2}{2m}=\frac{p_x^{\;2}}{2m}$$

∴
$$\frac{2m}{\hbar^2}E=\frac{p_x^{\;2}}{\hbar^2}$$

But
$$\frac{p_x}{\hbar}=\frac{p_x\,.2\pi}{h}=\frac{2\pi}{h/p_x}=\frac{2\pi}{\lambda}=k$$

where k is the propagation constant.

∴
$$\frac{2m}{\hbar^2}E=k^2$$

Putting the value of $\frac{2m}{\hbar^2}E$ in equation (*i*), we get

$$\frac{\partial^2\psi(x)}{\partial x^2}+k^2\psi(x)=0 \qquad ...(ii)$$

The general solution of equation (*ii*) is :

$$\psi(x) = A\sin kx + B\cos kx \qquad ...(iii)$$

where A and B are arbitrary constants.

Let us now prove that $\psi(x) = A\sin kx + B\cos kx$ is the correct solution of the equation $\frac{\partial^2\psi(x)}{\partial x^2}+k^2\psi(x)=0$.

Differentiating equation (*iii*) w.r.t. x, we get

$$\frac{\partial\psi(x)}{\partial x}=kA\cos kx-kB\sin kx$$

Differentiating again, we get

$$\frac{\partial^2\psi(x)}{\partial x^2}=-k^2A\sin kx-k^2B\cos kx$$

$$=-k^2(A\sin kx+B\cos kx)=-k^2\psi(x)$$

or
$$\frac{\partial^2\psi(x)}{\partial x^2}+k^2\psi(x)=0$$

Therefore equation (*iii*) is the solution of the differential equation (*ii*).

Let us now determine the values of A and B by applying boundary conditions.

For continuity of the wave function at the boundary, we have

$$\psi(x) = 0 \text{ when } x = 0 \text{ and when } x = l$$

substituting $\psi(x) = 0$ when $x = 0$ in equation (*iii*), we get

$$B = 0$$

Substituting the value of B in equation (*iii*), we get

$$\psi(x) = A \sin kx \qquad ...(iv)$$

Substituting $\psi(x) = 0$ when $x = l$ in (*iv*), we get

$$A \sin kl = 0$$

$\therefore \quad kl = n\pi$ where $n = 1, 2, 3, ...$

or
$$k = \frac{n\pi}{l} \qquad ...(v)$$

Zero and negative values of n are not permissible. If we take $n = 0$, this would mean that the momentum and hence the kinetic energy of the particle are zero. This would further mean that the particle is at rest at all the times. This is not physically acceptable. For $k = 0$, $\psi(x) = 0$ everywhere in the box. This would mean that the probability density $|\psi|^2 = 0$ *i.e.*, the particle is not in the box. The negative values of n repeat the corresponding values for positive n. So, they do not give independent solutions.

$$\therefore \quad \psi(x) = A \sin \frac{n\pi}{l} x \qquad ...(vi)$$

Let us now *normalise* the Eigen function $\psi(x)$ **for finding the value of A.** The probability of finding the particle anywhere inside the box is given by $\psi^*(x)\, \psi(x)\, dx = |\psi(x)|^2\, dx$. Its value is one. This is because there is only one particle and there is **no** chance of its being located outside the box.

$$\therefore \quad \int_0^l |\psi(x)|^2\, dx = 1 \quad \text{or} \quad A^2 \int_0^l \sin^2 \frac{n\pi x}{l} dx = 1$$

or
$$\frac{A^2}{2} \int_0^l 2\sin^2 \frac{n\pi x}{l}\, dx = 1$$

But
$$\int_0^l 2\sin^2 \frac{n\pi x}{l}\, dx = \int_0^l \left(1 - \frac{\cos 2\pi n x}{l}\right) dx = l$$

$$\therefore \quad \frac{A^2}{2} l = 1 \text{ or } A = \left(\frac{2}{l}\right)^{1/2}$$

From equation (*vi*),
$$\psi(x) = \left(\frac{2}{l}\right)^{1/2} \sin \frac{n\pi}{l} x \qquad ...(vii)$$

Eq. (*vii*) is the normalised wave function for a particle inside a box of finite width l.

From the wave equation (*vii*), the various **eigen functions** are

$$\psi_1 = \left(\frac{2}{l}\right)^{1/2} \sin \frac{\pi}{l} x,\ \psi_2 = \left(\frac{2}{l}\right)^{1/2} \sin \frac{2\pi}{l} x,\ ......\ \psi_n = \left(\frac{2}{l}\right)^{1/2} \sin \frac{n\pi}{l} x$$

with $p_1, p_2, ... p_n$ as eigen values of momentum, $k_1, k_2, ... k_n$ as eigen values of propagation constant and $E_1, E_2, ... E_n$ as eigen values for the total energy of the particle.

Let us now find the **eigen values of the propagation constant *k*.**

We know that $k = \frac{n\pi}{l}$ [Refer to equation (*v*)]

$\therefore$ Eigen values of k are $k_1 = \frac{\pi}{l}, k_2 = \frac{2\pi}{l} \ldots\ldots, k_n = \frac{n\pi}{l}$ ...(*viii*)

Let us now find the **eigen values of momentum.**

$$k = \frac{2\pi}{\lambda} = \frac{2\pi p}{h} = \frac{p}{\hbar} \quad \text{or} \quad p = \hbar k = \frac{n\pi\hbar}{l}$$

$\therefore$ The eigen values of the momentum p of the particle are :

$$p_1 = \frac{\pi\hbar}{l}, p_2 = \frac{2\pi\hbar}{l} \ldots\ldots, p_n = \frac{n\pi\hbar}{l} \qquad \text{...}(ix)$$

Let us now find the **eigen values of the energy** E of the particle.

If we consider the energy of the particle non-relativistically, then

$$E = \frac{p^2}{2m} = \frac{n^2\pi^2\hbar^2}{2ml^2}$$

So, the eigen values of the energy E of the particle are :

$$E_1 = \frac{\pi^2\hbar^2}{2ml^2}, E_2 = \frac{2^2\pi^2\hbar^2}{2ml^2} \ldots\ldots, E_n = \frac{n^2\pi^2\hbar^2}{2ml^2} \qquad \text{...}(x)$$

The results based on quantum-mechanical considerations **differ** from classical results in following respects.

(1) According to classical considerations, the free particle in a box can have any value of energy from zero to infinity. However according to quantum mechanics, only discrete values of energy (and hence momentum) are permissible *i.e.,* momentum and energy are quantised.

Quantisation of momentum and energy

The momentum of the particle can only have values which are integral multiples of

$$\frac{\pi\hbar}{l} = \frac{\pi h}{2\pi l} = \frac{h}{2l} \quad \text{or} \quad p_n = \frac{nh}{2l}$$

and no other values. So, the linear momentum of the free particle is quantised. The only values it can have are multiples of $\frac{h}{2l}$. We also find that the quantum of momentum does not depend upon the mass of the particle. In a box of length l, the momentum of any particle (proton, neutron etc.) will be integral multiple of $\frac{h}{2l}$. Thus, we conclude that the momentum of a particle inside the box is

(*i*) directly proportional to the number n,

(*ii*) inversely proportional to the length l of the box and

(*iii*) independent of the mass m of the particle.

Under non-relativistic conditions, the energy of the particle can *only* have values given by

$$E_n = \frac{n^2\pi^2\hbar^2}{2ml^2} \quad \text{where } n = 1, 2, 3$$

Clearly, the energy can have only discrete values. So, the energy of the free particle is quantised. The only values it can have is n^2 times $\frac{\pi^2\hbar^2}{2ml^2}$.

Thus, we conclude that the energy of a particle inside the box is

(*i*) directly proportional to the square of natural number *i.e.*, n^2,

(*ii*) inversely proportional to the square of the length of the box *i.e.*, $\frac{1}{l^2}$ and

(*iii*) inversely proportional to the mass m of the particle.

(2) According to quantum mechanics, the minimum energy that the particle can have is $E_1 = \frac{\pi^2\hbar^2}{2ml^2}$ and not zero, as required classically.

(3) Classically, the particle is equally likely to be found at any point within the box from 0 to l. But according to quantum mechanics, the probability of finding the particle is maximum at the *antinodes* and minimum at the *nodes*.

1.46 ENERGY LEVELS FOR A PARTICLE INSIDE A BOX OF FINITE WIDTH l

The normalised wave function for a particle in a box of finite width l is given by

$$\psi(x) = \left(\frac{2}{l}\right)^{1/2} \sin\frac{n\pi}{l}x$$

Here,
$$\frac{n\pi}{l} = k$$

where k is the propagation constant.

Momentum,
$$p = \frac{h}{\lambda} = \frac{h}{2\pi}\frac{2\pi}{\lambda} = \hbar k$$

$\therefore$
$$p = \frac{n\hbar\pi}{l}$$

Also kinetic energy
$$E = \frac{1}{2}mv^2 = \frac{m^2v^2}{2m} = \frac{p^2}{2m} = \frac{n^2\pi^2\hbar^2}{2ml^2}$$

Conclusion. For a free particle in a one-dimensional box *i.e.*, an infinite rectangular or square potential well, the energy of the particle can have only values given by

$$E_n = \frac{n^2\pi^2\hbar^2}{2ml^2} \text{ where } n = 1, 2, 3, ...$$

These discrete values of energy are known as energy states of *energy levels.*

Ground State

The minimum value of energy is for $n = 1$. It is given by

$$E_1 = \frac{\pi^2\hbar^2}{2ml^2}$$

This energy is called the *ground state energy* of the free particle. The state of the particle for $n = 1$ is called the *ground state.*

1.47 EXCITED STATES

The states of the particle for $n = 2, 3, 4$ etc. are called ***the excited states.*** For $n = 2$, we get the *first* excited state with energy $E_2 = 4\frac{\pi^2\hbar^2}{2ml^2} = 4E_1$.

For $n = 3$, we get the *second* excited state with energy $E_3 = 9\dfrac{\pi^2\hbar^2}{2\,ml^2} = 9E_1$ and so on.

1.48 ENERGY LEVEL DIAGRAM

From the above discussion we find that the particle can have only *discrete* values of energy given by E_1, E_2, E_3 .... *i.e.*, the energy of the particle is quantised. These discrete allowed values of energy E_1, E_2, E_3 .... etc. are called *energy levels* and are shown in Fig. 1.17.

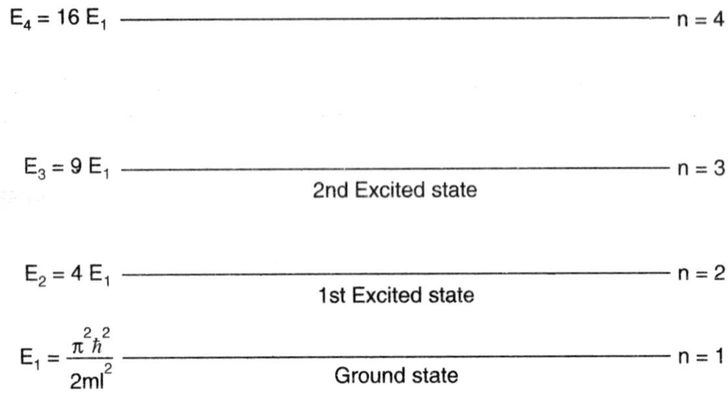

Fig. 1.17

1.49 PROBABILITY DENSITY OF A PARTICLE CONFINED TO ONE-DIMENSIONAL BOX

The probability of finding the particle in the one-dimensional space between the points x and $x + dx$ is given by $\psi\psi^*\,dx$. Here ψ is the wave function and ψ^* its complex conjugate.

The probability of finding the particle in a finite interval between x_1 and x_2 is given by

$$P = \int_{x_1}^{x_2} \psi\,\psi^*\,dx$$

In the case of a free particle in one-dimensional box, the value of normalised wave function $\psi(x)$ is given by the relation

$$\psi_n(x) = \left(\frac{2}{l}\right)^{1/2} \sin\frac{n\pi}{l}x$$

where l is the length of the box.

The probability of finding a particle in a distance dx about the mean position x is

$$P_n\,dx = \psi_n\,\psi_n^*.\,dx = \left(\frac{2}{l}\right)^{\frac{1}{2}} \sin\frac{n\pi x}{l}\left(\frac{2}{l}\right)^{1/2} \sin\frac{n\pi x}{l}\,dx$$

$$= \frac{2}{l}\sin^2\frac{n\pi}{l}x\,dx$$

Probability density $\quad P_n = \psi_n\psi_n^* = |\psi_n|^2 = \dfrac{2}{l}\sin^2\dfrac{n\pi}{l}x$

1.50 GRAPHS

For $n = 1, 2, 3$ and 4, the normalised wave functions ψ_1, ψ_2, ψ_3 and ψ_4 and corresponding probability densities $|\psi_1^2|$, $|\psi_2^2|$, $|\psi_3^2|$ and $|\psi_4^2|$ are plotted as shown in Fig. 1.18 (*a*) and (*b*). It is clear from the figure that ψ_n has *positive* as well as *negative* values whereas $|\psi_n^2|$ is always *positive.*

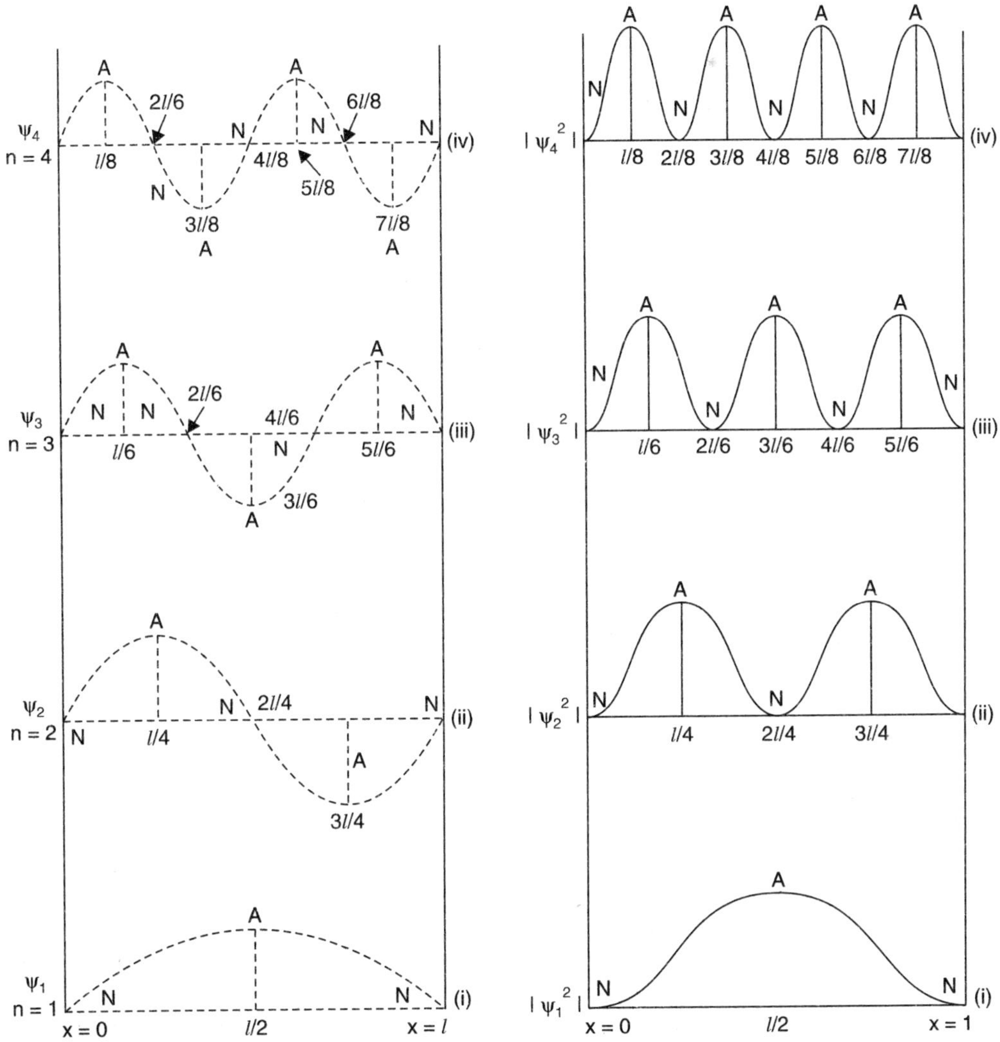

Normalised wave function
(a)

Probability density
(b)

Fig. 1.18

In the figure, A and N represent antinodes and nodes respectively.

1.51 NODES AND ANTINODES

The position of minimum (or zero) displacement is called a **node**. An **antinode** is the position of maximum displacement. A node corresponds to $\psi_n(x) = 0$.

For the nth quantum state, the nodes occur where

$$\sin \frac{n\pi x}{l} = 0 \text{ or } \frac{n\pi x}{l} = m\pi$$

where $m = 0, 1, 2, \ldots n$

$$\therefore \qquad x = m\frac{l}{n}$$

Clearly, the nodes are located at, $0, \frac{l}{n}, \frac{2l}{n} \ldots, \frac{nl}{n}$

An antinode corresponds to the maximum value of $\psi_n(x)$. Since the maximum value of $\sin \frac{n\pi}{l}x = 1$, therefore for an antinode

$$\psi_n(x) = \left(\frac{2}{l}\right)^{1/2}$$

For the nth quantum state, the antinodes occur where

$$\sin \frac{n\pi x}{l} = 1 \text{ or } \frac{n\pi x}{l} = (2m + 1)\frac{\pi}{2}$$

$$\therefore \qquad x = \frac{2m+1}{2n} l$$

Clearly, the antinodes are located at $x = \frac{l}{2n}, \frac{3l}{2n} \ldots \frac{(2n-1)}{2n} l$.

When $n = 1$, nodes occur at $x = 0$ and $x = l$. Antinodes occur at $x = \frac{l}{2}$.

When $n = 2$, nodes occur at $x = 0$; $x = \frac{l}{2}$; and $x = l$. Antinodes occur at $x = \frac{l}{4}$; $x = \frac{3l}{4}$

When $n = 3$, nodes occur at $x = 0$; $x = \frac{l}{3}$; $x = \frac{2l}{3}$; and $x = l$. Antinodes occur at $x = \frac{l}{6}$; $x = \frac{3l}{6}$; $x = \frac{5l}{6}$.

When $n = 4$, nodes occur at $x = 0$; $x = \frac{l}{4}$; $x = \frac{2l}{4}$; $x = \frac{3l}{4}$; and $x = l$. Antinodes occur at $x = \frac{l}{8}$; $x = \frac{3l}{8}$; $x = \frac{5l}{8}$; $x = \frac{7l}{8}$

In terms of probability density, an *antinode* is a point of *maximum* probability density and a *node* is a point of *zero* probability density. We can therefore conclude that for the nth quantum state, there are n *antinodes i.e.,* there are n equally probable positions in the box at which the particle is most likely to be found. Similarly, there are $n - 1$ nodes, each node lying in between two antinodes *i.e.,* there are $n - 1$ equally spaced positions where the probability of finding the particle is zero. The two positions $x = 0$ and $x = l$ are permanent positions of nodes.

In the ground state, $n = 1$. The position of maximum probability is given by $x = \frac{l}{2}$.

For the first excited state, $n = 2$. The probability is maximum for $x = \frac{l}{4}$ and $x = \frac{3l}{4}$ but the probability for $x = \frac{l}{2}$ is zero.

According to the classical theory, the probability of finding the particle is the same at every point in the box. According to the quantum theory, there is a greater probability of finding the particle at some points than at others.

The probability of locating the particle at $x = 0$ or $x = l$, *i.e.*, near the wall is always zero.

Maxima and Minima. For a particle in an infinite square well of width l, the wave function is given by

$$\psi_n = \left(\frac{2}{l}\right)^{1/2} \sin\left(\frac{2\pi}{l}x\right)$$

The probability density $P_n = \psi_n\psi_n^* = \frac{2}{l}\sin^2\left(\frac{n\pi}{l}x\right)$

At a maxima, the probability density has a maximum value. For maximum value of probability density,

$$\sin^2\left(\frac{n\pi}{l}x\right) = 1 \quad \text{or} \quad \frac{n\pi}{l}x = (2m+1)\frac{\pi}{2}$$

$\therefore$ For a maxima $\quad x = \frac{2m+1}{2n}l$ where $m = 0, 1, 2, \ldots. \frac{2n-1}{2}$

Thus maxima are located at $x = \frac{l}{2n}, \frac{3l}{2n}, \ldots. \left(\frac{2n-1}{2n}\right)l$

At a minima, the probability density has a minimum (or zero) value. For minimum (or zero) value of probability density,

$$\sin^2\left(\frac{n\pi}{l}x\right) = 0 \quad \text{or} \quad \frac{n\pi x}{l} = m\pi$$

$\therefore$ For a minima $\quad x = \frac{ml}{n}$, where $m = 0, 1, 2, \ldots, n$

Thus minima are located at $x = 0, \frac{l}{n}, \frac{2l}{n}, \ldots. \frac{nl}{n}$.

(*i*) **Ground state.** For ground state, $n = 1$.

$\therefore$ Maximum occurs at $\quad x = \frac{l}{2}$.

Minima occur at $\quad x = 0$ and $x = l$.

Thus there is *one* maxima and *two* minima.

(*ii*) **First excited state.** For the first excited state, $n = 2$

$\therefore$ Maxima occurs at $\quad x = \frac{l}{4}$ and $x = \frac{3l}{4}$

Minima occurs at $\quad x = 0$; $x = \frac{l}{2}$ and $x = l$.

Thus there are *two* maxima and *three* minima.

(*iii*) **Second excited state.** For second excited state, $n = 3$

$\therefore$ Maxima occur at $\quad x = \frac{l}{6}; \frac{3l}{6}$ and $x = \frac{5l}{6}$.

Minima occur at $\quad x = 0$; $x = \frac{l}{3}$; $x = \frac{2l}{3}$ and $x = l$

Thus there are *three* maxima and *four* minima.

Note that the number of maxima = n and number of minima is one more *i.e.*, $n + 1$, where n is quantum number. Further $x = 0$ and $x = l$ is always a minima.

1.52 A PARTICLE OF ZERO ENERGY SHALL NOT EXIST IN A BOX–QUANTUM MECHANICAL PROOF

The wave equation for a free particle in a one-dimensional box of length l is given by

$$\psi_n(x) = \left(\frac{2}{l}\right) \sin \frac{n\pi}{l} x \qquad ...(i)$$

The given values of energy are

$$E_n = \frac{n^2\pi^2\hbar^2}{2ml^2} \qquad ...(ii)$$

For $n = 0$, $E_0 = 0$ and $\psi_0(x) = 0$

For $E_0 = 0$, the probability of the particle being within the box is nil. In other words, there can be no particle in the box.

Conclusion. If the particle is in the box, n cannot be zero. In other words, a particle will not exist in a box if its energy is zero.

Zero point energy

The smallest possible value of n in relation (i) and (ii) is **one,** for which

$$\psi_1(x) = \left(\frac{2}{l}\right)^{1/2} \sin \frac{\pi x}{l} \text{ and } E_1 = \frac{\pi^2\hbar^2}{2ml^2}$$

Clearly, the minimum energy that the particle can have is $E_1 = \frac{\pi^2\hbar^2}{2ml^2}$. In other words, even when the temperature of the box is reduced to 0 K, the total energy will still be E_1 and will not be zero. $E_1 = \frac{\pi^2\hbar^2}{2ml^2}$ is called the *zero point energy* of the particle.

The zero point energy is

(i) inversely proportional to m, and

(ii) inversely proportional to the square of the length of the box.

If m is very large, *i.e.*, for a heavy particle or if l is very large *i.e.*, for a box of the laboratory size, zero point energy is *almost* zero. If $l \to \infty$, the particle behaves as a free particle without any box. Such a particle does not possess any zero point energy.

1.53 ORTHOGONAL WAVE FUNCTIONS

The two wave functions $\psi_m(x)$ and $\psi_n(x)$ are said to be orthonormal or orthogonal if

$$\int_{\text{Available region}} \psi_m^*(x)\, \psi_n(x)\, dx = 0$$

1.54 WAVE FUNCTIONS OF TWO DIFFERENT STATES FOR A PARTICLE IN ONE-DIMENSIONAL BOX

Let us now prove that for a particle in one-dimensional box, the wave functions of two different states are orthogonal (orthonormal).

The wave functions for two different states m and n of a free particle in a box are given by

$$\psi_m(x) = \left(\frac{2}{l}\right)^{1/2} \sin\frac{m\pi}{l}x \text{ and } \psi_n(x) = \left(\frac{2}{l}\right)^{1/2} \sin\frac{n\pi}{l}x$$

$$\therefore \quad \int_0^l \psi_m{}^*(x)\,\psi_n(x)\,dx = \frac{2}{l}\int_0^l \sin\frac{m\pi}{l}x \sin\frac{n\pi}{l}x$$

$$= \frac{2}{l}\frac{1}{2}\int_0^l \left[\frac{\cos(m-n)\pi x}{l} - \frac{\cos(m+n)\pi x}{l}\right]dx$$

$$= \frac{1}{l}\left[\frac{\sin(m-n)\pi x/l}{(m-n)\pi/l}\right]_0^l - \left[\frac{\sin(m+n)\pi x/l}{(m+n)\pi/l}\right]_0^l = 0$$

[Note that each function within the bracket is equal to zero].

Thus, the wave functions for two different states are orthogonal or orthonormal.

1.55 SCHRÖDINGER'S EQUATION FOR A PARTICLE IN A THREE DIMENSIONAL RIGID RECTANGULAR BOX

Consider a *free* particle of rest mass m confined to move non-relativistically in a rectangular box of sides l_x, l_y and l_z, the sides being rigid and parallel to the X, Y and Z co-ordinate axes respectively. Since the particle is *free* therefore the potential energy V(x, y, z) of the particle inside the box is zero. Thus, the total energy E of the particle is equal to the kinetic energy and is always a *positive* quantity. For non-relativistic motion

$$E = \frac{1}{2}mv^2 = \frac{m^2v^2}{2m} = \frac{p^2}{2m}$$

Here p is the magnitude of total momentum of the particle. Since the potential energy is independent of time therefore we need to employ Schrödinger's time independent, three dimensional wave equation, given by

$$\nabla^2\psi + \frac{2m}{\hbar^2}(E - V)\psi = 0 \qquad \text{or} \qquad \nabla^2\psi + \frac{2m}{\hbar^2}E\psi = 0$$

or $$\left(\frac{\partial^2\psi}{\partial x^2} + \frac{\partial^2\psi}{\partial y^2} + \frac{\partial^2\psi}{\partial z^2}\right) + \frac{p^2}{\hbar^2}\psi = 0 \qquad ...(i)$$

ψ is a function of x, y, z co-ordinates.

$$\therefore \qquad \psi(x, y, z) = \psi_x\,\psi_y\,\psi_z$$

where ψ_x, ψ_y, ψ_z are the functions respectively of x, y and z co-ordinates only.

$$\therefore \qquad \frac{\partial^2\psi}{\partial x^2} = \psi_y\,\psi_z\,\frac{\partial^2\psi_x}{\partial x^2}.$$

$$\frac{\partial^2\psi}{\partial y^2} = \psi_x\,\psi_z\,\frac{\partial^2\psi_y}{\partial y^2} \quad \text{and} \quad \frac{\partial^2\psi}{\partial z^2} = \psi_x\,\psi_y\,\frac{\partial^2\psi_z}{\partial z^2}$$

Also $$p^2 = p_x^2 + p_y^2 + p_z^2$$

where p_x, p_y and p_z are the x, y and z components of the total linear momentum $\vec{p}$ respectively.

Substituting in (i), we have

$$\psi_y\psi_z \frac{\partial^2\psi_x}{\partial x^2} + \psi_x\psi_z \frac{\partial^2\psi_y}{\partial y^2} + \psi_x\psi_y \frac{\partial^2\psi_z}{\partial z^2} + \frac{p_x^2 + p_y^2 + p_z^2}{\hbar^2} \psi_x\psi_y\psi_z = 0$$

or

$$\frac{1}{\psi_x}\frac{\partial^2\psi_x}{\partial x^2} + \frac{1}{\psi_y}\frac{\partial^2\psi_y}{\partial y^2} + \frac{1}{\psi_z}\frac{\partial^2\psi_z}{\partial z^2} + \frac{p_x^2 + p_y^2 + p_z^2}{\hbar^2} = 0 \quad ...(ii)$$

The three co-ordinates are independent of each other. So, we can separate equation (ii) into three independent equations. Each equation is a function of only *one* co-ordinate.

$$\therefore \quad \frac{\partial^2\psi_x}{\partial x^2} + \frac{p_x^2}{\hbar^2}\psi_x = 0; \frac{\partial^2\psi_y}{\partial y^2} + \frac{p_y^2}{\hbar^2}\psi_y = 0 \text{ and } \frac{\partial^2\psi_z}{\partial z^2} + \frac{p_z^2}{\hbar^2}\psi_z = 0$$

The respective solutions of these equations are

$$\psi_x = A_x \sin\frac{p_x}{\hbar}x + B_x \cos\frac{p_x}{\hbar}x \quad ...(iii)$$

$$\psi_y = A_y \sin\frac{p_y}{\hbar}y + B_y \cos\frac{p_y}{\hbar}y \quad ...(iv)$$

$$\psi_z = A_z \sin\frac{p_z}{\hbar}z + B_z \cos\frac{p_z}{\hbar}z \quad ...(v)$$

Using the boundary conditions, we have

$\psi_x = 0$ when $x = 0$ and $x = l_x$

$\psi_y = 0$ when $y = 0$ and $y = l_y$

$\psi_z = 0$ when $z = 0$ and $z = l_z$

Substituting $\psi_x = 0$ when $x = 0$ in equation (iii), we get $B_x = 0$.

Again substituting $B_x = 0$ and $\psi_x = 0$ when $x = l_x$, we have

$$\sin\frac{p_x}{\hbar}l_x = 0 \quad \text{or} \quad \frac{p_x}{\hbar}l_x = n_x\pi$$

where n_x can have only integral values 1, 2, 3 ... etc.

Similarly, substituting $\psi_y = 0$ when $y = 0$ and $y = l_y$ in equation (iv), we get

$$B_y = 0 \text{ and } \frac{p_y}{\hbar}l_y = n_y\pi$$

where n_y has only integral values 1, 2, 3 ... etc.

and substituting $\psi_z = 0$ when $z = 0$ and $z = l_z$ in equation (v), we get

$$B_z = 0 \text{ and } \frac{p_z}{\hbar}l_z = n_z\pi$$

where n_z has only integral values 1, 2, 3 ... etc.

Combining all these, we get

$$\psi = \psi_x\psi_y\psi_z = A_xA_yA_z \sin\frac{n_x\pi}{l_y}x \sin\frac{n_y\pi}{l_y}y \sin\frac{n_z\pi}{l_z}z$$

$$= A \sin\frac{n_x\pi}{l_x}x \sin\frac{n_y\pi}{l_y}y \sin\frac{n_z\pi}{l_z}z \quad ...(vi)$$

where $A = A_x A_y A_z$ and n_x, n_y and n_z are all independent integers.

The probability of finding the particle anywhere inside the box is equal to one because there is only one particle and there is no chance of it being located outside the box.

$$\therefore \quad \int_0^{l_x}\int_0^{l_y}\int_0^{l_z} |\psi|^2 \, dx\, dy\, dz$$

$$= A^2 \int_0^{l_x}\int_0^{l_y}\int_0^{l_z} \sin^2 \frac{n_x\pi}{l_x} x \sin^2 \frac{n_y\pi}{l_y} y \sin^2 \frac{n_z\pi}{l_z} z \, dx\, dy\, dz$$

$$= A^2 \frac{l_x}{2}\frac{l_y}{2}\frac{l_z}{2} = \frac{A^2}{8} l_x l_y l_z = 1 \qquad ...(vii)$$

$$\left[\because \int_0^{l_x} \sin^2 \frac{n_x\pi}{l_x} x = \frac{l_x}{2}; \int_0^{l_y} \sin^2 \frac{n_y\pi}{l_y} y + \frac{l_y}{2} \text{ and } \int_0^{l_z} \sin^2 \frac{n_z\pi}{l_z} z = \frac{l_z}{2}\right]$$

$\therefore$ Amplitude A from equation (*vii*) is given by

$$A = \sqrt{\frac{8}{l_x l_y l_z}} = \sqrt{\frac{8}{V}}$$

where $V = l_z l_y l_z$ = the volume of the box.

The eigen functions are

$$\psi(n_x, n_y, n_z) = \sqrt{\frac{8}{l_x l_y l_z}} \sin \frac{n_x\pi}{l_x} x \sin \frac{n_y\pi}{l_y} y \sin \frac{n_z\pi}{l_z} z$$

The momentum $p = [p_x^2 + p_y^2 + p_z^2]^{1/2}$. Therefore, the **eigen values of momentum are**

$$p(n_x, n_y, n_z) = \left[\left(\frac{n_x\pi\hbar}{l_x}\right)^2 + \left(\frac{n_y\pi\hbar}{l_y}\right)^2 + \left(\frac{n_z\pi\hbar}{l_z}\right)^2\right]^{\frac{1}{2}} = \pi\hbar\left[\frac{n_x^2}{l_x^2} + \frac{n_y^2}{l_y^2} + \frac{n_z^2}{l_z^2}\right]^{\frac{1}{2}}$$

The energy $E = \frac{p^2}{2m}$. Therefore, the **eigen values of energy** are

$$E(n_x, n_y, n_z) = \frac{\pi^2\hbar}{2m}\left[\frac{n_x^2}{l_x^2} + \frac{n_y^2}{l_y^2} + \frac{n_z^2}{l_z^2}\right]$$

Thus, we find that instead of only one quantum number required in the case of a particle in one dimensional box, we require three quantum numbers, one corresponding to each co-ordinate *x*, *y*, *z*.

If any of the three quantum number n_x, n_y, n_z is zero then, $\psi = 0$. This implies the absence of the particle. This is not physically acceptable. Therefore, none of the quantum numbers can have zero values. Thus, the *ground state* corresponds to $n_x = n_y = n_z = 1$

$$\therefore \quad \psi_{1,1,1} = \sqrt{\frac{8}{l_x l_y l_z}} \sin \frac{\pi x}{l_x} \sin \frac{\pi y}{l_y} \sin \frac{\pi z}{l_z}$$

The *zero point energy* is given by

$$E_{1,1,1} = \frac{\pi^2\hbar^2}{2m}\left[\frac{1}{l_x^2} + \frac{1}{l_y^2} + \frac{1}{l_z^2}\right]$$

The first excited state is obtained when one of the quantum numbers is 2 and the other two are one each.

$$\therefore \qquad E_{2,1,1} = \frac{\pi^2\hbar^2}{2m}\left[\frac{4}{l_x^2}+\frac{1}{l_y^2}+\frac{1}{l_z^2}\right]$$

$$E_{1,2,1} = \frac{\pi^2\hbar^2}{2m}\left[\frac{1}{l_x^2}+\frac{4}{l_y^2}+\frac{1}{l_z^2}\right]$$

$$E_{1,1,2} = \frac{\pi^2\hbar^2}{2m}\left[\frac{1}{l_x^2}+\frac{1}{l_y^2}+\frac{4}{l_z^2}\right]$$

Let us now suppose that the box is *cubical i.e.*, $l_x = l_y = l_z = l$.

The *eigen functions* are now given by

$$\psi(n_x, n_y, n_z) = \sqrt{\frac{8}{l^2}}\sin\frac{n_x\pi}{l}x\sin\frac{n_y\pi}{l}y\sin\frac{n_z\pi}{l}z$$

Note that $n_x = 1, 2, 3$ ... etc., $n_y = 1, 2, 3$ ... etc., and $n_z = 1, 2, 3$ ... etc., are independent integral quantum numbers.

The eigen values of momentum are

$$p(n_x, n_y, n_z) = \frac{\pi\hbar}{l}\left[n_x^2+n_y^2+n_z^2\right]^{1/2}$$

The eigen values of energy now are

$$E(n_x, n_y, n_z) = \frac{\pi^2\hbar^2}{2ml^2}\left[n_x^2+n_y^2+n_z^2\right]$$

The lowest state or ground state is given by $n_x = n_y = n_z = 1$. The corresponding values of momentum and energy are :

$$p_{1,1,1} = \frac{\pi\hbar}{l}\sqrt{3} \quad \text{and} \quad E_{1,1,1} = \frac{\pi^2\hbar^2}{2ml^2}3$$

The eigen state will be

$$\psi_{1,1,1} = \sqrt{\frac{8}{l^3}}\sin\frac{\pi x}{l}\sin\frac{\pi y}{l}\sin\frac{\pi z}{l}$$

The next higher state or first excited state will have 3 sets of quantum numbers 2, 1, 1 or 1, 2, 1 or 1, 1, 2. For all these three quantum states, the values of momentum will be given by

$$p_{2,1,1} = \frac{\pi\hbar}{l}\left[2^2+1^2+1^2\right]^{1/2} = \frac{\pi\hbar}{l}\sqrt{6},$$

$$p_{1,2,1} = \frac{\pi\hbar}{l}\left[1^2+2^2+1^2\right]^{1/2} = \frac{\pi\hbar}{l}\sqrt{6},$$

$$p_{1,1,2} = \frac{\pi\hbar}{l}\left[1^2+1^2+2^2\right]^{1/2} = \frac{\pi\hbar}{l}\sqrt{6},$$

Clearly,

$$p_{2,1,1} = p_{1,2,1} = p_{1,1,2} = \frac{\pi\hbar}{l}\sqrt{6}$$

Similarly, the values of energy for all the three quantum states will be given by

$$E_{2,1,1} = E_{1,2,1} = E_{1,1,2} = \frac{\pi^2\hbar^2}{2ml^2}6$$

The *eigen function* corresponding to the three sets of quantum number are, however, different.

$$\psi_{2,1,1} = \sqrt{\frac{8}{l^3}} \sin \frac{2\pi x}{l} \sin \frac{\pi y}{l} \sin \frac{\pi z}{l}$$

$$\psi_{1,2,1} = \sqrt{\frac{8}{l^3}} \sin \frac{\pi x}{l} \sin \frac{2\pi y}{l} \sin \frac{\pi z}{l}$$

and
$$\psi_{1,1,2} = \sqrt{\frac{8}{l^3}} \sin \frac{\pi x}{l} \sin \frac{\pi y}{l} \sin \frac{2\pi z}{l}$$

Conclusion. For a single value of energy or momentum ; there are three different quantum states.

The property of two or more quantum states of a particle, having different sets of quantum numbers and different eigen functions, to have the same value of momentum and energy, is called **degeneracy.**

A free particle in a *cubical* box is a case of degeneracy. The lowest state is not degenerate, as there is only *one* eigen function for $n_x = n_y = n_z = 1$. The next higher state is said to be *three fold* degenerate as three sets of quantum numbers (2, 1, 1), (1, 2, 1) and (1, 1, 2) and three eigen functions $\psi_{2,1,1}$; $\psi_{1,2,1}$; $\psi_{1,1,2}$ have the same value of energy and momentum.

The next higher state (2, 2, 1) has also *three fold* degeneracy, because the three different eigen function $\psi_{2,2,1}$; $\psi_{1,2,2}$; $\psi_{2,1,2}$ will have the same value of momentum given by

$$p_{2,2,1} = p_{1,2,2} = p_{2,1,2} = \frac{\pi\hbar}{l}[2^2 + 2^2 + 1]^{1/2} = \frac{\pi\hbar}{l} 3$$

and the same value of energy, given by

$$E_{2,2,1} = E_{1,2,2} = E_{2,1,2} = \frac{\pi^2\hbar^2}{2ml^2}[2^2 + 2^2 + 1^2] = \frac{\pi^2\hbar^2}{2ml^2} 9$$

The next higher state (2, 2, 2) is not degenerate, but if we proceed further we get three fold degeneracy, except when all the three quantum numbers are the same.

1.56 ONE-DIMENSIONAL HARMONIC OSCILLATOR

When a system vibrates about an equilibrium position such that (i) the magnitude of the force acting on it is proportional to its displacement from the equilibrium position and (ii) the direction of the force is opposite to that of displacement, it is called a harmonic oscillator.

Examples. A simple pendulum, an object floating in a liquid, a diatomic molecule and an atom in a crystal lattice. There are many more examples in the macroscopic and microscopic field.

Importance. The importance of a harmonic oscillator in both classical and quantum physics, lies in the fact that if any force, which is a function of displacement x, acts upon a particle or system, it can be shown to be proportional to x for *small* displacements and thus the particle acts as a simple harmonic oscillator.

Any force which is a function of x can be expressed in a Maclaurin's series about the equilibrium position $x = 0$ as

$$F(x) = F_{x=0} + \left(\frac{dF}{dx}\right)_{x=0} x + \frac{1}{2}\left(\frac{d^2F}{dx^2}\right)_{x=0} x^2 + \frac{1}{6}\left(\frac{d^3F}{dx^3}\right)_{x=0} x^3$$

Since $x = 0$ is the equilibrium position $F_{x=0} = 0$

Also when x is small x^2, x^3 etc are *very very* small and the terms containing x^2, x^3 etc can be neglected, as compared to the term containing x. The only term of significance, when x is small is, therefore

$$F(x) = \left(\frac{dF}{dx}\right)_{x=0} x$$

This is the statement of Hooke's law when $\left(\frac{dF}{dx}\right)_{x=0}$ is negative *i.e.*, it is a restoring force.

Thus the importance of a harmonic oscillator lies in the fact that all oscillations are simple harmonic in character, when their amplitude is sufficiently small.

1.57 ENERGY OF A HARMONIC OSCILLATOR (According to Classical Mechanics)

Consider a particle of mass m constrained to move along X-axis. If we take $x = 0$ as the equilibrium position, the restoring force F acting on the oscillating particle, at a displacement x, is given by

$$F \propto -x \text{ or } F = -kx \qquad ...(i)$$

The potential energy of the particle is related to the force by the equation

$$F = -\frac{dV}{dx} \qquad ...(ii)$$

Equating (*i*) and (*ii*), we get

$$\frac{-dV}{dx} = -kx \quad \text{or} \quad dV = k\,x\,dx$$

Integrating, $$V = \int_0^x k\,x\,dx = \frac{1}{2}kx^2 \qquad ...(iii)$$

Clearly, $$V \propto x^2.$$

So, the graph between potential energy V of the harmonic oscillator and displacement x is a parabola.

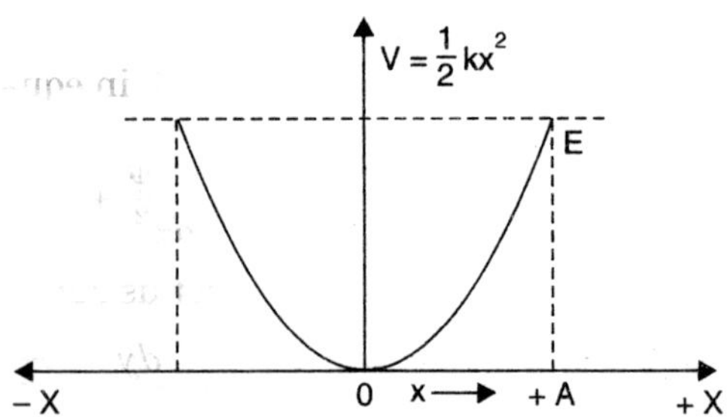

Fig. 1.19

According to classical mechanics

$$F = ma = \frac{d^2x}{dt^2} = -kx$$

or $$\frac{d^2x}{dt^2} + \frac{k}{m}x = 0 \qquad ...(iv)$$

This is the equation of motion of the particle.

Its solution is

$$x = A \sin(\omega t + \phi) \qquad ...(v)$$

Here A represents the amplitude of the oscillations, ϕ the initial phase and ω the *classical angular frequency* of oscillation, given by

$$\omega = \sqrt{\frac{k}{m}} \qquad ...(vi)$$

When the vibrating particle is at the mean position ($x = 0$), its total energy E is all kinetic. Potential energy is zero. As the particle moves away from mean position, its kinetic energy decreases and potential energy increases. However, the total energy E remains constant. At the extreme positions $x = \pm$ A, the kinetic energy is zero and the potential energy V = E, the total energy. The particle, therefore, cannot go beyond $x = \pm$ A, because then the potential energy will be greater than V and hence kinetic energy will have to be *negative*, which is a meaningless concept. Thus, at the points $x = \pm$ A the velocity becomes zero, changes sign and the particle reverses its direction. These points are called **classical turning points.** At the turning points,

Total energy E = potential energy V

$$V = \frac{1}{2}kx^2 = \frac{1}{2}kA^2 \text{ ; at } x = \pm A$$

The total energy E of the oscillator is proportional the square of the amplitude A. As a can have any value, the total energy can also have any value ranging from zero to ∞.

1.58 ENERGY OF A HARMONIC OSCILLATOR (According to Quantum Mechanics)

Schrödinger's one dimensional time independent equation is

$$\frac{d^2\psi}{dx^2} + \frac{2m}{\hbar^2}(E - V)\psi = 0 \quad ...(vii)$$

For a harmonic oscillator $V = \frac{1}{2}kx^2$

$$\therefore \quad \frac{d^2\psi}{dx^2} + \frac{2m}{\hbar^2}\left(E - \frac{1}{2}kx^2\right)\psi = 0 \quad ...(viii)$$

Substituting the value of $k = m\omega^2$ in equation (*viii*), we have

$$\frac{d^2\psi}{dx^2} + \frac{2m}{\hbar^2}\left(E - \frac{1}{2}m\omega^2x^2\right)\psi = 0 \quad ...(ix)$$

or

$$\frac{d^2\psi}{dx^2} + \left(\frac{2mE}{\hbar^2} - \frac{m^2\omega^2}{\hbar^2}x^2\right)\psi = 0 \quad ...(x)$$

Put $\frac{m\omega}{\hbar} = \alpha$ and $\frac{2mE}{\hbar^2} = \beta$ in equation (x), we get

$$\frac{d^2\psi}{dx^2} + (\beta^2 - \alpha^2x^2)\psi = 0 \quad ...(xi)$$

To solve this equation, let us substitute $\sqrt{\alpha}\, x = y$. It is a *dimensionless* constant.

When $\quad y = \sqrt{\alpha}\, x, \; \frac{dy}{dx} = \sqrt{\alpha}$

and

$$\frac{d\psi}{dx} = \frac{d\psi}{dy} \cdot \frac{dy}{dx} = \sqrt{\alpha}\,\frac{d\psi}{dy}$$

$$\frac{d^2\psi}{dx^2} = \frac{d}{dx}\left(\frac{d\psi}{dx}\right) = \frac{d}{dy}\left(\sqrt{\alpha}\,\frac{d\psi}{dy}\right)\frac{dy}{dx} = \alpha\frac{d^2\psi}{dy^2}$$

Putting this value of $\frac{d^2\psi}{dx^2}$ in equation (*xi*), we get

$$\alpha\frac{d^2\psi}{dy^2} + \left(\beta - \alpha^2\frac{y^2}{\alpha}\right)\psi = 0$$

or

$$\frac{d^2\psi}{dy^2} + \left(\frac{\beta}{\alpha} - y^2\right)\psi = 0 \quad ...(xii)$$

To have an *acceptable* solution of the above equation, let us first try to get an *asymptotic solution i.e.,* when $y >> \beta/\alpha$. In such a case, the term β/α can be neglected, as compared to the term y^2. Now, from equation (*xii*), we have

$$\frac{d^2\psi}{dy^2} - y^2\psi = 0 \qquad ...(xiii)$$

The solution of the above equation is given by $\psi = e^{\pm y^2/2}$ *i.e.,* there are two possible solutions $\psi = e^{+y^2/2}$ and $\psi = e^{-y^2/2}$. The quantity $|\psi|^2$ gives the probability of finding the particle somewhere along X-axis. So, it must decrease continuously to zero as $x \to -\infty$ and $x \to +\infty$. But the solution $\psi = e^{+y^2/2}$ increases rapidly with increasing y and hence with increasing x. Therefore, it is not an acceptable solution of the wave function.

So, we conclude that $\qquad \psi = e^{-y^2/2}; |y| \to \infty$

is the *only acceptable solution*.

The standard procedure for solving differential equation like equation (*xii*) valid for all values of y is to try a solution of the form

$$\psi = Ae^{-y^2/2}\,H(y)$$

where A is a constant and H(y) a polynomial in y, known as *Hermite polynomial.*

To determine the function H(y) we calculate

$$\frac{d\psi}{dy} = -Aye^{-y^2/2}H + Ae^{-y^2/2}\frac{dH}{dy}$$

and

$$\frac{d^2\psi}{dy^2} = -Ae^{-y^2/2}H + Ay^2e^{-y^2/2}H - Aye^{-y^2/2}\frac{dH}{dy}$$
$$- Aye^{-y^2/2}\frac{dH}{dy} + Ae^{-y^2/2}\frac{d^2H}{dy^2}$$

Putting this value of $\frac{d^2\psi}{dy^2}$ in equation (*xii*), we have

$$Ae^{-y^2/2}\left(-H + y^2H - 2y\frac{dH}{dy} + \frac{d^2H}{dy^2}\right) + \left(\frac{\beta}{\alpha} - y^2\right)Ae^{-y^2/2}H = 0$$

or

$$\frac{d^2H}{dy^2} - 2y\frac{dH}{dy} + \left(\frac{\beta}{\alpha} - 1\right)H = 0 \qquad ...(xiv)$$

Writing $(\beta/\alpha - 1) = 2n$ or $\beta/\alpha = 2n + 1$, equation (*xiv*) becomes

$$\frac{d^2H}{dy^2} - 2y\frac{dH}{dy} + 2nH = 0 \qquad ...(xv)$$

This is a standard mathematical equation. It is popularly known as *Hermite equation*. The solutions of equation (*xv*) are called *Hermite polynomials*. These are given by :

$$H_n(y) = (-1)^n e^{y^2}\frac{d^n}{dy^n}[e^{-y^2}]$$

Only those solution of equation (*xv*) are acceptable for all values of y for which

$$\beta/\alpha = 2n + 1$$

where n is an integral number 0, 1, 2, 3, etc. This number n is called *quantum number*.

Putting the value of $$\beta = \frac{2mE}{\hbar^2} \text{ and } \alpha = \frac{\omega m}{\hbar},$$

we get $$\beta/\alpha = \frac{2mE}{\hbar^2} \cdot \frac{\hbar}{\omega m} = \frac{2E}{\hbar\omega}$$

$$\therefore \quad 2E = \hbar\omega(\beta/\alpha) = \hbar\omega(2n+1)$$

or $$E = \hbar\omega\left(n + \frac{1}{2}\right)$$

In general $$E_n = \hbar\omega\left(n + \frac{1}{2}\right)$$

If ν is the *classical frequency*, then $\hbar\omega = \frac{h}{2\pi} \cdot 2\pi\nu = h\nu$

and $$E_n = h\nu\left(n + \frac{1}{2}\right) \qquad ...(xvi)$$

Equation (*xvi*) gives the eigen values (or permitted values) or total energy for a harmonic oscillator for $n = 0, 1, 2, 3$ ... These energies are discrete. Their corresponding energy levels are shown in Fig. 1.20.

1.59 ENERGY LEVEL DIAGRAM OF HARMONIC OSCILLATOR

The energy of a harmonic oscillator is quantised in steps of $h\nu$. The energy levels are *equally spaced* with the separation between adjacent levels being $h\nu$. When $n = 0$, $E_0 = \frac{1}{2}h\nu$. This is the lowest value of energy known as *zero point energy*. Thus we find that according to quantum theory, the energy of a harmonic oscillator is quantised in steps of $h\nu$, with lowest energy $\frac{1}{2}h\nu$. This is in sharp contrast to classical theory, where the energy of the oscillator is continuous, from zero to infinity.

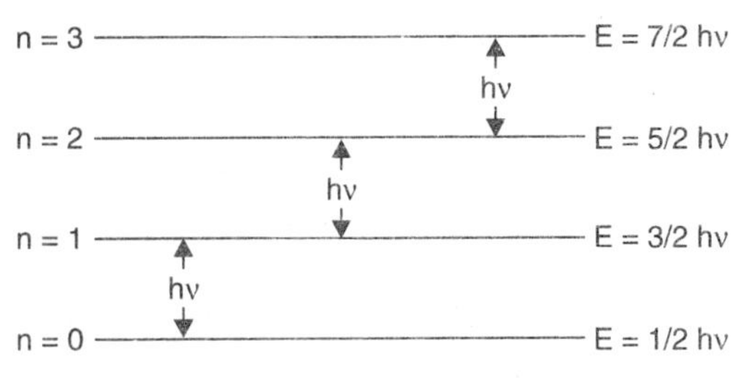

Fig. 1.20

Additional Information

- ***Quantisation of energy of a harmonic oscillator.*** The energy of a harmonic oscillator is given by

$$E_n = h\nu\left(n + \frac{1}{2}\right) \text{ where } n = 0, 1, 2, 3 \text{ ... etc.} \qquad ...(i)$$

Thus a harmonic oscillator cannot have any value of energy, as predicted by the classical theory, but only discrete values of energy, given by equation (*i*) *i.e.*, the energy is quantised in steps of $h\nu$.

■ ***Zero point energy.*** The energy of a harmonic oscillator has the smallest value for $n = 0$. Thus the minimum energy of the oscillator is

$$E_0 = \frac{1}{2}h\nu$$

This is known as *zero point energy*, because a harmonic oscillator in equilibrium with its surroundings would approach an energy $E = E_0 = \frac{1}{2}h\nu$ and not $E = 0$ as, the temperature approaches 0 K.

■ ***Zero point energy in conformity with uncertainty principle.*** The existence of zero point energy is required by the uncertainty principle, according to which

$$\Delta E\, \Delta t \geq \hbar/2$$

or $$\Delta E \geq \frac{\hbar}{2\Delta t} \text{ which is not zero.}$$

For a harmonic oscillator, the energy levels are equally spaced with an interval of $h\nu$ between successive levels, whereas for a particle in a box, the energy levels are not equally spaced. *The energy-interval* increases as we proceed from $n = 1$, onwards. For the ground state, $n = 1$, energy $E_1 = \frac{\pi^2\hbar^2}{2ml^2}$. For the first excited state for $n = 2$, it is $2^2 E_1 = 4E_1$; for the 2nd excited state, it is $3^2 E_1 = 9 E_1$ and for the 3rd excited state $16E_1$ and so on. Thus the interval between the first excited state and ground state is $3E_1$ and between third excited state and second excited state $7E_1$ and so on *i.e.*, the interval goes on increasing.

The lowest energy for a harmonic oscillator is for $n = 0$ and $= \frac{1}{2}h\nu$, whereas for a particle in a box, it is for $n = 1$ and $= \frac{\pi^2\hbar^2}{2ml^2}$.

1.60 GENERAL AND NORMALISED WAVE FUNCTIONS OF A HARMONIC OSCILLATOR

The general wave function of a harmonic oscillator is given by

$$\psi_n = Ae^{-y^2/2}H_n(y)$$

where A is a constant $H_n(y)$ is the *Hermite polynomial*, given by

$$H_n(y) = (-1)^n e^{y^2} \frac{d_n}{dy^n}(e^{-y^2}) \qquad \ldots(i)$$

and $$y = \sqrt{\alpha}\, x \text{ with } \alpha = m\omega/\hbar.$$

On normalising the wave function, we get

$$A = \left(\frac{\sqrt{\alpha}}{2^n n! \sqrt{\pi}}\right)^{1/2}$$

So, the normalised wave function is given by

$$\psi_n = \left(\frac{\sqrt{\alpha}}{2^n n!\sqrt{\pi}}\right)^{\frac{1}{2}} e^{-\alpha x^2/2}\, H_n\left(\sqrt{\alpha}x\right) \quad ...(ii)$$

or

$$\psi_n = \left(\frac{2m\nu}{\hbar}\right)^{1/4} \left(2^n n!\right)^{-1/2} H_n(y)\, e^{-y^2/2} \quad ...(iii)$$

for $n = 0, 1, 2, 3$ etc. because $\sqrt{\frac{\alpha}{\pi}} = \sqrt{\frac{m\omega}{\hbar\pi}} = \sqrt{\frac{2m\nu\pi}{\hbar\pi}} = \left(\frac{2m\nu}{\hbar}\right)^{1/2}$

First three normalised wave functions. The first three normalized wave function are (*i*) ground state for $n = 0$, (*ii*) first excited state for $n = 1$ and (*iii*) second excited state for $n = 2$.

1. Ground state wave function. For the ground state wave function $n = 0$ and $H_n(y)$ for $n = 0$ from equation (*i*) is given by

$$H_0(y) = (-1)^0 e^{y^2} e^{-y^2} = 1; \text{ Also } n! = 0! = 1$$

∴ From equation (*ii*), we get

$$\Psi_0 = \left(\frac{\alpha}{\pi}\right)^{\frac{1}{4}} e^{-\alpha x^2/2} \quad ...(iv)$$

The plot of the ground state wave function is shown in Fig. 1.21(*a*).

2. First excited state wave function. For the first excited state $n = 1$ and $H_n(y)$ for $n = 1$ from equation (*i*) is given by

$$H_1(y) = (-1)^1 e^{y^2} \frac{d}{dy}(e^{-y^2}) = -1e^{y^2}(-2ye^{-y^2})$$

$$= 2y = 2\sqrt{\alpha}\, x$$

∴ From equation (*ii*), we get

$$\psi_1 = \left(\frac{\sqrt{\alpha}}{2\sqrt{\pi}}\right)^{1/2} e^{-\alpha x^2/2} \times 2\sqrt{\alpha}\, x = \left(\frac{\alpha}{\pi}\right)^{1/4} \sqrt{2}\sqrt{\alpha}\, x\, e^{-\alpha x^2/2} \quad ...(v)$$

The plot of the first excited state wave function is shown in Fig. 1.21 (*b*).

3. Second excited state wave function. For the second excited state $n = 2$ and $H_n(y)$ for $n = 2$ from equation (*i*) is given by

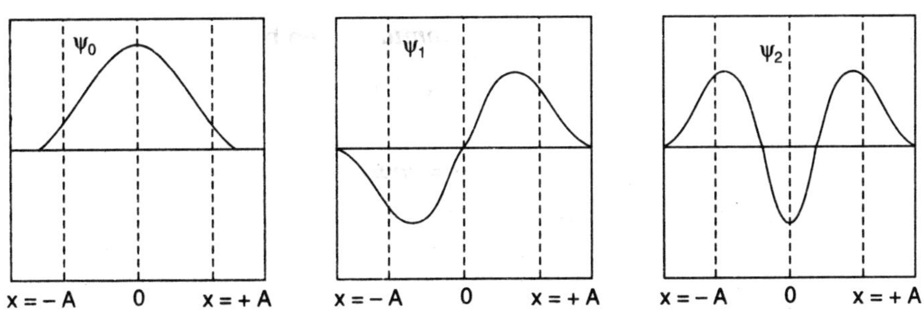

(*a*) (*b*) (*c*)

Fig. 1.21

$$H_2(y) = (-1)^2 e^{y^2} \frac{d^2}{dy^2}(e^{-y^2}) = e^{y^2} \frac{d}{dy}(-2ye^{-y^2})$$

$$= e^{y^2}(4y^2 - 2)\, e^{-y^2} = 2(2y^2 - 1) = 2(2\alpha x^2 - 1)$$

$\therefore$ From equation (*ii*), we get

$$\psi_2 = \left(\frac{\sqrt{\alpha}}{2^2 \times 2\sqrt{\pi}}\right)^{\frac{1}{2}} \times 2(2\alpha x^2 - 1)\, e^{-\alpha x^2/2}$$

$$= \left(\frac{\sqrt{\alpha}}{2\sqrt{\pi}}\right)^{\frac{1}{2}} (2\alpha x^2 - 1)\, e^{-\alpha x^2/2} \qquad ...(vi)$$

The plot of the second excited state wave function is shown in Fig. 1.21 (*c*).

The vertical lines show the limits at $x = +A$ and $x = -A$ between which a classical oscillator, with the same energy, would vibrate. Quantum mechanically, we find that the harmonic oscillator slightly exceeds the amplitude A with an exponentially decreasing probability and penetrates into classically forbidden regions.

NUMERICAL EXAMPLES BASED ON SCHRÖDINGER EQUATION

Example 1.59. *The quantum numbers for the particle in a box are taken to be positive. Give arguments for neglecting n = 0, –1, –2, –3 etc.*

(H.P.U., 1997)

Solution. Zero and negative values of *n* not permissible. The normalised wave function for a free particle inside a box of finite width l is given by

$$\psi(x) = \left(\frac{2}{l}\right)^{1/2} \sin \frac{n\pi}{l} x$$

where n is the quantum number having values $n = 1, 2, 3 ...$

Now $\frac{n\pi}{l} = k$ where k is *the propagation constant*

and $$k = \frac{2\pi}{\lambda} = \frac{p_x 2\pi}{h} = \frac{p_x}{\hbar}$$

$\therefore \quad p_x = \hbar k$

Also $$E = \frac{1}{2}mv^2 = \frac{m^2v^2}{2m} = \frac{p_x^2}{2m} = \frac{\hbar^2}{2m}k^2$$

We only take positive values of quantum number n. We do not take $n = 0$ because it would mean $k = 0$ and hence momentum p_x and kinetic energy E of the particle will be zero *i.e.,* the particle is at rest for all times which is not physically acceptable. For $k = 0$, $\psi(x) = 0$ everywhere in the box, which means that the probability density $|\psi|^2 = 0$ *i.e., the particle is not in the box,* which again is not possible.

The negative values of n simply repeat the corresponding values for positive n and do not give independent solutions.

Example 1.60. *State the values of momentum and energy of a particle in one dimensional box with impenetrable walls. Find their values for an electron in a box of length 1Å for n = 1 and n = 2 Mass of the electron = 9.1 × 10^{-31} kg and h = 6.63 × 10^{-34} Js.*

(G.N.D.U., 1997)

Solution. For a particle in one dimensional box of length l with impenetrable walls

Momentum $$p_n = n\frac{\hbar\pi}{l} \text{ where } n = 1, 2, 3, ...$$

and Energy $$E_n = \frac{p_n^2}{2m} = \frac{n^2\hbar^2\pi^2}{2ml^2}$$

Momentum for $n = 1$ and $n = 2$

For $n = 1$ $$p_1 = \frac{\hbar\pi}{l} = \frac{h}{2\pi}\frac{\pi}{l} = \frac{h}{2l}$$

$$= \frac{6.63 \times 10^{-34}}{2 \times 10^{-10}} = 3.315 \times 10^{-24} \text{ kg m s}^{-1}$$

For $n = 2$

$$p_2 = 2\frac{h}{2l} = \frac{h}{l} = 6.63 \times 10^{-24} \text{ kg m s}^{-1}$$

Energy for $n = 1$ and $n = 2$

For $n = 1$

$$E_1 = \frac{\hbar^2\pi^2}{2ml^2} = \frac{h^2\pi^2}{4\pi^2 . 2ml^2} = \frac{h^2}{8ml^2}$$

$$= \frac{(6.63 \times 10^{-34})^2}{8 \times 9.1 \times 10^{-31} \times 10^{-20}} \text{ joule}$$

$$= 60.4 \times 10^{-19} \text{ joule}$$

$$= \frac{60.4 \times 10^{-19}}{1.6 \times 10^{-19}} = 37.75 \text{ eV}$$

For $n = 2$

$$E_2 = 2^2 \frac{h^2}{8ml^2} = 4 \times 37.75 = \mathbf{151 \text{ eV}}.$$

Example 1.61. *Calculate the value of lowest energy of an electron in one dimensional force free region of length 4 Å.* **(H.P.U. 1997)**

Solution. For a particle in one dimensional force free region of length l, the energy is given by

$$E_n = \frac{p_n^2}{2m} = \frac{n^2\hbar^2\pi^2}{2ml^2} = \frac{n^2h^2\pi^2}{4\pi^2 2ml^2} = \frac{n^2h^2}{8ml^2}$$

The lowest energy state is for $n = 1$

$$\therefore \quad E_1 = \frac{h^2}{8ml^2}$$

$$= \frac{(6.63 \times 10^{-34})^2}{8 \times 9.1 \times 10^{-31} \times (4 \times 10^{-10})^2} \text{ J}$$

$$= 3.78 \times 10^{-19} \text{ J}$$

$$= \frac{3.78 \times 10^{-19}}{1.6 \times 10^{-19}} \text{ eV} = \mathbf{2.36 \text{ eV}}.$$

Example 1.62. *The lowest energy possible for a certain particle entrapped in a box is 40 eV. What are the next three higher energies the particle can have ?*

Solution. For a particle in a one dimensional box

$$E_n = \frac{n^2\hbar^2\pi^2}{2ml^2}$$

The lowest possible energy corresponds to $n = 1$ and is given by

$$E_1 = \frac{\hbar^2\pi^2}{2ml^2} = 40 \text{ eV}$$

The next higher energy state corresponds to $n = 2$ and is given by

$$E_2 = 2^2 \frac{\hbar^2\pi^2}{2ml^2} = 4 \times 40 \text{ eV} = \mathbf{160 \text{ eV}}$$

Similarly for $n = 3$

$$E_3 = 3^2 E_1 = 9 \times 40 \text{ eV} = \mathbf{360 \text{ eV}}$$

and for $n = 4$

$$E_4 = 4^2 E_1 = 16 \times 40 \text{ eV} = \mathbf{640 \text{ eV}}.$$

Example 1.63. *The minimum energy possible for a particle entrapped in a one dimensional box is 3.2×10^{-18} J. What are the next three energies in eV, the particle can have ?*

Solution. Energy of a particle trapped in a one dimensional box

$$E_n = \frac{n^2\hbar^2\pi^2}{2ml^2}$$

i.e., $E_n \propto n^2$ where for minimum energy $n = 1$

Now minimum energy of the particle in eV

$$= \frac{3.2 \times 10^{-18}}{1.6 \times 10^{-19}} = 20 \text{ eV}$$

$\therefore$ Next three values of energy are

20×2^2 eV = **80 eV**

20×3^2 eV = **180 eV**

20×4^2 eV = **320 eV.**

Example 1.64. *Find the probability that a particle in a one-dimensional box of length L can be found between 0.4 L to 0.6 L*

(*i*) *for the ground state*

(*ii*) *for the first excited state and*

(*iii*) *for the second excited state.*

Solution. The wave function for a free particle in a one dimensional box of length L is given by

$$\psi_n = \left(\frac{2}{L}\right)^{1/2} \sin\frac{n\pi}{L} x$$

The probability of finding a particle in a distance Δx about the mean position x is

$$P_n \Delta x = \psi_n \psi_n^* \Delta x$$

$$= \left(\frac{2}{L}\right)^{1/2} \sin\frac{n\pi x}{L} \left(\frac{2}{L}\right)^{1/2} \sin\frac{n\pi x}{L} \Delta x$$

$$= \frac{2}{L} \sin^2 \frac{n\pi}{L} x \, \Delta x$$

Here mean position

$$x = \frac{0.4\,\text{L} + 0.6\,\text{L}}{2} = 0.5\,\text{L} = \frac{\text{L}}{2}$$

and $\Delta x = 0.6\,\text{L} - 0.4\,\text{L} = 0.2\,\text{L}$

(*i*) For the ground state $n = 1$

$$\therefore \quad P_1 \Delta x = \frac{2}{\text{L}} \sin^2\left(\frac{\pi}{\text{L}}\frac{\text{L}}{2}\right) \times 0.2\,\text{L}$$

$$= 2\sin^2\frac{\pi}{2} \times 0.2 = \mathbf{0.4}$$

(*ii*) For the first excited state $n = 2$

$$\therefore \quad P_2 \Delta x = \frac{2}{\text{L}} \sin^2\left(\frac{2\pi}{\text{L}}\frac{\text{L}}{2}\right) \times 0.2\,\text{L}$$

$$= 2\sin^2\pi \times 0.2 = \textbf{Zero}$$

(*iii*) For the second excited state $n = 3$

$$\therefore \quad P_3 \Delta x = \frac{2}{\text{L}} \sin^2\left(\frac{3\pi}{\text{L}}\frac{\text{L}}{2}\right) \times 0.2\,\text{L}$$

$$= 2\sin^2\left(\frac{3\pi}{2}\right) \times 0.2 = \mathbf{0.4}$$

Thus for $n = 1, 3, 5$ (odd values) etc., the probability of finding the particle within a distance $\Delta x = 0.2$ L in the mean position $x = 0.5$ L is 0.4.

For $n = 2, 4, 6$ (even values) etc., the probability is zero.

■ **Example 1.65.** *A particle is moving in one dimensional potential box of infinite height and width 10 Å. Calculate the probability of finding the particle within an interval of 4 Å at the centre of the box when it is in the state of least energy.*

Solution. The state of least energy is the ground state for $n = 1$

Width of one dimensional potential box

$$\text{L} = 10\,\text{Å} = 10 \times 10^{-10}\,\text{m}$$

Position of the centre of the box

$$x = \frac{\text{L}}{2} = \frac{10 \times 10^{-10}}{2}\,\text{m} = 5 \times 10^{-10}\,\text{m}$$

Interval $\Delta x = 4\,\text{Å} = 4 \times 10^{-10}\,\text{m}$

$\therefore$ Probability of finding the particle within an interval Δx about the mean position x

$$P_n \Delta x = \frac{2}{\text{L}} \sin^2\left(\frac{n\pi x}{\text{L}}\right) \Delta x$$

$$= \frac{2}{10 \times 10^{-10}} \sin^2\left(\frac{\pi \times 5 \times 10^{-10}}{10 \times 10^{-10}}\right) 4 \times 10^{-10}$$

$$= 0.8 \sin^2\frac{\pi}{2} = \mathbf{0.8.}$$

■ **Example 1.66.** *Evaluate the first three energy levels of an electron enclosed in a box of width 10 Å. Compare it with those of glass marble of mass 1 g, contained in a box of width 20 cm. Can these levels of the marble be measured experimentally ?*

Solution. Energy of a particle in a one dimensional box

$$E_n = \frac{p_n^2}{2m} = \frac{n^2 \hbar^2 \pi^2}{2ml^2} = \frac{n^2 h^2}{8ml^2}$$

For **electron** $n = 1$; $l = 10\,\text{Å} = 10 \times 10^{-10}$ m ; $m = 9.1 \times 10^{-31}$ kg

$$E_1 = \frac{1 \times (6.63 \times 10^{-34})^2}{8 \times 9.1 \times 10^{-31} \times 100 \times 10^{-20}}\,\text{J}$$

$$= 0.604 \times 10^{-19}\,\text{J}$$

$$= \frac{0.604 \times 10^{-19}}{1.6 \times 10^{-19}}\,\text{eV} = 0.377\,\text{eV}$$

For $n = 2$;

$$E_2 = 2^2 \frac{h^2}{8ml^2} = 4 \times 0.377\,\text{eV} = 1.508\,\text{eV}$$

For $n = 3$;

$$E_3 = 3^2 \frac{h^2}{8ml^2} = 9 \times 0.377\,\text{eV} = 3.393\,\text{eV}$$

For glass **marble** $n = 1$; $l = 20$ cm $= 0.2$ m ; $m = 1\,\text{g} = 10^{-3}$ kg

$$E_1 = \frac{1 \times (6.63 \times 10^{-34})^2}{8 \times 10^{-3} \times (0.2)^2}\,\text{J}$$

$$= 1.374 \times 10^{-63}\,\text{J}$$

$$= \frac{1.374 \times 10^{-63}}{1.6 \times 10^{-19}}\,\text{eV} = 0.858 \times 10^{-44}\,\text{eV.}$$

For $n = 2$;

$$E_2 = 2^2 \frac{h^2}{8ml^2} = 4 \times 0.858 \times 10^{-44}\,\text{eV}$$

$$= 3.432 \times 10^{-44}\,\text{eV}$$

For $n = 3$;

$$E_3 = 3^2 \frac{h^2}{8ml^2} = 9 \times 0.858 \times 10^{-44}\,\text{eV}$$

$$= 7.722 \times 10^{-44}\,\text{eV}$$

These levels of the marble cannot be measured experimentally as the energy involved is very small and is almost zero.

Example 1.67. *Calculate the average value of p^2 for the wave function $\psi_x = \left(\frac{2}{l}\right)^{1/2} \sin\left(\frac{\pi x}{l}\right)$ in the region $0 < x < l$ and $\psi_x = 0$ outside the region.*

Solution. The average or expectation value of p^2 is given by

$$< p^2 > = \int_0^l \psi^* \hat{p}^2 \psi \, dx$$

$$= \int_0^l \psi^* (-i\hbar)^2 \frac{\partial^2}{\partial x^2} \psi \, dx$$

$$\int_0^l \left(\frac{2}{l}\right)^{1/2} \sin\left(\frac{\pi x}{l}\right) (-i\hbar)^2 \frac{\partial^2}{\partial x^2} \left(\frac{2}{l}\right)^{1/2} \sin\left(\frac{\pi x}{l}\right) dx$$

$$= -\hbar^2 \left(\frac{2}{l}\right) \int_0^l \sin\left(\frac{\pi x}{l}\right) \frac{\partial^2}{\partial x^2} \sin\left(\frac{\pi x}{l}\right) dx$$

$$= -\hbar^2 \left(\frac{2}{l}\right) \int_0^l \sin\frac{\pi x}{l} \left(-\frac{\pi^2}{l^2}\right) \sin\left(\frac{\pi x}{l}\right) dx$$

$$= \hbar^2 \frac{\pi^2}{l^2} \frac{2}{l} \int_0^l \sin^2\left(\frac{\pi x}{l}\right) dx$$

$$= \frac{2\hbar^2\pi^2}{l^3} \int_0^l \frac{1 - \cos\left(\frac{2\pi x}{l}\right)}{2} dx$$

$$= \frac{2\hbar^2\pi^2}{l^3} \times \frac{1}{2} \left[\int_0^l dx - \int_0^l \cos\left(\frac{2\pi x}{l}\right) dx\right]$$

Now $$\int_0^l \frac{\cos 2\pi x}{l} dx = \left[\frac{l}{2\pi} \sin\left(\frac{2\pi x}{l}\right)\right]_0^l = 0$$

and $$\int_0^l dx = [x]_0^l = l$$

$$\therefore \quad < p^2 > = \frac{2\hbar^2\pi^2}{l^3} \cdot \frac{l}{2} = \frac{\hbar^2\pi^2}{l^2} = \left(\frac{\hbar\pi}{l}\right)^2.$$

Example 1.68. *An electron is put in a cubical box of each side 1 Å. Find the values of its momentum and energy for the ground state and the first excited state.*

Solution. For the ground state

$$n_x = n_y = n_z = 1$$

$$\therefore \text{ Momentum } p_{1,1,1} = \frac{\pi\hbar}{l}\sqrt{3} = \frac{\pi h}{2\pi l}\sqrt{3}$$

$$= \frac{h}{2l}\sqrt{3} = \frac{6.63 \times 10^{-34}}{2 \times 10^{-10}}\sqrt{3} \text{ kg m s}^{-1}$$

$$= (3.315 \times 10^{-24})\sqrt{3} \text{ kg m s}^{-1}$$

$$= 5.74 \times 10^{-24} \text{ kg m s}^{-1}$$

$$\text{Energy } E_{1,1,1} = \frac{\pi^2\hbar^2}{2ml^2} 3$$

$$= \frac{\pi^2 h^2}{4\pi^2 ml^2} 3 = \frac{h^2}{8ml^2} 3$$

$$= \frac{(6.63 \times 10^{-34})^2 \times 3}{8 \times 9.1 \times 10^{-31} \times 10^{-20}} \text{ J}$$

$$= 60.4 \times 3 \times 10^{-19} = 181.2 \times 10^{-19} \text{ J}$$

$$= 113.25 \text{ eV}$$

For the first excited state, one of the quantum numbers is = 2.

$$\therefore \text{ Momentum } p_{2,1,1} = p_{1,2,1} = p_{1,1,2}$$

$$= \frac{\pi\hbar}{l}\sqrt{6}$$

$$= \frac{h}{2l}\sqrt{6} = 3.315 \times 10^{-24}\sqrt{6} \text{ kg m s}^{-1}$$

$$= 8.12 \times 10^{-24} \text{ kg m s}^{-1}$$

$$\text{Energy } E_{2,1,1} = E_{1,2,1} = E_{1,1,2}$$

$$= \frac{\pi^2\hbar^2}{2ml^2} 6 = \frac{h^2}{8ml^2} 6$$

$$= 60.4 \times 6 \times 10^{-19} \text{ J} = 362.4 \times 10^{-19} \text{ J}$$

$$= \mathbf{226.25 \text{ eV}.}$$

Example 1.69. *Think of the nucleus as a box with a size of 10^{-14} m across. Compute the energy of a neutron confined to the nucleus. ($h = 6.62 \times 10^{-34}$ Js, $m = 1.6 \times 10^{-27}$ kg).*

Solution. Take the nucleus to be a cubical box of size 10^{-14} m across.

$$\therefore \quad l_x = l_y = l_z = l = 10^{-14} \text{ m}$$

The neutron will have the lowest energy in the quantum state given by

$$n_x = n_y = n_z = 1$$

$$\text{Now energy } = E_{1,1,1}$$

$$= \frac{\pi^2\hbar^2}{2m}\left[\frac{n_x^2}{l_x^2} + \frac{n_y^2}{l_y^2} + \frac{n_z^2}{l_z^2}\right]$$

$$= \frac{\pi^2\hbar^2}{2m}\left[\frac{3}{l^2}\right]$$

$$= \frac{3\pi^2 h^2}{4\pi^2 \times 2ml^2} = \frac{3h^2}{8ml^2}$$

$$= \frac{3 \times [6.62 \times 10^{-34}]^2}{8 \times 1.6 \times 10^{-27} \times 10^{-28}}$$

$$= 10.27 \times 10^{-13} \text{ J} = 6.42 \text{ MeV}$$

For the first excited state, one of the quantum numbers is = 2

$\therefore$ Energy $= E_{2,1,1} = E_{1,2,1} = E_{1,1,2}$

$$= \frac{\pi\hbar^2}{2m}\left[\frac{1+4+1}{l^2}\right] = \frac{\pi^2\hbar^2}{2\,ml^2} \times 6$$

$$= \frac{h^2}{8\,ml^2} 6 = 12.84 \text{ MeV}$$

Thus the first excited state is three fold degenerate.

For the second excited state two of the quantum number are = 2

$\therefore$ Energy $= E_{2,2,1} = E_{2,1,2} = E_{1,2,2}$

$$= \frac{\pi\hbar^2}{2m}\left[\frac{4+4+1}{l^2}\right] = \frac{\pi^2\hbar^2}{2\,ml^2} \times 9$$

$$= \frac{h^2}{8\,ml^2} 9 = 19.26 \text{ MeV}$$

The second excited state is also three fold degenerate.

For the third excited state all the quantum number are = 2

$\therefore$ Energy $= E_{2,2,2}$

$$= \frac{\pi\hbar^2}{2m}\left[\frac{4+4+4}{l^2}\right] = \frac{\pi^2\hbar^2}{2\,ml^2} \times 12$$

$$= \frac{h^2}{8ml^2} 12 = 25.68 \text{ MeV}$$

The third excited state is not degenerate.

Example 1.70. *A particle of mass 1 mg is attached to a spring of spring constant 10^{-3} Nm^{-1}. Calculate its zero point energy and classical value of amplitude of zero point vibration.*

Solution. Mass of the spring

$$m = 1 \text{ mg} = 10^{-6} \text{ kg}$$

Spring constant $s = 10^{-3}$ N m^{-1}

Frequency of vibration $\nu = \frac{1}{2\pi}\sqrt{\frac{s}{m}}$

$$= \frac{1}{2\pi}\sqrt{\frac{10^{-3}}{10^{-6}}} \text{ s}^{-1} = 5.03 \text{ s}^{-1}$$

$\therefore$ Zero point energy $E_0 = \frac{1}{2} h\nu$

$$= \frac{1}{2} \times 6.63 \times 10^{-34} \times 5.03 \text{ J}$$

$$= 16.67 \times 10^{-34} \text{ J} = \frac{16.67 \times 10^{-34}}{1.6 \times 10^{-19}} \text{ eV}$$

$$= 10.4 \times 10^{-15} \text{ eV}$$

In classical physics total energy of the oscillator is given by

$$E = \frac{1}{2} ma^2\omega^2$$

But $\omega^2 = \frac{s}{m}$ $\quad\therefore\quad E = \frac{1}{2} ma^2 \frac{s}{m} = \frac{1}{2} sa^2$

or $\quad a = \sqrt{\frac{2E}{s}} = \sqrt{\frac{2 \times 16.67 \times 10^{-34}}{10^{-3}}}$

$$= 1.826 \times 10^{-15} \text{ m.}$$

Example 1.71. *Show that the quantity $y = \sqrt{\alpha}\, x$ where $\alpha = \frac{\omega m}{\hbar}$ as used in a harmonic oscillator is dimensionless.*

Solution. The dimensions of $\omega = \frac{1}{T}$ and $\hbar = M^1 L^2 T^{-1}$ (Work × time) $= M^1 L^2 T^2 T^1 = M^1 L^2 T^{-1}$

$\therefore$ Dimensions of $\alpha = \frac{1}{T} \frac{M}{M^1 L^2 T^{-1}} = \frac{1}{L^2}$

and dimension of $\sqrt{\alpha} = \frac{1}{L}$. Hence dimensions of $y = \sqrt{\alpha}x$ are $\frac{1}{L}$. L = a dimensionless quantity.

Example 1.72. *Find the energy of an electron moving in one dimension in an infinitely high potential box of width 1.0 Å. Given : mass of electron = 9.11 × 10^{-31} kg and h = 6.63 × 10^{-34} Js.*

Solution. $E_n = \frac{n^2h^2}{8\,ml^2}$

Substituting values,

$$E_n = \frac{n^2(6.63 \times 10^{-34})^2}{8 \times 9.11 \times 10^{-31} \times (10^{-10})^2} \text{ J}$$

$$= (6.031 \times 10^{-18})\, n^2 \text{ J}$$

$$= 37.694n^2 \text{ eV}$$

$$\approx \mathbf{38n^2\ eV}, \quad n = 1, 2, 3, \ldots\ldots$$

Clearly, the energy of electron confined in one-dimensional box is quantised *i.e.,* discrete.

EXERCISES

50. Calculate the probability of finding a simple harmonic oscillator within classical limits if the oscillator is in its normal state. **[Ans.** 84%]

51. A particle is in motion along a line between $x = 0$ and $x = a$ with zero potential energy. At points for which $x < 0$ and $x > a$, the potential energy is infinite. The wave function for the particle in nth state is given by

$$\psi_n = A \sin \frac{n\pi x}{a}.$$

52. Find the expression for the normalised wave function. $\left[\textbf{Ans. } \sqrt{\frac{2}{a}} \sin \frac{n\pi x}{a}\right]$

53. An electron is bound in a one dimensional potential box which has a width of 2.5×10^{-10} m. Assuming the height of the box to be infinite, calculate the lowest two permitted energy values of the electron.

($h = 6.63 \times 10^{-34}$ J-s, $m_e = 9.1 \times 10^{-31}$ kg)

$\left[\textbf{Hint. } E_n = \frac{n^2h^2}{8\,mL^2}, n = 1, 2\right]$ **[Ans.** 6.03 eV, 24.12 eV]

SHORT QUESTIONS

1. Why the de-Broglie wave associated with a moving car is not visible ?
2. Name one instrument based on de-Broglie hypothesis.
3. What is the rest mass of a photon ?
4. What is the approximate energy of a photon of visible light of maximum wavelength ?
5. What is the approximate energy of a photon of visible light of minimum wavelength ?
6. What is the momentum of a photon of wavelength λ ?
7. Are matter waves electromagnetic ?
8. Write down the relation between the energy and momentum of a photon.
9. What is the frequency of a photon whose energy is 66.3 eV ? Given : $h = 6.63 \times 10^{-34}$ Js.
10. What is the basic principle of electron microscope ?
11. An electron and a photon have got the same K.E. Which of the two has greater de-Broglie wavelength ?
12. An electron and a photon have got the same total energy. Which of the two has greater de-Broglie wavelength ?
13. Prove that de-Broglie wavelength λ of electrons of energy E is given by $\lambda = \frac{h}{\sqrt{2mE}}$.
14. An electron and a proton have the same amount of kinetic energy. Which of the two possesses greater wavelength ?
15. What is the dimensional formula of $\frac{h}{mv}$?
16. What useful information is derived from electron diffraction experiments ?
17. What is the momentum of a photon of frequency ν ?
18. For a given kinetic energy, which of the following has the smallest de-Broglie wavelength ; electron, proton, α-particle ?

19. Show that the wavelength of electromagnetic radiation is equal to de-Broglie wavelength of its quantum (photon).

20. "A photon of energy 6×10^{-20} J has linear momentum 2×10^{-30} kg m s^{-1}". Is this statement correct ?

21. "de-Broglie hypothesis supports the Bohr's model of stationary orbits." Comment on this statement.

22. Is there any difference between the wavelength of the radiation and the de-Broglie wavelength of a photon of that radiation ?

23. What is the difference between light waves and matter waves ?

24. Which photon is more energetic—violet or red ?

25. What led de-Broglie to think that the material particles may also show wave nature ?

26. Why a fast neutron beam needs to be thermalised with the environment, before it can be used for neutron diffraction experiments ?

27. Why is the wave nature of matter not noticeable in our daily observations ?

28. Crystal diffraction can be studied by electron diffraction as well as by neutron diffraction. What should be the ratio of the velocities of electron and neutron for the de-Broglie wavelength to be the same ?

29. How can you connect de-Broglie wave concept with Bohr's postulate : "Angular momentum is an integral multiple of $\frac{h}{2\pi}$" ?

30. In an accelerator experiment on high energy collisions of electrons with positrons, a certain event is interpreted as annihilation of an electron-positron pair of total energy 10.2 BeV into two γ-rays of equal energy. What is the wavelength associated with each γ-ray ? Given : 1 BeV = 10^9 eV.

31. The wavelength of a probe is roughly a measure of the size of a structure that it can probe in some detail. The quark structure of protons and neutrons appears at the minute length scale of 10^{-15} m or less. This structure was first probed in early 1970's using high energy electron beams produced by a linear Accelerator at Stanford, USA. Guess what might have been the order of energy of these electron beams ? (Rest mass energy of electron = 0.511 MeV).

32. The extent of localisation of a particle is determined roughly by its de-Broglie wavelength. If an electron is localised within the nucleus (of size about 10^{-14} m) of an atom, what is its energy ? Compare this energy with the typical binding energies (of the order of a few MeV) in a nucleus, and hence argue why electrons cannot reside in a nucleus.

33. The energy and momentum of an electron are related to the frequency and wavelength of the associated matter wave by the relations : $E = h\nu$, $p = \frac{h}{\lambda}$.

    But while the value of λ is physically significant, the value of ν (and therefore the value of the phase speed $\nu\lambda$) has no physical significance. Why ?

ANSWERS

1. $\lambda \propto \frac{1}{m}$. Since m is very large therefore λ is very small.Consequently, the de-Broglie wave is not visible.

2. Electron microscope. **3.** Zero. **4.** 1.7 eV.

5. 3.1 eV. **6.** $\frac{h}{\lambda}$.

7. No. This is because electromagnetic waves are produced by accelerated charge. On the other hand, the de-Broglie wave is independent of the charge of a particle.

8. $p = \frac{E}{c}$.

9. $E = h\nu, \nu = \frac{E}{h} = \frac{66.3 \times 1.6 \times 10^{-19}}{6.63 \times 10^{-34}}$ Hz $= 16 \times 10^{15}$ Hz $= 1.6 \times 10^{16}$ Hz.

10. It is based on de-Broglie hypothesis. A beam of accelerated electrons behaves like wave. This wave can be handled by electric and magnetic fields in exactly the same way as electromagnetic waves can be handled by lenses.

11. Try yourself.

12. Try yourself.

13. $\frac{1}{2} mv^2 = eV$ or $v = \sqrt{\frac{2eV}{m}}$

$$\lambda = \frac{h}{mv} = \frac{h}{m}\sqrt{\frac{m}{2eV}} = \frac{h}{\sqrt{2meV}}$$

14. $E = \frac{1}{2} m v^2 = \frac{m^2v^2}{2m} = \frac{p^2}{2m} = \frac{h^2}{2m\lambda^2}$ $\therefore$ $E \propto \frac{1}{m\lambda^2}$

But E is constant in the given problem.

$\therefore$ $m\lambda^2 = \text{constant}$ $\Rightarrow$ $\lambda \propto \frac{1}{\sqrt{m}}$

Since m of electron is less than that of proton therefore λ of electron is greater than that of proton.

15. We know that $\lambda = \frac{h}{mv}$. So, the dimensional formula of $\frac{h}{mv}$ is the same as that of λ *i.e.*, [M°L T°].

16. Particles possess wave nature.

17. Momentum of photon $= \frac{h\nu}{c}$.

18. α-particle.

19. $\lambda = \frac{h}{p} = \frac{h}{\frac{h\nu}{c}} = \frac{c}{\nu}$

20. No. Momentum of photon, $p = \frac{h\nu}{c}$

$\therefore$ $p = \frac{6 \times 10^{-20}}{3 \times 10^8}$ kg m s^{-1} $= 2 \times 10^{-28}$ kg m s^{-1}.

21. We know that $\lambda = \frac{h}{mv}$

Again, if n electrons are moving in an orbit of radius r, then $2\pi r = n\lambda$ or $\lambda = \frac{2\pi r}{n}$

$\therefore$ $\frac{h}{mv} = \frac{2\pi r}{n}$ or $mvr = n\frac{h}{2\pi}$,

which is in accordance with Bohr's concept of stationary orbits. So, the given statement is correct.

22. The momentum p of a photon of frequency $\nu = \frac{h\nu}{c} = \frac{h\nu}{\nu\lambda} = \frac{h}{\lambda}$ or $\lambda = \frac{h}{p}$ ...(1)

de-Broglie wavelength λ' of a photon $= \frac{h}{mv} = \frac{h}{p}$ ...(2)

It follows from (1) and (2) that $\lambda = \lambda'$.

So, there is no difference between the wavelength of the radiation and the de-Broglie wavelength of a photon of that radiation.

23. The velocity of light waves in vacuum is a constant quantity. On the other hand, the velocity of matter waves in vacuum depends upon their wavelength.

24. Since the frequency of violet is more than that of red therefore the photon of violet light is more energetic than the photon of red light.

25. Nature loves symmetry. Since matter and energy are inter-convertible therefore matter and energy are two aspects of the same entity.

26. The wavelength of fast neutrons is very small as compared to inter-atomic spacing. So, these are not useful for diffraction experiments. On the other hand, the wavelength of thermal neutrons is of the order of inter-atomic spacing. So, thermal neutrons are useful for diffraction experiments. Thus, a fast neutron beam needs to be thermalised with the environment before it can be used for neutron diffraction experiments.

27. This is because the value of de-Broglie wavelength is very small as compared to the size of the body.

28. $\lambda = \dfrac{h}{mv}$; Clearly, mv should be constant.

$$\therefore \qquad \frac{\text{velocity of electron}}{\text{velocity of neutron}} = \frac{\text{mass of neutron}}{\text{mass of electron}}.$$

29. According to de-Broglie wave concept, only that orbit is possible for an electron which contains an integral multiple of wavelengths.

$\therefore$ $2\pi r = n\lambda$, where r is the radius of the orbit and λ is the de-Broglie wavelength of the electron.

Now, $$2\pi r = n\left(\frac{h}{mv}\right) \quad \text{or} \quad mvr = n\,\frac{h}{2\pi}$$ which is Bohr's postulate.

30. $$\lambda = \frac{hc}{E} = \frac{6.63\times10^{-34}\times3\times10^{8}}{5.1\times10^{9}\times1.6\times10^{-19}}\ \text{m} = 2.437\times10^{-16}\ \text{m}.$$

31. Use $$p = \frac{h}{\lambda} = \frac{6.63\times10^{-34}\ \text{Js}}{10^{-15}\ \text{m}} = 6.63\times10^{-19}\ \text{kg m s}^{-1}.$$

Use the relativistic formula for energy :

$$E^2 = c^2p^2 + m_0^{\,2}c^4 = 9\times(6.63)^2\times10^{-22} + (0.511\times1.6)^2\times10^{-26} \simeq 9\times(6.63)^2\times10^{-22},$$

the second term (rest mass energy) being negligible. Therefore, $E = 1.989\times10^{-10}$ J = 1.24 BeV. Thus, electron energies from the accelerator must have been of the order of a few BeV.

32. For $\lambda = 10^{-14}$ m, E = 124 MeV, this is much too large compared to the binding energy that Coulomb force can provide within the nucleus. Therefore, electrons localised within a nucleus are far too energetic to stay bound within. This is why electrons do not reside in a nucleus.

33. The absolute value of energy E (but not momentum p) of any particle is arbitrary to within an additive constant. Hence while λ is physically significant, absolute value of ν of a matter wave of an electron has no direct physical meaning. The phase speed $\nu\lambda$ is likewise not physically significant.

The group speed given by $$\frac{d\nu}{d(1/\lambda)} = \frac{dE}{dp} = \frac{d}{dp}\left(\frac{p^2}{2m}\right) = \frac{p}{m}$$ is physically meaningful.

MULTIPLE CHOICE QUESTIONS

1. A photon behaves as if it had a mass equal to
(*a*) $h\nu c^2$ (*b*) $c^2/h\nu$ (*c*) $\nu c^2/h$ (*d*) $h\nu/c^2$.

2. The following particles are moving with same speed. The smallest matter wave will be associated with :
(*a*) α-particle (*b*) proton (*c*) electron (*d*) deuteron.

3. The energy of a photon corresponding to the visible light of maximum wavelength is
(*a*) 1 eV (*b*) 1.8 eV (*c*) 3.1 eV (*d*) 7 eV.

4. The de-Broglie wavelength of a 0.08 eV neutron is :
Given : mass of neutron = 1.66×10^{-27} kg
(*a*) 1.01×10^{-14} m (*b*) 1.01 Å (*c*) 2.02 m (*d*) 1.01×10^{9} m.

5. The de-Broglie wavelength of 2 kg mass whose velocity is 25 m s^{-1} is
(*a*) 1.33×10^{-25} Å (*b*) 1.33×10^{-25} m (*c*) 1.33 cm (*d*) 1.33 Å.

6. A photon and an electron have the same wavelength. If E and E′ be their respective energies, then
(*a*) $E' < E$ (*b*) $E' > E$ (*c*) $E' = E$ (*d*) $E' = 4E$.

7. Which of the following is not the property of photons ?
(*a*) momentum (*b*) energy (*c*) frequency (*d*) rest mass

8. If E_1, E_2 and E_3 are the respective kinetic energies of an electron, an alpha particle and a proton each having the same de-Broglie wavelength, then
(*a*) $E_1 = E_2 = E_3$ (*b*) $E_2 > E_3 > E_1$ (*c*) $E_1 > E_3 > E_2$ (*d*) $E_1 > E_2 > E_3$.

9. A photon and an electron have the same wavelength. Then the velocity of photon is :
(*a*) less than that of the electron (*b*) greater than that of the electron
(*c*) equal to that of the electron (*d*) infinite.

10. The energy of a photon is 20 eV. Its momentum is :
Given : $h = 6.625 \times 10^{-34}$ Js.
(*a*) 5.33×10^{-27} kg m s^{-1} (*b*) 10.66×10^{-25} kg m s^{-1}
(*c*) 10.66×10^{-27} kg m s^{-1} (*d*) 5.33×10^{-30} kg m s^{-1}.

11. The de-Broglie wavelength of helium atoms (mass 6.65×10^{-27} kg) having average velocity 1.635×10^{3} m s^{-1} is :
(*a*) 6.61×10^{-4} m (*b*) 4×10^{-24} m (*c*) 6.1×10^{-9} m (*d*) 6.1×10^{-11} m.

12. When a photon with a wavelength of 450 nm strikes the surface of metallic sodium, an electron having an energy of 2.1 eV is ejected. The momentum of the photon is :
(*a*) 4.4×10^{-27} kg m s^{-1} (*b*) 1.1×10^{-25} kg m s^{-1}
(*c*) 1.5×10^{-27} kg m s^{-1} (*d*) 3.7×10^{-28} kg m s^{-1}.

13. In the above question, the work function of photo-emitter is :
(*a*) 0.7 eV (*b*) 1.4 eV (*c*) 2.1 eV (*d*) 2.8 eV.

14. Kinetic energy of neutron whose de-Broglie wavelength is 0.7 Å is :
Given : mass of neutron = 1.66×10^{-27} kg
(*a*) 0.167 eV (*b*) 0.334 eV (*c*) 0.167 J (*d*) 0.167 erg.

15. The fact that the photon carries momentum was demonstrated through
(*a*) the famous Bohr's theory of atom (*b*) the wonderful Thomson's e/m experiment
(*c*) the amazing Compton effect (*d*) the interesting Doppler effect.

16. The de-Broglie wavelength of an electron whose speed is half that of light is :
(*a*) 3.6×10^{-12} m (*b*) 4.2×10^{-12} m (*c*) 8.4×10^{-12} m (*d*) 6.12×10^{-12} m.

17. A photon of frequency ν has a momentum associated with it. If c is the velocity of light, then this momentum is

(a) $h\nu/c^2$ (b) $h\nu/c$ (c) ν/c (d) ν/c^2.

18. We may state that

(i) the energy E of a photon of frequency ν is $E = h\nu$, where h is Planck's constant.

(ii) the momentum p of a photon is $p = h/\lambda$ where λ is the wavelength of the photon.

From the above statements, one may conclude that the wave velocity of light is equal to

(a) 3×10^{-8} m s^{-1} (b) the ratio E/p

(c) the product of E and p (d) the ratio $(E/p)^2$.

19. The de-Broglie wavelength of an electron is 1.66×10^{-10} m. Its energy will be nearly :

(a) 1.08 Me V (b) 0.5 Me V (c) ≈ 54 eV (d) 55 Me V.

20. What is the wavelength of an electron moving with a velocity of 500 km s^{-1} ?

(a) 2.9 nm (b) 3.0 nm (c) 0.5 nm (d) 1.45 nm.

21. The energy of a particle in a one dimensional closed box of length l is

(a) directly proportional to l (b) directly proportional to l^2

(c) inversely proportional to l (d) inversely proportional to l^2.

22. The uncertainty principle is applicable to :

(a) the events in macroscopic world only but not the events in microscopic world

(b) the events in microscopic world only but not to the events in macroscopic world

(c) the events in both the macroscopic and microscopic world.

23. Two wave functions $\psi_m(x)$ and $\psi_n(x)$ are orthogonal if

(a) $\int \psi_m(x)\, \psi_n(x)\, dx = 0$ (b) $\int \psi_m{}^*(x)\, \psi_n(x)\, dx = 0$

(c) $\int \psi_m(x)\, \psi_n(x)\, dx = 1$ (d) $\int \psi_m{}^*(x)\, \psi_n(x)\, dx = 1$.

24. The energy of a particle in a one dimensional closed box is

(a) directly proportional to quantum number (b) directly proportional to n^2

(c) inversely proportional to n (d) inversely proportional to n^2.

25. The number of nodes in the ground state wave function of a particle in an infinite square well is

(a) 1 (b) 2 (c) 3 (d) 4.

26. For a particle trapped in a box of length l the average value of momentum $< p >$ is

(a) $\frac{h}{l}$ (b) $\frac{h}{2l}$ (c) 0 (d) 1.

ANSWERS

1. (d)	2. (a)	3. (b)	4. (a)	5. (a)	6. (b)	7. (d)	8. (c)
9. (b)	10. (c)	11. (d)	12. (c)	13. (a)	14. (a)	15. (c)	16. (b)
17. (b)	18. (b)	19. (c)	20. (d)	21. (d)	22. (c)	23. (b)	24. (b)
25. (b)	26. (c).						

LONG ANSWER QUESTIONS

1. What are de-Broglie matter waves ? *(UPTU, B.Tech. I, Q. Bank, 2001)*

2. Show that the phase velocity of de-Broglie wave is greater than the velocity of light, but the group velocity is equal to the velocity of particle with which the waves are associated.

(UPTU, B.Tech, I Sem. 2001)

3. What are de-Broglie waves ? How do they help in the interpretation of Bohr's quantisation rule ? *(UPTU, B.Tech. I, Q. Bank, 2001)*
4. Deduce an expression for de-Broglie wavelength of helium atom having energy at temperature T°K. *(UPTU, B.Tech. I, Q. Bank, 2001)*
5. Derive the formula for de-Broglie wavelength of a particle in terms of kinetic energy and its rest energy m_0c^2. *(UPTU, B.Tech. I, Q. Bank, 2001)*
6. Find the de-Broglie wavelength of an electron of energy V eV. *(UPTU, B.Tech. I, Q. Bank, 2001)*
7. A particle of charge q and mass m is accelerated from rest through a potential difference V. Find its de-Broglie wavelength. *(UPTU, B.Tech, I. Sem. 2001)*
8. Derive de-Broglie wave equation for a material particle.
9. Derive an expression for the de-Broglie wavelength of an electron moving under a potential of V volt.
10. Describe Davisson and Germer experiment to depict the wave nature of electron.
11. Describe Davisson and Germer experiment on diffraction of electrons.
12. Describe Thomson's method on diffraction of electrons.
13. Describe one practical application of de-Broglie waves.
14. Write a note on the de-Broglie waves associated with atoms and molecules.
15. Write a note on dual nature of matter.
16. What do you understand by dual nature of radiation ?
17. Outline schematically the experiment of Davisson and Germer to demonstrate the wave nature of matter.
18. Distinguish between the phase velocity (V_p) and the group velocity (V_g) of a wave paper. Prove that $V_pV_g = c^2$. *(UPTU, B.Tech, II Sem. 2001)*
19. Discuss the dual nature of matter and Waves. *(UPTU, B.Tech. I, Q. Bank, 2001)*
20. What is uncertainity principle ? Apply it to prove the non-existence of the electron in the nucleus. *(UPTU, B.Tech. I, Q. Bank, 2001)*
21. Describe Division-Germer experiment to demonstrate the wave nature of particle. *(UPTU, B.Tech. I, Q. Bank, 2001)*
22. State and explain Heisenberg's uncertainty principle ? Using this principle show that the electron cannot reside in an atomic nucleus. *(UPTU, B.Tech. II Sem. 2001)*
23. Illustrate Heisenberg's uncertainty principle by diffraction of a beam of electrons by a narrow slit.
24. Prove position momentum uncertainty principle using particle approach.
25. Derive Schrödinger's time independent wave equation. What is the physical significance of stultification 'ψ' used in this equation ? What condition must it fulfil ? *(UPTU, B.Tech. II Sem. 2001)*
26. A particle of mass m is confined to a one dimensional box of length l. Derive expression for (*i*) wave function (*ii*) probability density of the particle. Show these on separate graphs.
27. What are nodes and antinodes ? Where do they occur ? Explain their presence in terms of probability density.
28. Obtain energy levels for a particle inside a box of finite width l.
29. How many maxima and minima are there in the (*i*) ground state (*ii*) first excited state and (*iii*) second excited state wave function of a particle in a square well ?
30. Prove quantum mechanically that a particle will not exist in a box if its energy is zero. Discuss zero point energy.
31. What are orthogonal wave functions ? For a particle in one dimensional box, show that the wave functions for two different states are orthogonal (ortho-normal).
32. Show that the linear momentum and energy of a particle in a one dimensional box is quantised.

33. Derive time dependent and time independent Schrödinger wave equation.
(UPTU, B.Tech. I, Q. Bank, 2001)

34. Derive time independent Schrödinger wave equation. *(Model Q. Paper, B.Tech. I, 2001)*

35. Derive time dependent Schrödinger wave equation. *(UPTU, B.Tech. I Sem. 2001)*

36. (*a*) Show quantitatively that energy of an harmonic oscillator is quantised in steps of $h\nu$ and approaches $\frac{1}{2}h\nu$ as temperature approaches 0 K. How do you explain the existence of zero point energy?

(*b*) Draw energy level diagram of a harmonic oscillator and compare with those of particle in a box.

37. (*a*) Write Schrödinger's equation for a particle in a rigid box and solve it. Find the eigen functions when a particle is kept in a rectangular box of dimensions $l_x \times l_y \times l_z$. Find the eigen values of momentum and energy.

(*b*) Show that for a single value of energy or momentum, different quantum states are possible. Explain the term degeneracy.

38. Using uncertainty principle show that, protons neutrons and α-particles can exist within the nucleus.

39. Apply Heisenberg's uncertainty principle to explain the following :

(*i*) Non-existence of electrons within the nucleus.

(*ii*) Existence of protons, neutrons and α-particles.

(*iii*) Existence of finite zero-point energy.

(*iv*) Binding energy of an electron in a hydrogen atom is of the order of 15 eV.

(*v*) Size of the elementary cell in phase space.

(*vi*) The radiation emitted from an excited atom does not appear with a precise frequency.

2

X-rays

2.01 INTRODUCTION

X-rays were accidently discovered by a German scientist Rontgen in 1895. He was working with a cathode ray discharge tube when he noticed that a fluorescent screen (coated with barium platinocyanide) which happened to be near the discharge tube began to glow with brilliant greenish light. The screen continued to glow even after the discharge tube was completely covered with thick black paper to stop all visible light. Considering that some unknown rays were coming out of the tube which could penetrate thick black paper and cause fluorescence, he called them X-rays. These rays were later named as Rontgen rays after the name of the discoverer.

2.02 PRODUCTION OF X-rays

Principle

The underlying principle of nearly all methods for the production of X-rays is that when fast moving electrons (cathode rays) are stopped by a target of high atomic weight, a part of the energy of the cathode rays appears in the form of X-rays. Dr. William Coolidge, in 1913, designed a tube for the production of X-rays. This tube is known as Coolidge tube or vacuum tube or Modern X-ray tube.

Apparatus

This tube consists of an evacuated discharge tube, to one side of which a tungsten filament F is fused. For getting a copious supply of electrons, the tungsten filament is coated with BaO. The filament is heated with the help of a low power D.C. circuit consisting of a battery B, rheostat Rh and an ammeter A. The electrons emitted by the filament are focussed on the anticathode or anode A with the help of a hollow molybdenum cylinder S surrounding the filament. S is maintained at negative potential and called *cylindrical shield*.

The anode is a tungsten block (high atomic weight) held at an angle of 45° to the horizontal. It is embedded in a copper block. The copper block is connected to a long copper rod R to conduct away the heat produced. The outer portion of the rod R is cooled by a cooling pipe.

The anode is connected to the positive and cathode to the negative of a high tension power supply.

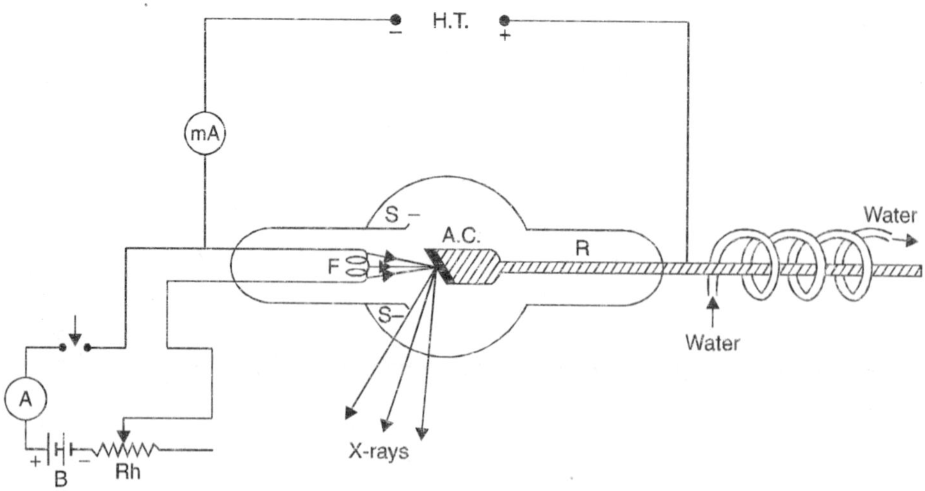

Fig. 2.1

Working

The filament F is heated by passing a suitable current through it. The electrons emitted from the filament are focussed at a point on the anticathode with the help of cylindrical shield. The negative potential of the shield exerts a lateral force of repulsion from all sides. So, the electrons are forced to converge to a point in the middle of the anode. On account of extremely high potential difference between the cathode and the anode, the electrons arrive at the anode with a large amount of kinetic energy. On striking the anode, the electrons are stopped. Nearly 98% of the energy of the incident electrons is converted into heat. The remaining energy appears in the form of X-rays.

Modern X-ray tube has the advantage of independent control of intensity and quality. The intensity of X-rays depends upon the number of electrons emitted from the filament. This depends upon the electric current flowing through the filament. So, by controlling the current with the help of rheostat (in low power D.C. circuit), we can control the intensity of X-rays. The quality of X-rays depends upon their energy which, in turn, depends upon the energy of the incident electrons. The energy of the incident electrons depends upon the potential difference between the anode and the cathode. This potential difference may be called accelerating potential. Thus, by controlling accelerating potential we can control the quality of X-rays. If X-rays have low energy, then they have low penetrating power and are called *soft X-rays*. If X-rays have high energy, then they have more penetrating power and are called *hard X-rays*.

2.03 WAVELENGTH OF X-RAYS

If V be the potential difference between the cathode and anode, then an electron acquires energy eV.

If ν be the frequency of X-rays, then the energy of one X-ray photon is $h\nu$.

Here, h is Planck's constant (6.624×10^{-34} Js).

Now, $$h\nu = eV \quad \text{or} \quad \nu = \frac{eV}{h}$$

But $$^{*}c = \nu\lambda \quad \text{or } \nu = \frac{c}{\lambda}$$

$$\therefore \qquad \frac{c}{\lambda} = \frac{eV}{h} \quad \text{or} \quad \boxed{\lambda = \frac{ch}{eV}}$$

2.04 PROPERTIES

Some of the properties of X-rays are given below :

(*i*) X-rays are electromagnetic radiations like light and thus these also possess the following properties (*a*) rectilinear propagation and casting shadows of obstacles (*b*) travelling with the speed of light (*c*) reflection (*d*) refraction (*e*) diffraction (*f*) interference.

(*ii*) The wavelength of X-rays is very short. It ranges from 0.1 Å to 100 Å (1Å = 10^{-10} m).

(*iii*) These are not deviated by electric or magnetic fields.

(*iv*) These cause fluorescence in certain chemical compounds. A plate of barium-platinocynide glows bright greenish yellow in X-rays.

(*v*) These ionise the gas through which these pass.

(*vi*) These penetrate through solids. Materials which are opaque (like wood etc.) to visible light are transparent for X-rays. The absorption of X-rays in matter is regulated by the relation $I = I_0 e^{-\mu x}$, where I_0 is the original intensity and I is the intensity after absorption through a material of absorbing power μ and thickness x.

(*vii*) When these strike a metal, these give rise to *secondary X-rays* issuing out of that metal.

(*viii*) These have harmful effect on human tissues. So, a continued exposure to these rays must be avoided.

(*ix*) When these fall on certain Group I metals like caesium, these liberate electrons from the surfaces of these metals. This is called photoelectric effect.

(*x*) These have a continuous spectrum as also characteristic spectrum depending upon the nature of the target.

2.05 NATURE OF X-RAYS

Just like light rays, X-rays are also electromagnetic waves. However, the wavelength of visible light is nearly 10^3 times more than the wavelength of X-rays. While the wavelength of visible light ranges from 3900 Å, to 7800 Å, the wavelength of X-rays generally lies between 1 Å and 100 Å. Since wavelength is inversely proportional to frequency therefore the frequency of X-rays is nearly 10^3 times more than the frequency of visible light. Again, since the energy of a photon is proportional to frequency therefore X-ray photons are much stronger than the photons of visible light.

2.06 X-RAY SPECTRA

When a fast moving electron beam strikes a target, an X-ray spectrum is produced. The spectrum so produced consists of a series of intense lines (corresponding to definite wavelengths) superimposed on a continuous X-ray emission background. The individual sharp or peak spectral lines constitute the *characteristic spectrum*. This spectrum depends upon the nature of the target material. These lines occur in small groups. The group in the short wavelength region is K-series (Fig. 2.3). The next group in the direction of increasing wavelength is called L-series and so on.

The continuous background is the *continuous spectrum* which is independent of the nature of the target but whose intensity is proportional to the atomic number of the material of the target. The continuous spectrum, on the analogy of visible radiation of light, is sometimes called the white spectrum.

On changing the nature of the material, the nature of the characteristic spectrum changes, the continuous spectrum remains as such though its intensity suffers a change. If we keep the same target but lower the operating voltage, the intensity of both types of spectra is lowered. But while the characteristic spectrum remains the same, the maximum wavelength (or shortest frequency) position in the case of the continuous spectrum, λ_m increases with the decrease in voltage.

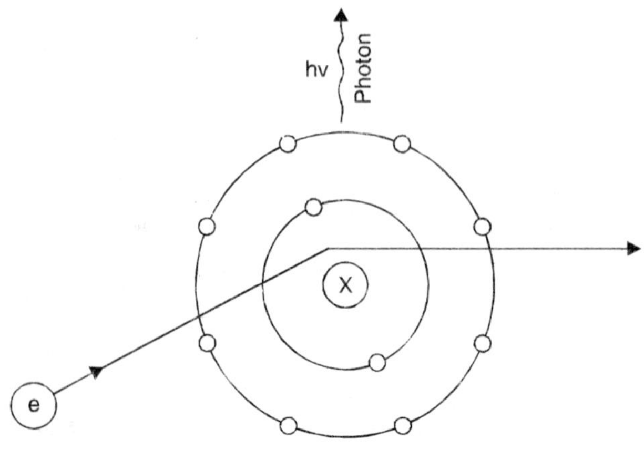

Fig. 2.2

***(i)** Continuous spectrum.* The origin of the continuous spectrum is due to the retardation of fast moving electrons as they move closer to the nuclei of atoms of the target. As a fast moving electron passes through an atom of the target, it is deflected from its path on account of attractive force exerted by the nucleus. During deflection (in the electric field of nucleus), a photon of frequency ν is emitted (Fig. 2.2). This frequency lies in the X-ray region of the electromagnetic spectrum. The closer an electron comes to the nucleus of the target atom, larger is the loss of its velocity or energy. Consequently, higher is the frequency of the emitted photon. Since a large number of electrons are there and they suffer different deviations therefore *radiations of all the possible wavelengths or frequencies with a finite range radiation are emitted. This gives rise to the continuous X-ray spectrum. It starts at a fixed minimum* wavelength $\lambda_{min.}$, increases in intensity with increasing wavelength and then falls off gradually as shown in Fig. 2.3.

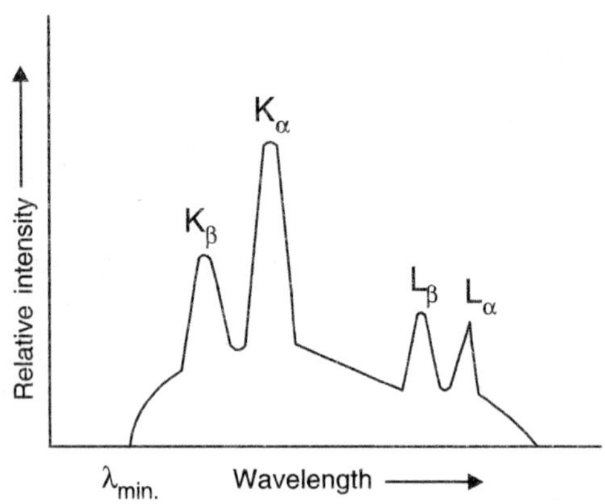

Fig. 2.3

Let us now consider the case of an electron which hits the target with maximum speed. Suppose this electron is deflected by the atom to such an extent that it completely stops. In this case, the X-ray emitted will have maximum frequency (or minimum wavelength).

Now, $$\frac{1}{2}mv^2_{max.} = h\nu_{max.} = \frac{hc}{\lambda_{min}} \qquad \left[\because \ \nu = \frac{c}{\lambda}\right]$$

But $$\tfrac{1}{2}\,mv^2_{max.} = eV$$

where V is the accelerating potential of the cathode beam.

$$\therefore \qquad \frac{hc}{\lambda_{min}} = eV \quad \text{or} \quad \lambda_{min} = \frac{hc}{eV}$$

Putting the values of h, c and e, we get

$$\lambda_{min} = \frac{12396}{V} \text{ Å}$$

$$\therefore \qquad \boxed{\lambda_{min} \propto \frac{1}{V}}$$

So, the $\lambda_{min.}$ of continuous X-ray spectrum is inversely proportional to the accelerating potential of the cathode beam. The radiations of the continuous X-ray spectra are called *Bremsstrahlung* which, in German, means 'braking radiation'. This is a fairly correct description of continuous radiation because it is actually produced by the 'slowing down' of the incident beam of electrons.

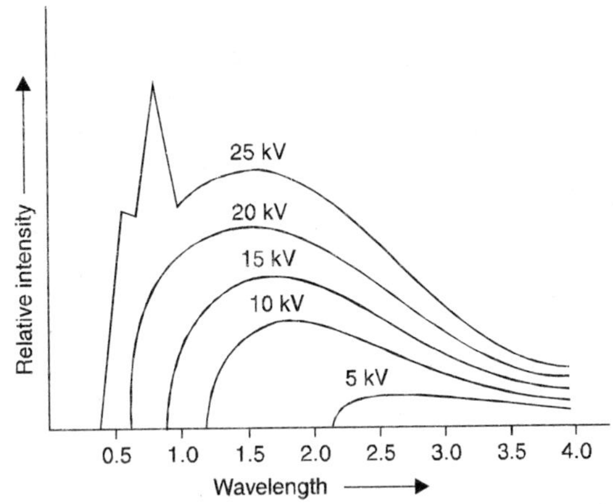

Fig. 2.4

In Fig. 2.4, graphs have been drawn between wavelength and relative intensity for different values of accelerating potential (potential difference between cathode and target) in the case of molybdenum target. It is clear from the figure 2.4, that on increasing accelerating potential, the relative intensity increases, however the $\lambda_{min.}$ decreases. At lower acceleration potentials, we notice only the continuous spectra. However, at an accelerating potential of 25 kV, K_α and K_β characteristic lines are also noticed.

(ii) *Characteristics of sharp line X-ray spectrum*. According to Bohr's theory of atom, each atom has a nucleus around which a number of electrons are revolving in various shells K, L, M, N etc. as dictated by quantum conditions. Sometimes, it so happens that a high speed electron impinging on the target (Fig. 2.5 (*a*)) penetrates through the outer shells of an atom and strikes so hard against an electron in a K shell that the electron of K shell is knocked out. This creates a vacant space in K shell.

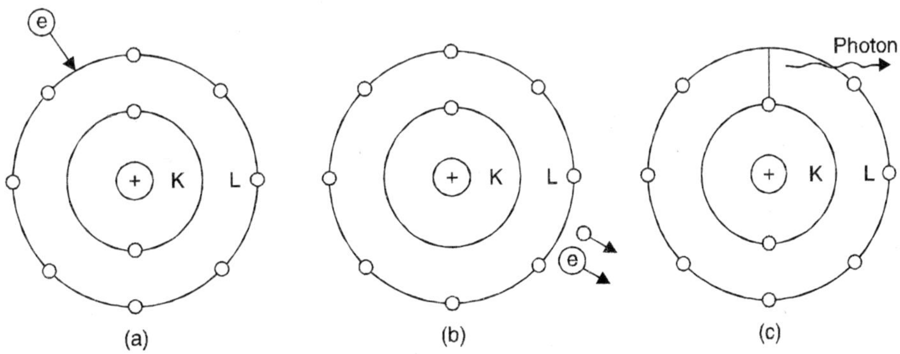

Fig. 2.5

But we know that the vacancy cannot exist for more than 10^{-8} second. So, almost immediately, an electron from some upper shell, say L shell, makes a quantum jump to the empty K shell. The difference of energies of the two shells is radiated out as electromagnetic radiation. The frequency of this radiation lies in the X-ray region of the electromagnetic spectrum and is given by

$$\nu = RZ^2 \left(\frac{1}{n_1^2} - \frac{1}{n_2^2} \right)$$

where Z is the atomic number of the target material and R is the Rydberg constant. For K shell, $n_1 = 1$. For L shell, $n_2 = 1$.

Thus, we get one characteristic line. The vacancy that was caused in L shell is filled by quantum jump of an electron from, say, M shell to L shell. In this way, a radiation of L-series is given out. This process of 'filling a vacancy' continues till the outermost orbit is reached. Thus, we get a series of characteristic spectral lines.

K-series of spectral lines has maximum energy and is, therefore, most penetrating. It is obtained by the quantum jumps of electrons from L, M and N orbits to the innermost K orbit. The respective X-ray lines are K_α, K_β and K_γ (Fig. 2.6).

Similarly, the L-series of characteristic spectral lines L_α, L_β and L_γ are obtained by the quantum jumps of electrons from the outer shells M, N, O to the L-shell respectively. In the same manner, we can get other series M, N, O etc. The characteristic spectral lines can be used to identify the target material because they have a characteristic property of the target material.

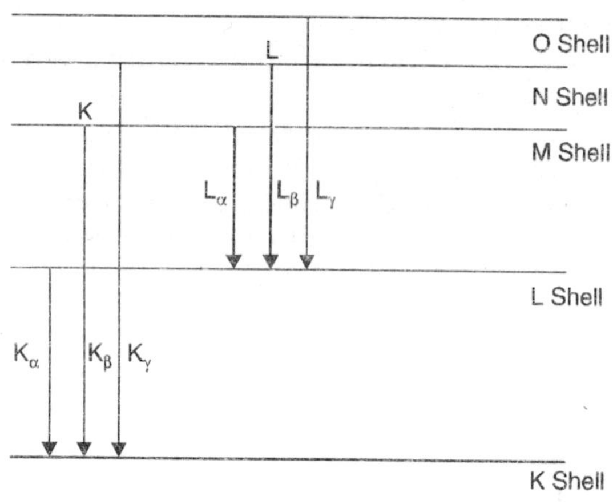

Fig. 2.6

2.07 MOSELEY'S LAW

Moseley examined the characteristic X-ray spectra of thirty five different elements in 1913. He observed that if the square root of the frequency ν of the most intense spectral line *i.e.*, K_α line is plotted against atomic number Z of the target, we get a straight line graph (Fig. 2.7) for every element. This graph is known as Moseley diagram. It is clear from the graph that

$$\nu \propto (Z - b)^2$$

where b is the intercept on the atomic number axis.

Now, $\nu = \alpha (Z - b)^2$ ...(1)

where α is a constant depending upon the spectral line. As an example, the value of α for K_α line is $\frac{3}{4}$ RC where R is the Rydberg constant and c is the speed of light. Equation (1) is the mathematical statement of Moseley's law. This law is applicable to other series also but the values of α and b are different.

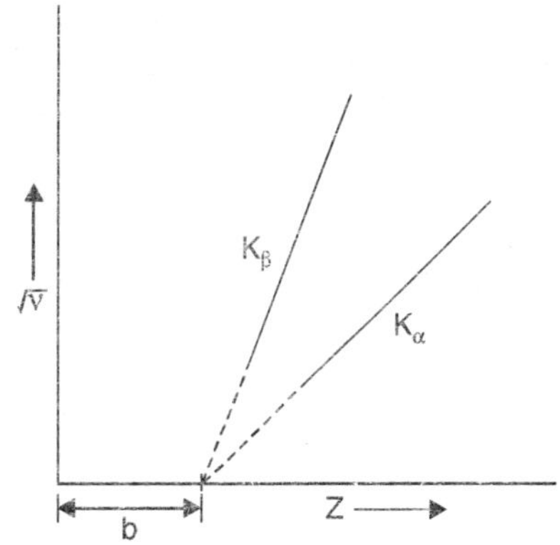

Fig. 2.7

2.08 ABSORPTION OF X-RAYS

X-rays are very penetrating indeed. But as these pass through some thickness of material, their intensity is decreased. Thus, the material has partly absorbed the X-rays. X-rays having greater penetrating power are called *hard* rays, while those having less penetrating power are called *soft* rays. In general, we know that K series radiations are about 300 times harder than corresponding L series from the same target atom.

The *intensity* of the X-ray beam at any given point is defined as the *amount of energy carried per second across a unit area held normally to the direction of flow of energy*. Let I_0 be the initial intensity of X-ray beam incident normally on a material. Let the intensity become I after travelling through thickness x of the material. If dI be the decrease in the intensity of beam in transversing a distance dx in the material, then

$$\text{rate of decrease of intensity} = -\frac{dI}{dx}$$

Now, $$-\frac{dI}{dx} \propto I \quad \text{or} \quad -\frac{dI}{dx} = \mu I$$

where μ is the absorption coefficient of the material.

Again, $$\frac{dI}{I} = -\mu\, dx$$

Integrating, $$\int_{I_0}^{I} \frac{1}{I} dI = \int_0^x -\mu dx$$

or $$\Big[\log_e I\Big]_{I_0}^{I} = -\mu \Big[x\Big]_0^x$$

or $$\log_e I - \log_e I_0 = -\mu (x - 0) = -\mu x$$

or $$\log_e \frac{I}{I_0} = -\mu x \quad \text{or} \quad \frac{I}{I_0} = e^{-\mu x}$$

or $$\boxed{I = I_0 e^{-\mu x}}$$

It is clear from the above relation that there is an *exponential fall* in the intensity of beam.

Absorption coefficient. Its value depends upon the wavelength λ of X-rays and the atomic number Z of the material. It is given by

$$\mu = c Z^4 \lambda^3 \rho$$

where c is a constant and ρ is the density of the absorbing material. It may be pointed that absorption coefficient exhibits sudden discontinuities for certain values of λ. However, on the whole, the variation of μ with x is smooth.

2.09 PRACTICAL APPLICATIONS OF X-RAYS

Applications and Uses of X-rays

X-rays, practically soon after their discovery in 1895, were put to a diverse range of applications even before more sophisticated methods for its production, like Coolidge tube were invented to produce X-rays of high quality and quantity. This single discovery is a good example of the utility of science in daily life. Some of the important applications are listed below.

1. In medical field

X-rays are used both for diagnosis and cure. *Radiography* is the branch of science used for diagnosis. X-rays can pass through light bodies like clothes, flesh etc. and are absorbed by thick substances like bones, metals etc. So the presence (or absence) of any fracture in the bones can be detected. Likewise the presence or absence of a metal body inside a human system can be detected. The stomach is studied by giving a meal of barium sulphate (which is opaque to X-rays), so that its progress within the system is studied and the source of trouble located. Likewise, iodine is used to study the urinary system in a body. Thus, ulcers, ruptures etc. are diagnosed. Diseased lungs can also be studied with the help of X-rays.

Radiotherapy is the process of treatment by the use of X-rays. The fact is that X-rays are harmful to the tissues—but more harmful to the diseased tissues. Thus, the diseased part having ulcer, cancer etc. is exposed to X-rays. The tumour or the diseased tissue dies quickly and thus the patient may get cured in due course of time.

2. In engineering and industry

Here, we make use of the penetrating power of X-rays. These rays are used in detecting flaws in heavy girders ; flaws, cracks and air bubbles in certain finished goods, the presence of pearls in oysters, testing the uniformity of the insulating material. Detection of faults in radio valves, tennis balls, metal, castings and welded joints is also carried out with the help of X-rays.

3. In detective departments

Here also the property of penetration of X-rays is utilised in making the detection of the presence of smuggled gold, at airports, sea-ports etc. The contents of parcels, the false bottoms of suitcases carrying smuggled watches etc., is all carried by mobile X-ray units.

4. In research

X-rays are widely employed for research and obtaining deeper knowledge into the crystal structure and various other allied effects of secondary radiations. By knowing the X-rays scattered by elements, we know their inner structure and chemical composition etc.

2.10 X-RAY DIFFRACTION

Diffraction is a phenomenon exhibited by waves. But for observing the phenomenon of diffraction, the size of the obstacle or slit should be comparable with the wavelength of the radiations. X-rays have wavelengths of the order of 10^{-10} m. Therefore diffraction could not be produced by the available grating of

> ***X-rays are electromagnetic waves of very short wavelength of the order of 10^{-8} cm.***
>
> ***Attempts to show diffraction of X-rays with the help of gratings failed. Diffraction pattern with a grating can be obtained only if the grating element is comparable with the wavelength of incident radiations.***

approximately ruled with 15000 lines per inch and it is not possible to rule a grating having a grating element of the order of 10^{-10} m.

Max Von Laue in 1913 suggested that the diffraction of X-rays may be observed with natural crystal in which atoms are regularly arranged at equal distances known as lattice. The spacing between two lattice planes is of the order of 10^{-10}m. Therefore the natural crystal may act as three dimensional diffraction grating for X-rays. This was later confirmed by experiment conducted by Max Von Laue and his associates.

Experimental Arrangement

In 1913 Max Von Laue with his associates W. Friedrich and P. Knipping succeeded in diffracting X-rays. The experimental arrangement consisted of a source of X-rays, *i.e.,* a X-rays tube. These X-rays were passed through the fine holes in two lead screens S_1 and S_2 as shown in Fig. 2.8. A fine beam of X-rays emerging out of them was made incident normally over the surface of a Zinc Sulphide (Zinc Blende) crystal C. The emerging X-rays were made to fall on a photograhic plate PP. The photographic plate was exposed for quite a long time. (In actual experiment it was 40 minute exposure). After developing, it was found that there was a big bright central spot surrounded by a number of symmetrically arranged faint spots called Laue spots. Laue knew that he had met with success, as these spots could only be due to diffraction from the atoms of crystal. Careful examination showed that each spot corresponds to a particular angle of diffraction.

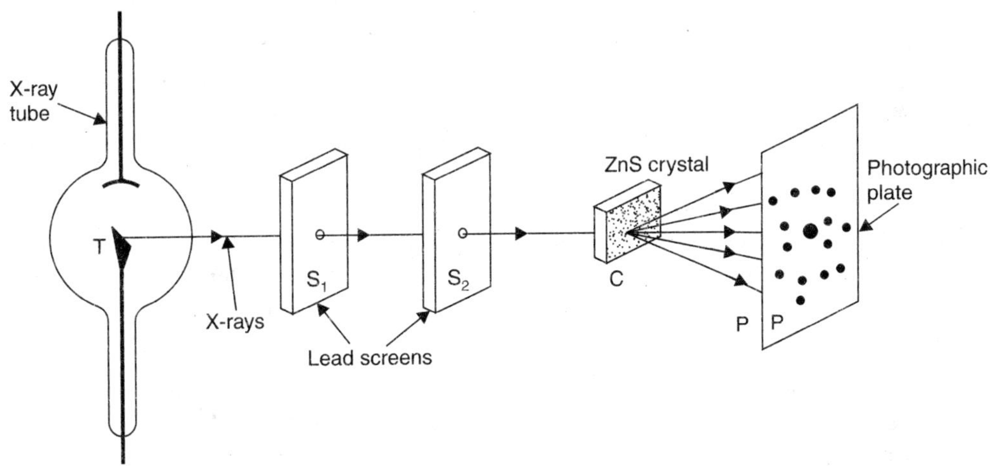

Fig. 2.8

Conclusions

The experiment confirmed the following :

(i) The X-rays are electromagnetic radiations of very short wavelength of the order of the distance between two atomic layers of the crystal.

(ii) Diffraction of X-rays was possible and the Laue's spots were due to the diffraction of X-rays while passing through crystal lattice.

(iii) The atoms in a crystal are arranged in a regular three dimensional arrays. In crystal the interatomic distance is of the order of 10^{-8} cm to 10^{-9} cm, *i.e.,* of the order of wavelength of X-rays. Thus a crystal acts as a three dimensional grating.

Explanation

Atoms in crystal act as secondary sources of disturbances. Thus diffracted X-rays spread out in all directions. These diffracted X-rays interfere and the result is determined by their relative phases. For certain directions and for particular wavelength, constructive interference will take place. Thus, diffraction pattern is formed. This explains the formation of Laue spots.

Soon after the Laue experiment W.L. Bragg explained the diffraction phenomenon in case of X-rays using the arrangement called Bragg spectrograph.

Crystal Structure

The regular arrangement of space positions of atoms in a crystal is called crystal array.

The unit cell of a lattice is the smallest block or geometric figure of crystal repeated again and again to form lattice structure.

According to Bragg, three dimensional crystal can be imagined to be divided into a set of parallel planes called Bragg planes. The combination of atoms on the planes is such that each set has a definite interplaner separation. These separations or spacings are called Bragg spacings.

In case of crystals where unit cell is a cube (simple cubic structure) in which all the atoms are of the same kind, the side of unit cell is equal to the Bragg spacing.

2.11 DIFFRACTION FROM A SINGLE PLANE OF ATOMS IN A CRYSTAL

Consider an incident plane wavefront AB of X-rays incident on a surface having row of atoms in a crystal plane (Fig. 2.9). The wavelets from this wavefront after falling on the atoms in the crystal plane get diffracted in different direction.

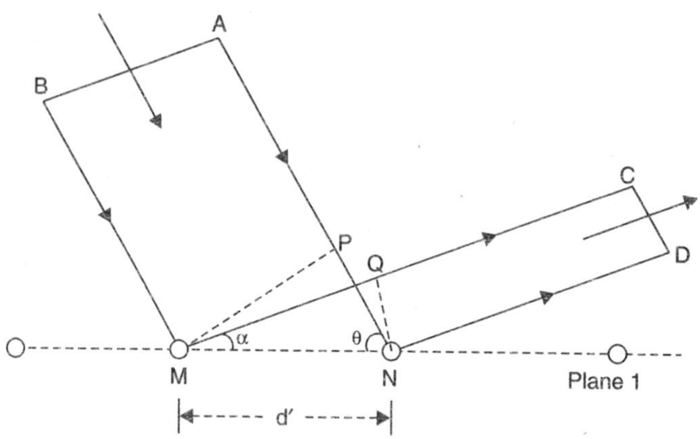

Fig. 2.9

Let d' is distance between adjacent atoms in surface plane, angle between incident X-rays and row of atoms in surface plane be θ. Also if α is angle between diffracted X-rays and surface plane and CD is diffracted plane wavefront.

Then in order to find path difference between diffracted beams of X-rays from adjacent atoms M and N draw MP and NQ perpendicular on AN and MC respectively.

Then, path difference = MQ – PN.

Now in ΔPMN, $\dfrac{PN}{MN} = \cos\theta$

or $$PN = MN \cos\theta = d' \cos\theta$$

In $$\Delta QMN, \frac{MQ}{MN} = \cos\alpha$$

$$MQ = MN \cos\alpha = d' \cos\alpha$$

$\therefore$ $$\text{Path difference} = d' \cos\alpha - d' \cos\theta$$

For constructive interference this path difference must be an integral multiple of wavelength λ,

i.e., $$d' \cos\alpha - d' \cos\theta = n\lambda.$$

2.12 BRAGG'S LAW

After Laue's experiment William Bragg devised another technique to study the diffraction of X-rays. He considered the parallel set of lattice planes for the scattering of X-rays. These X-rays are scattered by the atoms of two adjacent planes in all direction. The constructive interference will take place when the angle of incidence is equal to angle of scattering and incident ray, scattered ray and normal to the lattice plane should lie in one plane.

Consider a narrow monochromatic X-ray beam falling on the cleavage planes of a crystal at a glancing angle θ. The ray AB is reflected or scattered along BE. The other ray CD reflected or scattered from the second cleavage plane along DF. For the sake of simplicity, let us assume that the angle which the incident beam of X-rays makes with the plane is equal to the angle which diffracted beam of X-rays makes with the plane.

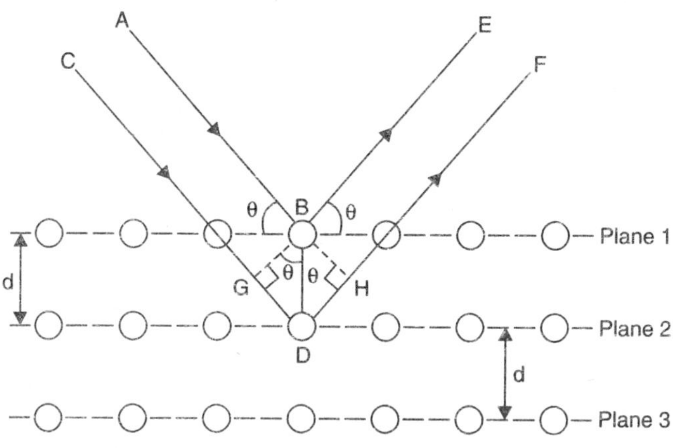

Fig. 2.10

The reflected rays BE and DF will reinforce and interfere with each other. From B, drop perpendiculars BG and BH on CD and DF respectively. Now, BG is the incident plane wavefront of X-rays while BH is the reflected wavefront. Path difference between rays, $p = GD + DH$.

Let d (= BD) be the distance between the successive cleavage planes of the crystal.

Then, $GD = DH = BD \sin\theta = d \sin\theta$

$\therefore$ Path difference, $p = d \sin\theta + d \sin\theta = 2d \sin\theta$

Applying the condition of constructive interference,

$$\boxed{2d \sin\theta = n\lambda}$$

where $n = 1, 2, 3, \ldots$

This equation called Bragg's equation is the mathematical statement of Bragg's law. Here, n is the order of the spectrum, d is called the lattice constant and θ is called Bragg angle.

If, in Bragg's equation, $n = 1$ and glancing angle θ_1, we have

$$2d \sin \theta_1 = \lambda$$

which is the condition for first order reflection. For $n = 2$, we get

$$2d \sin \theta_2 = 2\lambda$$

which gives the condition for second order reflection of the same wavelength at a larger glancing angle θ_2.

It is clear from Bragg's equation that if the wavelength of X-rays, *i.e.,* λ is known, the distance between two consecutive atomic layers, d can be calculated and vise versa.

Additional Information

$$2d \sin \theta = n\lambda$$

This is called Bragg's diffraction formula or Bragg's equation. Within the crystal there are many crystal planes from where Bragg's diffraction can result. But the certain set of planes the diffraction is more intense.

Elements with high atomic number diffract incident X-rays more effectively than those with low atomic numbers.

KNOWLEDGE PLUS

Sir William Bragg and Sir Lawrence Bragg (father and son respectively) devised a method for determining the wavelength of X-rays. This method is based on the fact that the cleavage planes, in some crystals, are parallel and the atoms in these are also at equal distances from each other. A narrow beam of X-rays is made to fall obliquely at the plane surface of a crystal at an angle θ with its surface. This is called *glancing angle*. It may be clearly noted that this is not the angle of reflection. In fact it is complimentary to the angle of reflection as understood in the case of light.

W.L. Bragg derived the condition under which the diffraction of X-rays takes place while passing through a crystal.

He imagined three dimensional crystal lattice to be divided up into a number of different sets of parallel planes so that all the lattice points lay on one of the planes and no point lay outside them. Each diffraction spot was produced by the interference between the rays reflected from a particular set of parallel planes within the crystal.

It is now easy to derive a condition which should be satisfied so that the rays scattered by the various atoms in a crystal may undergo constructive interference to produce an intense spot on the photographic plate. Consider a set of parallel atomic planes whose spacing is d and let a narrow monochromatic beam of X-rays fall upon these parallel planes at a glancing angle θ. Each atom in a given layer according to Huygen's principle, becomes the centre of expanding wavelets whose envelope gives rise to reflected wavefront. As in the case of optical reflection, the reflected wavefront will have maximum intensity at an angle θ to this plane which is equal to the glancing angle. Each parallel layer in the given set gives rise to the reflected wavefront. *The condition for constructive interference between the reflected wavefronts is that the path difference between the reflected wavefronts from one layer and that from the next must be equal to an exact wavelength or an integral multiple of it.*

2.13 BRAGG'S X-RAYS SPECTROMETER

For detailed study of crystal structure, Bragg design a spectrometer whose schematic diagramme is shown in figure (2.11). In this spectrometer there is a turn table which is capable of rotation about a vertical axis passing through its centre. The rotation can be read on a circular graduated scale V_1. X-rays from X-ray tube are made incident through slits S_1 and S_2 of lead screen on crystal C mounted on turn table. The reflected beam then passes through slit and enters in the ionization chamber. This ionization chamber filled with ethyle bromide and mounted on a special movable arm about same axis of crystal. The position of ionization chamber can be read through another scale V_2. The turn table and ionization chamber are coupled together in such a way that when turn table rotates through an angle of θ, the ionization chamber turns automatically by an angle of 2θ so that beam always reflect into ionization chamber. The X-rays entering the chamber ionize the gas which causes a current and measured by an electrometer.

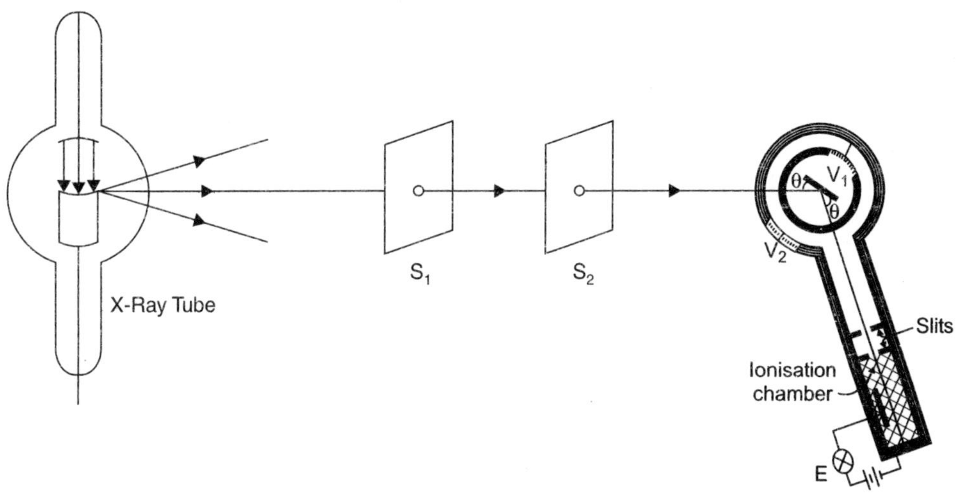

Fig. 2.11

Now the glancing angle θ is gradually increases by rotating the turn table and ionization current is recorded. A graph then plotted between ionization current and glancing angle θ. For certain value of θ, the ionization current increases abruptly. The number of peaks A_1, A_2, A_3... in figure 2.12 are the corresponding glancing angle θ which satisfy Bragg's equation

$$2d \sin \theta = n\lambda.$$

for first order spectrum

$$n = 1 \text{ and } 2\, d \sin \theta_1 = \lambda$$

for second order $\quad n = 2 \text{ and } 2\, d \sin \theta_2 = 2\lambda$

for third order $\quad n = 3 \text{ and } 2\, d \sin \theta_3 = 3\lambda$

$$\therefore \sin \theta_1 : \sin \theta_2 : \sin \theta_3 = 1 : 2 : 3$$

hence glancing angle for different order can be directly measured from graph.

If X-ray beam consist of two wavelength, λ_1 and λ_2 then peaks like A_1B_1 ; A_2B_2 ; A_3B_3 are obtained. In figure 2.13 peaks $A_1A_2A_3$ are for wavelength λ_1 and $B_1B_2B_2$ are for λ_2. The Bragg's equation is satisfied for both wavelengths and then

$\sin \theta_1 : \sin \theta_2 : \sin \theta_3 = 1 : 2 : 3$ for λ_1 . and

$\sin \theta_1' : \sin \theta_2' : \sin \theta_3' = 1 : 2 : 3$ for λ_2.

where θ and θ' are the glancing angle for wavelength λ_1 and λ_2 respectively.

Therefore from measured value of θ and known value of d and order, the wavelength of X-ray can be determined. There are some observation from the graph.

(i) The intensity of the reflected X-rays decreased with the increase of order of spectrum.

(ii) The ionization current does not fall to zero for any value of glancing angle θ. Its value is maximum for certain angle θ. This shows that a continuous spectrum exist for all angles which is superimposed by a line spectrum:

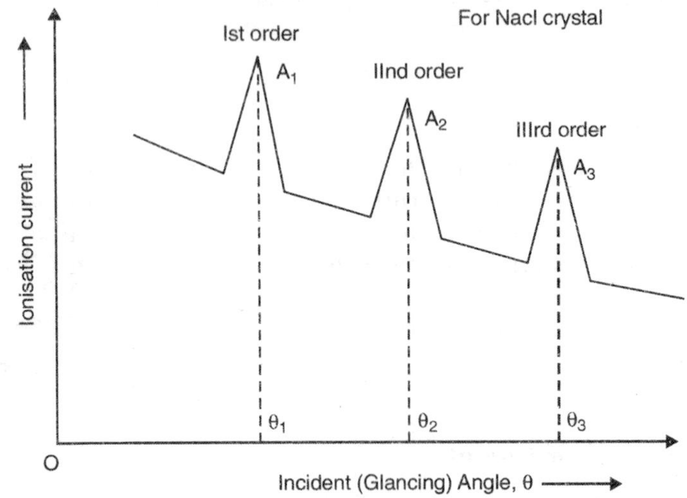

Fig. 2.12

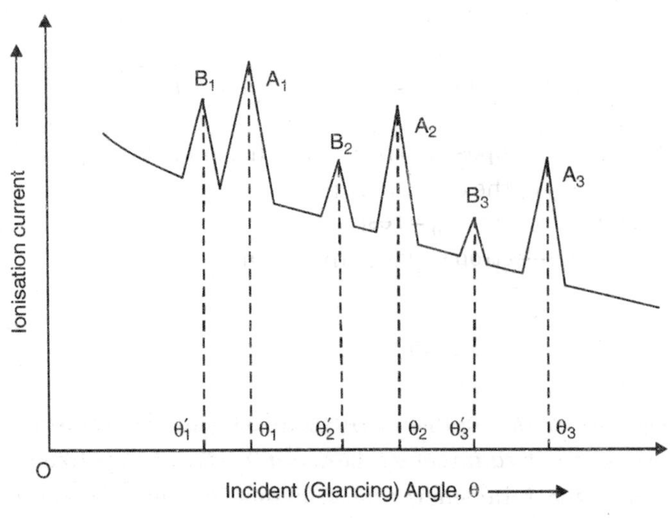

Fig. 2.13

2.14 COMPTON EFFECT

In 1921, Prof. A.H. Compton discovered that when a monochromatic beam of high frequency radiation such as X-rays is scattered by a substance such as carbon, the scattered radiation has two components. One component has the same wavelength (or frequency) as that of the incident radiation. While the other component is observed to have wavelength slightly greater than that of the incident radiation. The radiation of unchanged wavelength is called the unmodified radiation, the radiation of greater wavelength is called the modified radiation.

The phenomenon according to which the wavelength of the radiations scattered by an element is greater than of incident radiation is called compton effect.

The difference in the wavelength of the scattered and incident radiations is called **Compton shift.** Thus if λ is the wavelength of the incident radiation and λ' that of the scattered radiation, then

Compton shift, $$\Delta\lambda = \lambda' - \lambda$$

Cause

The scattering process can be considered to be an elastic collision between a photon of incident radiation and an electron of the scattering material. During the collision, the photon transfers a part of its energy to the electron and in turn gets scattered at a certain angle. The electron after absorbing some energy, flies away in a certain direction and the scattered photon carries less energy, its frequency decreases. In accordance with Planck's relation $E = h\nu = \frac{hc}{\lambda}$, the wavelength increases. Thus the wavelength of scattered X-rays is longer than that of the incident radiations.

Mathematical expression for Compton shift

Following **conditions** are required for Compton shift to take place.

(*i*) A free electron is assumed to be present in the scattering material.

(*ii*) The free electron is supposed to be at rest for the duration of the collision.

(*iii*) The collision between an electron and a photon is perfectly elastic and the total energy and momentum remain conserved.

(*iv*) The energy of the photon is much greater than the threshold energy of the free electron colliding with it.

(*v*) The beam of X-rays is supposed to be a monochromatic beam.

(*vi*) The rest mass energy of the electron is to be taken into account according to Einstein's mass energy relation, $E = m_0c^2$ where m_0 = rest mass of the electron and c = velocity of light.

(*vii*) After the collision, the electron moves with a certain velocity v given by Einstein's equation for relativistic variation of mass, *i.e.*,

$$m = \frac{m_0}{\sqrt{1 - v^2/c^2}}.$$

(*viii*) *A very small part of the energy of incident photon gets transferred to the recoiled electron. The rest of the energy is radiated as scattered X-rays of slightly greater wavelength.*

In such a collision, the total energy and momentum remain conserved. As shown in Fig. 2.14, let

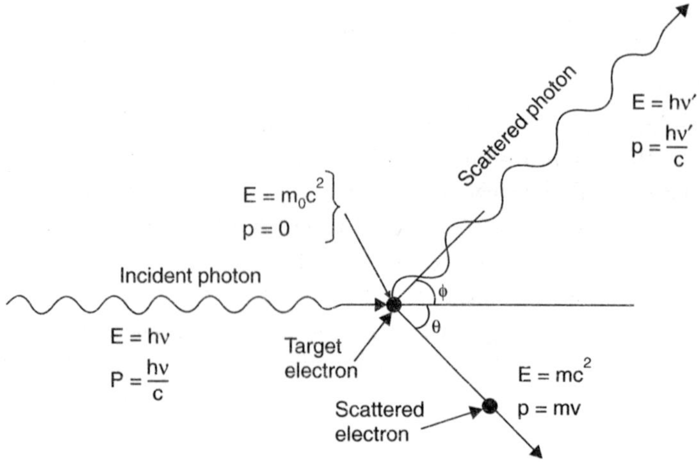

Fig. 2.14. X-ray photon being scattered by an electron.

λ = the wavelength of the photon

λ' = the wavelength of the scattered photon

ν = frequency of the incident photon

ν' = frequency of the scattered photon

ϕ = angle made by the scattered photon with the direction of incident photon

θ = angle made by the recoiled electron with the direction of the incident photon.

m_0 = rest mass of the electron,

m = mass of the recoil electron moving with a velocity v

and c = velocity of light

For calculating Compton shift, electron is considered to be at rest having rest mass energy equal to m_0c^2.

From the principle of conservation of energy, we have

$$h\nu + m_0c^2 = h\nu' + mc^2$$

or

$$mc^2 = m_0c^2 + h(\nu - \nu')$$

Squaring both sides,

$$m^2c^4 = m_0^2c^4 + h^2(\nu^2 + \nu'^2 - 2\nu\nu') + 2m_0 \cdot c^2h(\nu - \nu') \quad ...(1)$$

Unlike energy, momentum is a vector quantity, involving both magnitude and direction. The momentum must, therefore, remain conserved both along the original direction of motion of the photon and also along the direction perpendicular to it in the plane containing the electron and the scattered photon.

Applying the principle of conservation of momentum in the original direction of the incident photon, we have

$$\frac{{}^*h\nu}{c} + 0 = \frac{h\nu'}{c} \cos\phi + mv\cos\theta \quad ...(2)$$

Applying this law along perpendicular direction, we get

*The energy of a photon of mass m is given by $E = mc^2$. $\therefore$ $mc = E/c = h\nu/c$ which gives the momentum of a photon.

$$0 = \frac{h\nu'}{c} \sin \phi - mv \sin \theta \qquad ...(3)$$

From Eq. (2), $mvc \cos \theta = h\nu - h\nu' \cos \phi$ ...(4)

From Eq. (3), $mvc \sin \theta = h\nu' \sin \phi$ ...(5)

Squaring and adding equations (4) and (5), we have

$$m^2v^2c^2(\cos^2 \theta + \sin^2 \theta) = h^2\nu^2 + h^2\nu'^2 \cos^2 \phi - 2h^2 \nu\nu' \cos \phi + h^2\nu'^2 \sin^2 \phi$$

or
$$m^2v^2c^2 = h^2\nu^2 + h^2\nu'^2 - 2h^2\nu\nu' \cos \phi$$

or
$$m^2v^2c^2 = h^2(\nu^2 + \nu'^2 - 2\nu\nu' \cos \phi) \qquad ...(6)$$

Subtracting (6) from (1), we get

$$m^2c^2(c^2 - v^2) = m_0^2c^4 - 2\nu\nu' h^2(1 - \cos \phi) + 2m_0c^2h(\nu - \nu')$$

Substituting
$$m = \frac{m_0}{\sqrt{1 - v^2/c^2}}$$

the above equation becomes

$$m_0^2c^4 = m_0^2c^4 - 2\nu\nu' h^2(1 - \cos \phi) + 2m_0c^2h(\nu - \nu')^*$$

or
$$2\nu\nu' h^2(1 - \cos \phi) = 2m_0c^2h(\nu - \nu')$$

or
$$\frac{\nu - \nu'}{\nu\nu'} = \frac{h}{m_0c^2} (1 - \cos \phi)$$

or
$$\frac{1}{\nu'} - \frac{1}{\nu} = \frac{h}{m_0c^2} (1 - \cos \phi)$$

$$\frac{c}{\nu'} - \frac{c}{\nu} = \frac{h}{m_0c} (1 - \cos \phi) \qquad ...(7)$$

If λ and λ' are the wavelengths corresponding to the frequencies ν and ν', we have $\frac{c}{\nu} = \lambda$ and $\frac{c}{\nu'} = \lambda'$.

Thus the equation (7) becomes

$$\boxed{\lambda' - \lambda = \Delta\lambda = \frac{h}{m_0c} (1 - \cos \phi)} \qquad ...(8)$$

This equation was first derived by Compton and it gives the increase $\Delta\lambda$, (called *Compton shift*), in wavelength of an X-ray photon when scattered through an angle ϕ by a particle of rest mass m_0.

2.15 DIRECTION OF RECOIL ELECTRON

To find the direction of recoil electron divide eq. (5) by eq. (4)

$$\tan\theta = \frac{h\nu' \sin\phi}{h\nu - h\nu' \cos\phi}$$

$$= \frac{\nu' \sin\phi}{\nu - \nu' \cos\phi} \qquad ...(9)$$

using eq. (7)

$$\frac{1}{\nu'} = \frac{1}{\nu} + \frac{h}{m_0c^2}(1-\cos\phi)$$

$$= \frac{1}{\nu} + \frac{h}{m_0c^2}\left(2\sin^2\frac{\phi}{2}\right)$$

$$= \frac{1+\left(h\nu/m_0c^2\right)2\sin^2\frac{\phi}{2}}{\nu}$$

$$\nu' = \frac{\nu}{1+\left(h\nu/m_0c^2\right)2\sin^2\frac{\phi}{2}} \qquad ...(10)$$

putting the value of ν' from eq. (10) in eq. (9) we get

$$\tan\theta = \frac{\nu\sin\phi/\left[1+\alpha.2\sin^2\frac{\phi}{2}\right]}{\nu-\left[\frac{\nu}{1+\alpha.2\sin^2\frac{\phi}{2}}\right]\cos\phi} \qquad \left[\alpha = \frac{h\nu}{m_0c^2}\right]$$

$$= \frac{\sin\phi}{1+2\,.\,\alpha\sin^2\frac{\phi}{2}-\cos\phi}$$

$$= \frac{\sin\phi}{(1-\cos\phi)+2\alpha\sin^2\frac{\phi}{2}}$$

$$= \frac{2\sin\frac{\phi}{2}\cos\frac{\phi}{2}}{2\sin^2\frac{\phi}{2}+2\alpha\sin^2\frac{\phi}{2}}$$

$$= \frac{\cos\frac{\phi}{2}}{\sin\frac{\phi}{2}(1+\alpha)} = \frac{\cot\frac{\phi}{2}}{1+\alpha}$$

$$\tan\theta = \frac{\cot\frac{\phi}{2}}{\left[1+\left(h\nu/m_0c^2\right)\right]} \qquad ...(11)$$

Therefore recoil angle θ depends on scattering angle φ.

When $\phi = 0$ $\theta = 90°$ and

if $\phi = 180°$ $\theta = 0°$.

hence the electron can recoil only upto 90°.

2.16 KINETIC ENERGY OF RECOIL ELECTRON

The incident photon of energy $h\nu$ deliver part of its energy to electron which recoil at an angle of θ. Therefore, kinetic energy of electron is the difference of energy of photon before and after collision

$$KE = h\nu - h\nu'$$

puting the value of ν′ from eq. (10)

$$KE = h\nu - h\frac{\nu}{\left[1 + 2\alpha\sin^2\frac{\phi}{2}\right]}$$

$$= h\nu\frac{\left[2\alpha\sin^2\frac{\phi}{2}\right]}{\left[1 + 2\alpha\sin^2\frac{\phi}{2}\right]} \qquad ...(12)$$

The electron possess maximum KE when $\sin\frac{\phi}{2} = 1$

or $$\frac{\phi}{2} = \frac{\pi}{2} \text{ or } \phi = \pi$$

hence $$(KE)_{max} = \frac{h\nu\,(2\alpha)}{(1+2\alpha)}$$

$$= \frac{2h^2\nu^2}{m_0c^2\left(1+\frac{2h\nu}{m_0c^2}\right)} \qquad ...(13)$$

Eq. (13) shows the maximum kinetic energy that a recoil electron can have.

2.17 FACTORS ON WHICH THE COMPTON SHIFT DEPENDS

The compton shift Δλ is given by

$$\lambda' - \lambda = \Delta\lambda = \frac{h}{m_0c}(1 - \cos\phi)$$

(*a*) It is clear from the expression that *Compton shift depends only on φ, the angle of scattering, the values h, m_0 and c being constants.*

For $\phi = 0$, that is in the direction of incident photon, $\Delta\lambda = 0$.

For $\phi = \frac{\pi}{2}$, that is in the direction perpendicular to that of incident photon, $\Delta\lambda = \frac{h}{m_0 c}$.

This value of the Compton shift is termed as **Compton wavelength** of electron. It is a constant having value equal to 0.024 A or 2.4×10^{-12} m.

Fro $\phi = \pi$, that is in a direction *opposite to that of the incident* photon, *Compton shift is maximum* and is equal to $\Delta\lambda = 2\frac{h}{m_0 c}$, which is twice the Compton wavelength of the electron.

(*b*) From the expression (8) it is clear that the *Compton shift does not depend on the nature of the material* used for the purpose of scattering.

(*c*) Also *Compton shift is independent of the wavelength of the incident photon,* as λ does not occur in the relation (8).

These results were confirmed experimentally by Compton.

Experimental demonstration of Compton effect. Fig. 2.15 shows the schematic representation of the experimentally set up used to study Compton effect. A beam of X-rays of a single, known wavelength is obtained from a source and, after confining it through narrow slits, is made to fall upon a scattering substance. The radiations scattered at some angle ϕ are made to fall upon the crystal of an X-ray spectrometer and the wavelengths of the scattered radiations determined as already explained. The spectrometer must be capable of moving around the scattering substance in an arc, so as to measure the wavelength of the scattered radiations at any scattering angle ϕ.

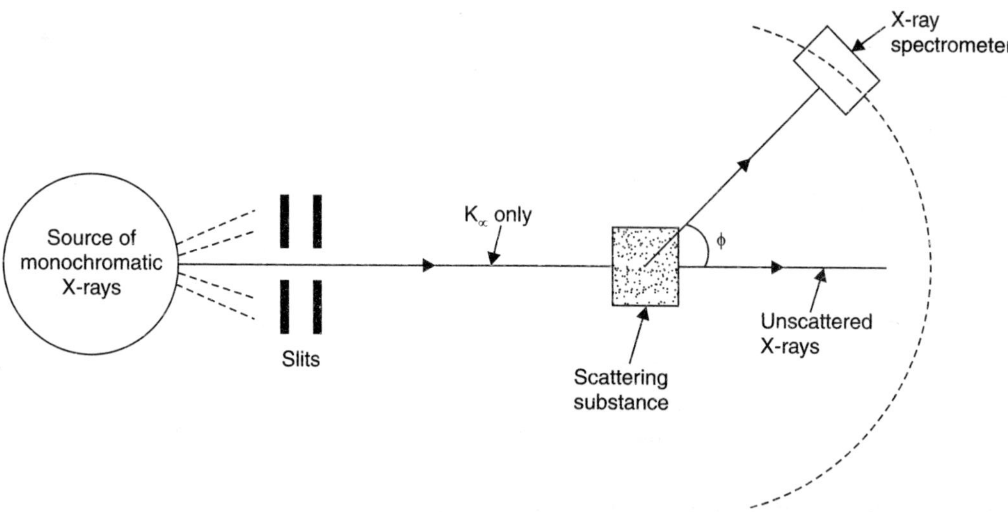

Fig. 2.15

The results of the wavelength displacement for the particular wavelength of X-rays known as Kα line from molybdenum target scattered by carbon for four different values of scattering angles have been shown in Figure 2.16. At each angle, the scattered beam consists of two distinct lines ; the unmodified line has the same wavelength as the incident radiation whereas the modified line has a longer wavelength, depending upon the angle of scattering. The wavelengths of the modified lines at different scattering angles are found to be in good agreement with those calculated by Eq. (8).

Significance of Compton scattering

1. It provides a convincing proof of the particle nature of the electromagnetic radiations.

2. As the Compton shift has been computed on the basis of

(*a*) Planck's quantum hypothesis

and (*b*) the relativistic relations

$$E = mc^2 \quad \text{and} \quad m = m_0\left[1 - \frac{v^2}{c^2}\right]^{-1/2}$$

it provides an indirect method for the verification of quantum hypothesis and theory of relativity.

2.18 COMPTON EFFECT WITH VISIBLE LIGHT

From eq. (8) the maximum compton shift is observed at $\theta = 180°$ and which is

$$(\Delta\lambda)_{max} = \frac{h}{m_0c^2}(1 - \cos 180°) = \frac{2h}{m_0c^2}$$

$$= 2 \times 0.0242 \text{ Å}$$

$$= .05 \text{ Å}$$

This shift is maximum whatever the wavelength of incident radiation used. Percentage of wavelength shift for higher wavelength is smaller than lower wavelength. Since the visible light are of higher wavelength (≈ 5000Å) the shift is very small (.001%). For example

% shift for X-rays (of 1Å)

$$\% \text{ shift} = \frac{.05 \times 100}{1} = 5\%$$

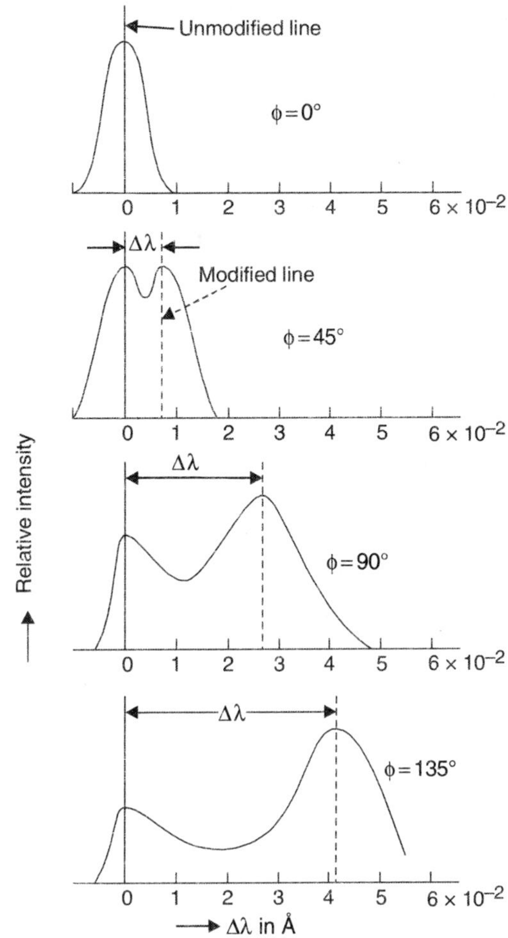

Fig. 2.16

% shift for visible light (of 5000Å)

$$\% \text{ shift} = \frac{.05 \times 100}{5000} = 0.001\%$$

Since the % shift is very small hence it is not observable for visible light.

2.19 EXPLANATION OF UNMODIFIED RADIATION

Compton also explained the unmodified radiation. The atom of target have free as well as bound electrons. When an incident photon instead of striking with free electron, it strike with electron which tightly bound with the nucleus, then instead of recoiling electron whole atom is recoils and then in eq. (8) m_0 will be considered mass of atom in place of mass of electron. The mass of atom is several thousand times the mass of electron which results the change in wavelength a very small value and is negligible for any angle of scattering. Therefore when a photon collides with bound electron, its wavelength does not change ($\Delta\lambda = 0$) and hence radiation obtained are unmodified radiation.

Arthur Holly Compton (1892–1962), a native of Ohio, was educated at College of Wooster and Princeton While at Washington University in St Louis he found that X-rays increase in wavelength when scattered, which he explained in 1923 on the basis of the quantum theory of light. This work convinced remaining doubters of the reality of photons.

After receiving the Nobel Prize in 1927. Compton, now at the University of Chicago, studied cosmic rays and helped establish that they are fast charged particles (today known to be atomic nuclei, largely protons) that circulate in space and are not high energy gamma rays as many had thought. He did this by showing that cosmic-ray intensity varies with lautude, which makes sense only if they are ions whose paths are influenced by the earth's magnetic field. During World War II Compton was one of the leader in the development of the atomic bomb.

NUMERICAL EXAMPLES BASED ON X-RAYS

■ **Example 2.01.** *Calculate the wavelength of a K_α X-ray for gold atoms. Given : Z (for gold) = 79.*

Solution. When an electron jumps from second orbit to first orbit, K_α X-ray spectrum line is emitted.

$$h\nu = k^2 \frac{2\pi^2 Z^2 me^4}{h^2}\left(\frac{1}{n_1^2} - \frac{1}{n_2^2}\right)$$

or $$\nu = k^2 \frac{2\pi^2 Z^2 me^4}{h^3}\left(\frac{1}{1^2} - \frac{1}{2^2}\right)$$

or $$\nu = \frac{k^2 \times 2\pi^2 z^2 me^4}{h^3}\left(1 - \frac{1}{4}\right)$$

or $$\nu = \frac{(9 \times 10^9)^2 \times 2 \times (3.142)^2 \times (79)^2 \times 9 \times 10^{-31} \times (1.6 \times 10^{-19})^4 \times 3}{(6.63 \times 10^{-34})^3 \times 4} \text{ Hz}$$

$$= 1.512 \times 10^{19} \text{ Hz}$$

or $$\frac{c}{\lambda} = 1.512 \times 10^{19} \text{ Hz}$$

or $$\lambda = \frac{3 \times 10^8}{1.512 \times 10^{19}} \text{ m}$$

or $$\boldsymbol{\lambda = 1.99 \times 10^{-11}} \textbf{ m.}$$

■ **Example 2.02.** *The angle of reflection for monochromatic X-rays from a crystal whose atomic spacing is 2.5 Å is 15°. Calculate the wavelength of X-rays.*

Solution. $\theta = 15°$, $d = 2.5$ Å $= 2.5 \times 10^{-10}$ m, l = ? $n = 1$

$$2d \sin\theta = n\lambda$$

$$\lambda = \frac{2d \sin\theta}{n}$$

or $$\lambda = \frac{2 \times 2.5 \times 10^{-10} \times \sin 15°}{1} \text{ m}$$

or $$\lambda = 5 \times 10^{-10} \times 0.2588\ m$$

$$= 1.294 \times 10^{-10}\ m = \textbf{1.294 Å.}$$

■ **Example 2.03.** *Calculate the maximum frequency of the continuous X-rays from an X-ray tube whose operating voltage is 50,000 volt. Given : e = 1.6 × 10^{-19} C, h = 6.63 × 10^{-34} Js.*

Solution. ν = ?, V = 50,000 volt

$$\text{Energy} = eV = 1.6 \times 10^{-19} \times 50{,}000 \text{ J}$$

Again, $E = h\nu$ or $\nu = \frac{E}{h}$

$$\therefore \quad \nu = \frac{1.6 \times 10^{-19} \times 50{,}000}{6.63 \times 10^{-34}} \text{ Hz}$$

$$= \mathbf{1.21 \times 10^{19}} \textbf{ Hz.}$$

Example 2.04. *Monochromatic X-rays (λ = 1 Å) are reduced to $\frac{1}{3}$ of their original intensity in passing through a gold foil (Z = 79) of 3 mm thickness. Calculate the absorption coefficient for the X-rays. What are the dimensions of the absorption coefficeint ?*

Solution. $\frac{I}{I_0} = \frac{1}{3}$, $x = 3$ mm = 0.3 cm

But $$\frac{I}{I_0} = e^{-\mu x}$$

or $$\frac{1}{3} = e^{-\mu (0.3)}$$

or $$e^{\mu (0.3)} = 3$$

$$\log_e e^{0.3\mu} = \log_e 3$$

or $$0.3\mu = 2.3026 \times 0.4771$$

or $$\mu = \frac{2.3026 \times 0.471}{0.3} \text{ cm}^{-1}$$

$$= \mathbf{3.662\ cm^{-1}.}$$

The dimensional formula of μ is $[M^0L^{-1}T^0]$.

Example 2.05. *The wavelength of a K_α line is 1.54 Å for copper. Calculate the ionisation potential of K shell electron in copper.*

Solution. The ionisation of K shell electron is equal to the energy required to move an electron from K shell to L shell. This is equal to the energy E radiated with the emission of K_α line.

$$E = h\nu = \frac{hc}{\lambda} = \frac{6.63 \times 10^{-34} \times 3 \times 10^8}{1.54 \times 10^{-10}} \text{ J}$$

$$= \mathbf{12.9 \times 10^{-16}\ J.}$$

Example 2.06. *An X-ray tube operates at 18 kV. Find the maximum speed of electron striking the anticathode. Given : mass of electron = 9×10^{-28} g, charge of electron = 4.8×10^{-10} esu.*

Solution. V = 18 kV = 18000 volt

$$= \frac{18000}{300} \text{ esu of potential}$$

$$= 60 \text{ esu of potential}$$

$$m = 9 \times 10^{-28} g,\ e = 4.8 \times 10^{-10} \text{ esu}$$

If the electrons strike the anticathode with maximum speed v,

$$\frac{1}{2} mv^2 = eV \quad \text{or} \quad v^2 = \frac{2\,eV}{m}$$

or $$v = \sqrt{\frac{2\,eV}{m}}$$

$$= \sqrt{\frac{2 \times 4.8 \times 10^{-10} \times 60}{9 \times 10^{-28}}} \text{ cm s}^{-1}$$

$$= \mathbf{8 \times 10^9\ cm\ s^{-1}.}$$

Example 2.07. *In Bragg's reflection of X-rays, a reflection was found at 30° glancing angle with lattice planes of spacing 1.87 Å. If this is a second order reflection, calculate the wavelength of X-rays.* **[UPTU, B.Tech. I, Q. Bank 2001]**

Solution. $2d \sin \theta = n\lambda$ or $\lambda = \frac{2d \sin \theta}{n}$ (Bragg's equation)

Now, $\theta = 30°$, $n = 2$ and $d = 1.87$ Å

$$\therefore \quad \lambda = \frac{2 \times 1.87 \times \sin 30°}{2} \text{ Å}$$

$$= 1.87 \times 0.5 \text{ Å}$$

$$= \mathbf{0.935\ Å.}$$

Example 2.08. *A set of lattice planes reflects X-rays of wavelength 1.32 Å at a glancing angle of 9°30′. Calculate the possible spacing of this set of planes.* **[Model Q. Paper, B. Tech. I, 2001]**

Solution. $2d \sin \theta = n\lambda$

Here, $\lambda = 1.32$ Å, $\theta = 9.5°$

$$d = \frac{n\lambda}{2 \sin \theta} = \frac{n \times 1.32}{2 \times \sin 9.5°} \text{ Å}$$

$$= \frac{n \times 0.66}{0.165} \text{ Å} = \mathbf{4n\ Å.}$$

Putting $n = 1, 2, 3, \ldots$, we get possible spacing of the set of planes.

Example 2.09. *An X-ray beam of wavelength 0.97 Å is obtained in the third order after reflection at 60° from the crystal plane. Another beam is obtained in the first order after reflection at 30° from the same crystal plane. Find the wavelength of the second X-ray beam.* **[UPTU, B. Tech. I, Q. Bank 2001]**

Solution. $2d \sin \theta = n\lambda$ (Bragg's law)

For the first beam, $\theta = 60°$, $n = 3$, $\lambda = 0.97$ Å

$$\therefore \quad 2d \sin 60° = 3 \times 0.97 \qquad \ldots(1)$$

If λ' is the wavelength of the second beam, then

$$2d \sin \theta' = n'\lambda'$$

Here, $\theta' = 30°$ and $n' = 1$

$\therefore \quad 2d \sin 30° = 1 \times \lambda' \quad ...(2)$

Dividing (2) by (1), we get

$$\frac{\lambda'}{3 \times 0.97} = \frac{\sin 30°}{\sin 60°}$$

or
$$\lambda' = 3 \times 0.97 \times \frac{\frac{1}{2}}{\frac{\sqrt{3}}{2}}$$

$$= \sqrt{3} \times 0.97 \text{ Å}$$
$$= 1.732 \times 0.97 \text{ Å}$$
$$= \mathbf{1.68\ Å.}$$

■ **Example 2.10.** *The first order reflection from the plane of NaCl is obtained at an angle $2\theta = 20°$ with the incident beam. If $d_0 = 2.82$ Å, calculate the wavelength of X-ray used.*

[UPTU, B. Tech. I, Q. Bank 2001]

Solution. $2d_0 \sin \theta = n\lambda$

or
$$\lambda = \frac{2d_0 \sin \theta}{n}$$

Here, $d_0 = 2.82$ Å, $2\theta = 20°$

or $\theta = 10°, n = 1$

$$\therefore \quad \lambda = \frac{2 \times 2.82 \times \sin 10°}{1} \text{ Å}$$
$$= 2 \times 2.82 \times 0.1736 \text{ Å}$$
$$= \mathbf{0.979\ Å.}$$

NUMERICAL EXAMPLES BASED ON COMPTON EFFECT

■ **Example 2.11.** *A beam of gamma radiation having photon energy 510 keV is incident on a foil of aluminium. Calculate the wavelength of the radiation at 90°.* **[UPTU, B. Tech. I, 2002]**

Solution. $E = h\nu = \frac{hc}{\lambda}$

or
$$\lambda = \frac{hc}{E}$$

Here $h = 6.63 \times 10^{-34}$ Js, $c = 3 \times 10^8$ m s^{-1}, $E = 510$ keV $= 510 \times 10^3 \times 1.6 \times 10^{-19}$ J

Substituting values,

$$\lambda = \frac{6.63 \times 10^{-34} \times 3 \times 10^8}{510 \times 10^3 \times 1.6 \times 10^{-19}} \text{ m}$$
$$= 2.4375 \times 10^{-12} \text{ m}$$

Compton shift, $\Delta\lambda = \lambda' - \lambda$

$$= \frac{h}{m_0 c}(1 - \cos \phi)$$

When $\phi = 90°, \lambda' = \lambda + \frac{h}{m_0 c}$

$$= 243.75 \times 10^{-10} + \frac{6.63 \times 10^{-34}}{9.1 \times 10^{-31} \times 3 \times 10^8}$$

On simplification, $\lambda' = \mathbf{4.866 \times 10^{-12}}$ **m.**

■ **Example 2.12.** *Calculate the Compton shift for X-rays which undergo a Compton scattering through angle of $r/3$. Given the Compton wavelength of the electron is 2.4×10^{-12} m.*

Solution. Compton shift $\Delta\lambda$ is given by

$$\Delta\lambda = \frac{h}{m_0 c}[1 - \cos \phi]$$

$$\frac{h}{m_0 c} = \text{Compton wavelength of electron}$$
$$= 2.4 \times 10^{-12} \text{ m}$$

and
$$\phi = \frac{r}{3}, \quad \therefore \quad \cos \phi = \frac{1}{2}$$

Hence
$$\Delta\lambda = 2.4 \times 10^{-12}\left(1 - \frac{1}{2}\right)$$

or
$$\Delta\lambda = \mathbf{1.2 \times 10^{-12}\ m.}$$

■ **Example 2.13.** *X-rays of wavelength 1.0×10^{-11} m are allowed to fall on graphite scatterer. The wavelength of the scattered radiation in a certain direction is found to be 1.11×10^{-11}m. Find the direction of the scattered photon. Given $h = 6.6 \times 10^{-34}$ J s, $m_0 = 9 \times 10^{-31}$ kg and $c = 3 \times 10^8$ m s^{-1}.*

Solution. Compton shift is given by

$$\Delta\lambda = \lambda' - \lambda = \frac{h}{m_0 c}(1 - \cos \phi)$$

where ϕ is the angle of scattering

Here $\lambda' = 1.11 \times 10^{-11}$ m, $\lambda = 1 \times 10^{-11}$ m

Hence $1.11 \times 10^{-11} - 1 \times 10^{-11}$

$$= \frac{6.6 \times 10^{34}}{9 \times 10^{-31} \times 3 \times 10^{8}} (1 - \cos \phi)$$

$$\text{or } 0.11 \times 10^{-11} = \frac{6.6}{9 \times 3} \times 10^{-11} (1 - \cos \phi)$$

$$\text{or } \quad 1 - \cos \phi = \frac{0.11 \times 10^{-11} \times 9 \times 3}{6.6 \times 10^{-11}}$$

$$= \frac{0.11 \times 9 \times 3}{6.6} = 0.45$$

$$\cos \phi = 1 - 0.45 = 0.55$$

$$\phi = \mathbf{56° \; 38'}.$$

■ **Example 2.14.** *X-rays of wavelength 10.0 pm are scattered from a target. (a) Find the wavelength of the X-rays scattered through 45°. (b) Find the maximum wavelength present in the scattered X-rays. (c) Find the maximum kinetic energy of the recoil electrons.*

Solution.

(*a*) $\lambda' - \lambda = \lambda_C(1 - \cos \phi)$, and so

$$\lambda' = \lambda + \lambda_C(1 - \cos 45°)$$
$$= 10.0 \text{ pm} + 0.293\lambda_C$$
$$= \mathbf{10.7 \text{ pm}}.$$

(*b*) $\lambda' - \lambda$ is a maximum when $(1 - \cos \phi) = 2$, in which case

$$\lambda' = \lambda + 2\lambda_C = 10.0 \text{ pm} + 4.9 \text{ pm} = \mathbf{14.9 \text{ pm}}.$$

(*c*) The maximum recoil kinetic energy is equal to the difference between the energies of the incident and scattered photons, so

$$KE_{max.} = h(\nu - \nu') = hc\left(\frac{1}{\lambda} - \frac{1}{\lambda'}\right)$$

where λ' is given in (*b*). Hence

$$KE_{max.} = \frac{(6.626 \times 10^{-34} \text{ J.s})(3.00 \times 10^{8} \text{ m/s})}{10^{-12} \text{ m/pm}}$$

$$\left(\frac{1}{10.0 \text{ pm}} - \frac{1}{14.9 \text{ pm}}\right)$$

$$= \mathbf{6.54 \times 10^{-15} \text{ J}}$$

which is equal to **40.8 keV.**

EXERCISES

1. Electrons bombarding the anode of a Coolidge tube produce X-rays of wavelength 1 Å. Find the energy of each electron at the moment of impact. Given : $h = 6.62 \times 10^{-27}$ erg s, $e = 4.8 \times 10^{-10}$ esu. **[Ans.** 12412.5 eV]
2. Calculate the potential difference to be applied to the X-ray tube to obtain the shortest wavelength of 0.5 Å. Given : $e = 4.8 \times 10^{-10}$ esu, 1 esu of potential = 300 volt, $h = 6.62 \times 10^{-27}$ erg s and $c = 3 \times 10^{10}$ cm s^{-1}. **[Ans.** 24825 volt]
3. An X-ray tube operates at 30 kV and emits a continuous spectrum with a short wavelength limit $\lambda = 0.414$ Å. Calculate the Planck's constant. Given : $e = 4.8 \times 10^{-10}$ esu, $c = 3 \times 10^{10}$ cm s^{-1}. **[Ans.** 6.624×10^{-27} erg s]
4. The absorption coefficient of aluminium of X-rays of certain wavelength is 0.143 m^{-1}. Find the thickness of aluminium absorber which will reduce the intensity of the X-ray beam to one-tenth of its incident value. **[Ans.** 1.61×10^{-3} m]
5. A certain crystal reflects monochromatic X-rays strongly, when the Bragg glancing angle (first order) is 15°. What are the glancing angles for the second and third order spectrum ? **[Ans.** 31° – 10′, 50° – 56′]
6. Determine the various orders of reflection possible from a crystal face, whose lattice constant is 1.44 Å, if the wavelength of the incident X-ray beam is 1.28 Å. **[Ans.** $\theta_1 = 26° - 23'$, $\theta_2 = 62° - 45'$]
7. The space between the principal planes of NaCl crystal is 2.82 Å. The first order Bragg's reflection occurs at angle of 10°. What is the wavelength of X-ray used ? **[Ans.** 0.98 Å]

8. Electrons are accelerated in television tubes through potential differences of about 10 kV. Find the highest frequency of the electromagnetic waves emitted when these electrons strike the screen of the tube. What kind of waves are these ? [**Ans.** 2.4×10^{18} Hz ; X-rays]
9. The distance between adjacent atomic planes in calcite ($CaCO_3$) is 0.300 nm. Find the smallest angle of Bragg scattering for 0.030-nm X-rays. [**Ans.** 2.9°]
10. A photon whose energy equals the rest energy of the electron undergoes a Compton collision with an electron. If the electron moves off at an angle of 40° with the original photon direction, what is the energy of the scattered photon ? [**Ans.** 335 keV]
11. What is the frequency of an X-ray photon whose momentum is 1.1×10^{-23} kg. m/s ? [**Ans.** 5.0×10^{18} Hz]
12. In some situations, the X-rays scattered by a crystal are assumed to undergo no change in wavelength. Show that this assumption is reasonable by calculating the Compton wavelength of a Na atom and comparing it with the typical X-ray wavelength of 0.1 nm. [**Ans.** $\lambda_C = 5.8 \times 10^{-8}$ nm << 0.1 nm]
13. An X-ray photon of initial frequency 3.0×10^{19} Hz collides with an electron and is scattered through 90°. Find its new frequency. [**Ans.** 2.4×10^{19} Hz.]
14. A photon of frequency ν is scattered by an electron initially at rest. Verify that the maximum kinetic energy of the recoil electron is $KE_{max.} = (2h^2\nu^2/mc^2)/(1 + 2h\nu/mc^2)$.
15. A beam of X-rays is scattered by a target. At 45° from the beam direction the scattered X-rays have a wavelength of 2.2 pm. What is the wavelength of the X-rays in the direct beam ? [**Ans.** 1.5 pm]
16. At what scattering angle will incident 100-keV X-rays leave a target with an energy of 90 keV ? [**Ans.** 64°]

SHORT QUESTIONS

1. How is penetrating power of X-rays related to their wavelength ?
2. What would be the velocity of X-rays having wavelength of 1 Å ?
3. What is the difference between the nature of X-rays and cathode rays ?
4. Which of the soft and hard X-rays have a longer wavelength ? Which of the two have a higher penetrating power ?
5. Why Compton effect is not observed with visible light ?

ANSWERS

1. The penetrating power of X-rays increases with the decrease in wavelength.
2. 3×10^8 m s^{-1}.
3. While X-rays are electromagnetic waves, cathode rays are a beam of fast moving electrons.
4. While the soft rays have longer wavelength, the hand rays have higher penetrating power.
5. Compton effect can most readily be detected for radiation of wavelength not greater than a few angstrom. In the case of visible light (4000 Å to 8000 Å), the maximum wavelength change is nearly 0.001% of incident wavelength. This is not detectable. So, Compton effect is not observed for visible light waves.

MULTIPLE CHOICE QUESTIONS

1. The structure of solids is investigated by using
   (*a*) Cosmic Rays (*b*) X-Rays (*c*) Infra-red Radiation (*d*) γ-Rays.

2. The penetrating power of X-rays can be increased by
   (*a*) increasing the current in the filament
   (*b*) decreasing the current in the filament
   (*c*) increasing the potential difference between the cathode and the anode
   (*d*) decreasing the potential difference between the cathode and anode.
3. X-ray region lies between
   (*a*) short radio waves and visible region
   (*b*) visible and ultraviolet region
   (*c*) gamma rays and ultra-violet region
   (*d*) short radio waves and long radio waves.
4. The wavelength of K_a X-rays produced by an X-rays tube is 0.76 Å. The atomic number of the anticathode material is
   (*a*) 82 (*b*) 41
   (*c*) 20 (*d*) 10.
5. Compare momentum of 10^5 eV X-ray photon (p_x) with that of a 10^5 eV electron (p_e)
   (*a*) $\frac{p_e}{p_x} = \frac{1}{2}$ (*b*) $\frac{p_e}{p_x} = \frac{16}{5}$
   (*c*) $\frac{p_e}{p_x} = \frac{1}{5}$ (*d*) $\frac{p_e}{p_x} = \frac{5}{1}$.
6. Molybdenum is used as a target element for production of X-rays because it is
   (*a*) a heavy element and can easily absorb high velocity electrons
   (*b*) a heavy element with a high melting point
   (*c*) an element having high thermal conductivity
   (*d*) heavy and can easily deflect electrons.
7. The frequencies of X-rays, γ-rays and ultravoilet rays are represented by *a*, *b* and *c*. Then
   (*a*) $a > b, b > c$ (*b*) $a < b, b > c$
   (*c*) $a < b, b < c$ (*d*) $a > b, b < c$.
8. The shortest wavelength of X-rays emitted from X-ray tube depends on
   (*a*) the current in the tube (*b*) the voltage applied to the tube
   (*c*) the nature of the glass in the tube (*d*) the atomic number of the target material.
9. The wavelength of K_a line from an element of atomic number 51 is λ. From another element the wavelength of K_a line is 4λ. What is the atomic number of second element ?
   (*a*) 25 (*b*) 26
   (*c*) 99 (*d*) 100.
10. The continuous X-rays spectrum produced by an X-rays machine at constant voltage has which of the following :
    (*a*) a maximum wavelength (*b*) minimum wavelength
    (*c*) a single wavelength (*d*) a minimum frequency.
11. In bcc structure of lattice constant '*a*', the minimum distance between atoms is
    (*a*) $\frac{\sqrt{3}}{2}a$ (*b*) $\sqrt{2}a$
    (*c*) $\frac{a}{\sqrt{2}}$ (*d*) $\frac{a}{2}$.

12. A beam of electrons accelerated through 50 kV strikes against a target. Which of the following X-ray frequency will be missing ?

(*a*) 4×10^8 Hz (*b*) 6×10^8 Hz

(*c*) 12×10^{18} Hz (*d*) 14×10^{18} Hz.

13. The laser produces

(*a*) a beam of fast moving neutron rays

(*b*) a beam of monochromatic incoherent light

(*c*) highly penetrating X-rays

(*d*) a beam of monochromatic coherent light.

14. The wavelength of limiting line of Lyman series is 911 Å. The atomic number of the element which emits minimum wavelength of 0.7 Å of X-rays will be

(*a*) 31 (*b*) 33

(*c*) 35 (*d*) 37.

15. In X-ray tube the accelerating potential applied at the anode is V_a volts. The minimum wavelength of the emitted X-rays will be

(*a*) eV_a/h (*b*) $h/e V_a$

(*c*) eV_a/ch (*d*) $h c/e V_a$.

16. A potential difference of 42,000 volts is used in an X-ray tube to accelerate electrons. The maximum frequency of the X-radiations produced is

(*a*) 10^{19} Hz (*b*) 10^{18} Hz

(*c*) 10^{16} Hz (*d*) 10^{20} Hz.

17. If the frequencies of K_α, K_β, and L_α, X-rays lines of a substance are ν_{K_α}, ν_{K_β} and ν_{L_α} respectively, then

(*a*) $\nu_{K_\alpha} + \nu_{K_\beta} = \nu_{L_\alpha}$ (*b*) $\nu_{K_\alpha} - \nu_{K_\beta} = \nu_{L_\alpha}$

(*c*) $\nu_{K_\alpha} + \nu_{L_\alpha} = \nu_{K_\beta}$ (*d*) None of these.

18. X-rays passing through a strong uniform magnetic field

(*a*) get deflected in the direction of the field

(*b*) get deflected in the direction opposite to that of the field

(*c*) get deflected in the direction perpendicular to that of the field

(*d*) do not get deflected at all.

19. The wavelength of K_α line for an element of atomic number 43 is λ. Then the wavelength of K_α line for an element of atomic number 29 is

(*a*) (43/29)λ (*b*) (42/28)λ

(*c*) (9/4)λ (*d*) (4/9)λ.

20. The potential difference applied to an X-ray tube is 5 kV and the current through it is 3.2 mA. Then the number of electrons striking the target per second is

(*a*) 2×10^{16} (*b*) 5×10^{16}.

21. The frequency of K_α line of a source of atomic number Z is proportional to

(*a*) 1/Z (*b*) Z

(*c*) Z^2 (*d*) $(Z-1)^2$.

22. The electrons accelerated through 40 kV are incident on a molebdenum target. Which of the following wavelength will be absent from the continuous X-ray spectrum ?

(*a*) 0.15 nm (*b*) 0.10 nm

(*c*) 0.05 nm (*d*) 0.031 nm.

23. The wavelength λ of K_α ray line of an anticathode element of atomic number Z is nearly proportional to

(*a*) Z^2 (*b*) $(Z-1)^2$

(*c*) $1/(Z-1)$ (*d*) $1/(Z-1)^2$.

24. X-rays are produced in a tube operating at 10^5 V. The speed of X-rays produced is

(*a*) 3×10^8 m/s (*b*) 2.8×10^8 m/s

(*c*) 3.1×10^8 m/s (*d*) 3×10^{10} m/s.

25. The wavelength of characteristic X-ray K_α line emitted by hydrogen like atom is 0.32 Å. The wavelength of K_β line emitted by the same element is

(*a*) 0.18 Å (*b*) 0.48 Å

(*c*) 0.27 Å (*d*) 0.38 Å.

26. The minimum wavelength of X-rays produced by accelerating potential V is λ. What will be the minimum wavelength when the accelerating potential is 2 V ?

(*a*) $\sqrt{(2\lambda)}$ (*b*) 2λ

(*c*) $\lambda\sqrt{2}$ (*d*) $\lambda/2$.

27. Which of the following is accompanied by the characteristic X-ray emission ?

(*a*) α-particle emission (*b*) electron emission

(*c*) positron emission (*d*) K-electron capture.

28. X-rays beam of intensity I_0 passes through an absorption plate of thickness *d*. If absorption coefficient of material of plate is μ, the correct statement regarding the transmitted intensity I of X-ray is

(*a*) $I = I_0(1 - e^{-\mu d})$ (*b*) $I = I_0 e^{-\mu d}$

(*c*) $I = I_0(1 - e^{-\mu/d})$ (*d*) $I = I_0 e^{-\mu/d}$.

29. X-rays

(*a*) are deflected in magnetic field

(*b*) are deflected in electric field

(*c*) remain undeflected by both the fields

(*d*) deflected in both the fields.

30. The maximum frequency of X-rays is determined by

(*a*) filament current

(*b*) filament potential

(*c*) current from high tension

(*d*) potential difference between cathode and anticathode.

31. The maximum frequency of X-rays is determined by

(*a*) current in filament

(*b*) potential difference across filament

(*c*) potential difference across cathode and anticathode

(*d*) current from high tension.

32. Intensity of X-rays of wavelength 1 Å reduces to one fourth when passed through a 6.93 mm thick aluminium sheet. The coefficient of absorption of aluminium will be

(*a*) 0.198 mm^{-1} (*b*) 0.2 mm^{-1}

(*c*) 0.386 mm^{-1} (*d*) 0.227 mm^{-1}.

33. Fig. 2.17 shows the intensity-wavelength relation of X-rays coming from two different Coolidge tubes. The solid curve represents the relation for tube A in which the potential difference between the target and the filament is V_A and the atomic number of the target material is Z_A. These quantities are V_B and Z_B for the other tube.

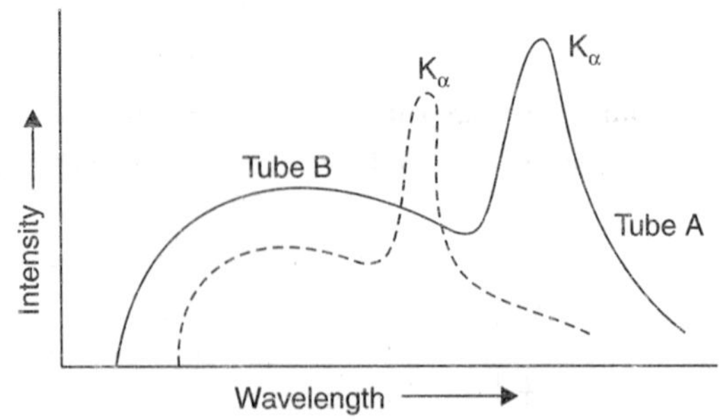

Fig. 2.17

Then

(*a*) $V_A > V_B$, $Z_A > Z_B$ (*b*) $V_A > V_B$, $Z_A < Z_B$
(*c*) $V_A < V_B$, $Z_A > Z_B$ (*d*) $V_A < V_B$, $Z_A < Z_B$.

34. The minimum wavelength of X-rays emitted from an X-ray tube operation at a voltage of 10^4 volts is roughly equal to

(*a*) 1 Å (*b*) 1.5 Å
(*c*) 2 Å (*d*) 2.5 Å.

35. The largest distance between the interatomic planes of a crystal is 10^{-9} m. The upper limit for the wavelength of X-ray which can usually be studied with the crystal is

(*a*) 1 Å (*b*) 10 Å
(*c*) 20 Å (*d*) 50 Å.

36. For the production of X-rays, the target should be made of

(*a*) steel (*b*) copper
(*c*) tungesten (*d*) aluminium.

37. The wavelength of X-rays is of order of

(*a*) micron (10^{-6} m) (*b*) metre
(*c*) centrimetre (*d*) angstrom (10^{-10} m).

38. The wavelength of most energetic X-ray emitted when emitted when a metal target is bombarded by 40 keV electrons, is approximately

(*a*) 300 Å (*b*) 10 Å
(*c*) 4 Å (*d*) 0.31 Å.

39. Electrons with energy 80 keV are incident on the tungesten target of an X-ray tube. K-shell electrons of tungsten have – 72.5 keV energy. X-rays emitted by the tube contains only

(*a*) a continuous X-ray spectrum with a minimum wavelength of 0.155 Å
(*b*) a continuous X-ray spectrum with all wavelengths
(*c*) the characteristic X-ray spectrum of tungsten
(*d*) a continuous X-ray spectrum with a minimum wavelength of 0.155 Å and the characteristic X-ray spectrum of tungsten.

40. The X-ray beam coming from an X-ray tube will be

(*a*) monochromatic
(*b*) having all wavelength smaller than certain maximum wavelength

(*c*) having all wavelength larger than certain minimum wavelength

(*d*) having all wavelength lying between a minimum and a maximum wavelength.

41. The co-efficient of absorption for aluminium for soft X-rays is 1.73 per cm. The fraction of X-rays transmitted through 1.331 mm thick aluminium plate will be

(*a*) 0.01 Å (*b*) 0.1 Å

(*c*) 1 Å (*d*) 10 Å.

42. The intensity of X-rays from a Coolidge tube is plotted against wavelength λ as shown in Fig. 2.18. The minimum wavelength found is λ_C and the wavelength of K_α line is λ_K. As the accelerating voltage is increased

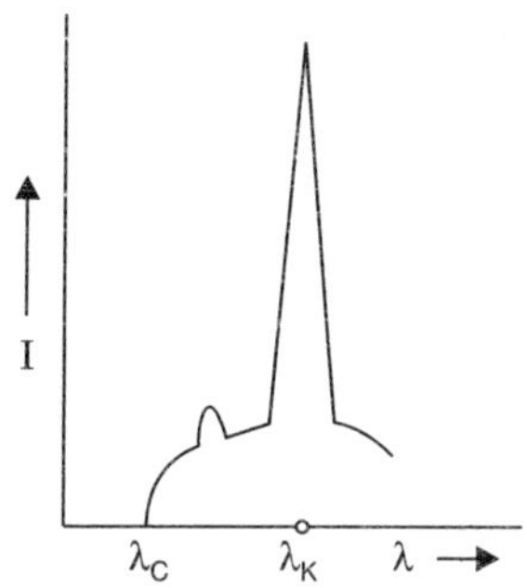

Fig. 2.18

(*a*) $\lambda_K - \lambda_C$ increases (*b*) $\lambda_K - \lambda_C$ decreases

(*c*) λ_K increases (*d*) λ_K decreases.

43. The binding energy of the inner most electron in tungsten is 40 keV. To produce characteristic X-rays using a tungsten target in X-ray tube the potential difference V between the cathode an anticathode should be

(*a*) V < 40 kV (*b*) V ≤ 40 kV

(*c*) V > 40 kV (*d*) V ≥ 40 kV.

44. The shortest wavelength produced in an X-ray tube operated at 0.5 million volt is

(*a*) dependent on the target material

(*b*) 0.025 Å approximately

(*c*) double of the shortest wavelength produced when the tube operating at 1 million volts

(*d*) dependent only on the target material and not on the operating voltage.

45. X-rays are known to be electromagnetic radiations. Therefore the X-ray photon has

(*a*) electric charge

(*b*) magnetic moment

(*c*) both electric charge and magnetic moment

(*d*) neither electric charge nor magnetic moment.

46. In α-rays diffraction experiment distance between atomic lattice planes of diffraction grating is 2.8 × 10^{-10} m, then the maximum wavelength of α-rays (in metre) is

(*a*) 2.5 (*b*) 5.6 × 10^{-10}

(*c*) 4 × 10^{-10} (*d*) 4.6 × 10^{-10}.

ANSWERS

1. (*b*)	**2.** (*c*)	**3.** (*c*)	**4.** (*b*)	**5.** (*b*)	**6.** (*b*)	**7.** (*b*)	**8.** (*b*)
9. (*b*)	**10.** (*b*)	**11.** (*a*)	**12.** (*c*)	**13.** (*d*)	**14.** (*d*)	**15.** (*d*)	**16.** (*a*)

17. (*c*)	**18.** (*d*)	**19.** (*c*)	**20.** (*a*)	**21.** (*d*)	**22.** (*d*)	**23.** (*d*)	**24.** (*c*)
25. (*c*)	**26.** (*d*)	**27.** (*d*)	**28.** (*b*)	**29.** (*c*)	**30.** (*b*)	**31.** (*c*)	**32.** (*b*)
33. (*b*)	**34.** (*a*)	**35.** (*c*)	**36.** (*a*)	**37.** (*d*)	**38.** (*d*)	**39.** (*c*)	**40.** (*c*)
41. (*d*)	**42.** (*d*)	**43.** (*c*)	**44.** (*b*), (*c*)	**45.** (*d*)	**46.** (*b*)		

LONG ANSWER QUESTIONS

1. What are continuous and characteristic X-rays ? How is the continuous X-ray spectrum and its short wavelength limit explained ? *(UPTU, B. Tech. I, Q. Bank, 2001)*

2. What is the origin of continuous X-ray spectra ? Explain why the continuous spectrum has a sharp limit on the short λ side.

3. What are continuous and characteristic X-rays ? Discuss their salient features.

How are continuous X-rays produced ? What is the significance of the short wavelength limit and how it is related with the voltage applied across the X-ray tube ?

4. What is Moselay's law ? How can it be explained on the basis of Bohr's theory ? What is its importance ? *(UPTU, B. Tech. I, Q. Bank, 2001)*

5. Explain how X-rays are produced. State the properties of X-rays.
(UPTU, B. Tech. I, Q. Bank, 2001)

6. How are continuous and characteristic X-rays produced ? How can you control intensity and penetrating power of X-rays ? Discuss some industrial and engineering applications.
(UPTU, B. Tech. I Sem. 2001)

7. Find out the relation between minimum wavelength and accelerating potential with respect to continuous X-ray emitted in a X-ray tube. *(UPTU, B. Tech. I, Q. Bank, 2001)*

8. Draw an energy level diagram to show the emission of K-series of X-rays. Discuss the importance of excitation and absorption limit in X-ray spectra. *(UPTU, B. Tech. I, Q. Bank, 2001)*

9. What are characteristic X-rays ? Explain their production. Describe an X-ray energy level diagram for an atom. *(UPTU, B. Tech. I, Q. Bank, 2001)*

10. Discuss Laue's principle of X-ray diffraction and obtain the diffraction condition for a simple cubic lattice. *(UPTU, B. Tech. I, Q. Bank, 2001)*

11. How does the frequency of characteristic X-rays depend upon atomic number of element ?
(UPTU, B. Tech. I, Q. Bank, 2001)

12. In what respect does the X-ray spectra of elements differ from their optical spectra ?
(UPTU, B. Tech. I, Q. Bank, 2001)

13. Derive Bragg's equation for reflection of X-rays by crystal planes.
(UPTU, B. Tech. I, Q. Bank, 2001)

14. What information does the sharp absorption edges in X-ray spectra give regarding the atomic structure. *(UPTU, B. Tech. I, Q. Bank, 2001)*

15. The optical spectra of elements show periodic characteristic with increasing Z while X-ray characteristic spectra show continuous order through all Z-values. Explain.
(UPTU, B. Tech. I, Q. Bank, 2001)

16. Describe Bragg's spectrometer and explain how it is used to study the structure of crystals.
(UPTU, B. Tech. I, Q. Bank, 2001)

17. Discuss Compton effect and derive an expression for the compton shift. Why are both the modified and the unmodified lines simultaneously present for non-zero scattering angles ?
(UPTU, B. Tech. II Sem. 2001)

UNIT II

DIELECTRICS AND MAGNETIC PROPERTIES OF MATERIALS

This unit contains the following chapters:

3

Dielectrics

3.01 INTRODUCTION

Dielectrics are nonconducting insulator materials such as glass, ceramic, polymers, wood and paper etc. The electrical conductivity of these materials is almost zero and the resistivity ranging from 10^{10} to 10^{20} ohm-meter. It was Faraday who first discovered that the capacitance of a capacitor increases if the space between conductor is filled with a dielectric material. The materials get polarized when an electric field is applied across the material. Electrical and electronic circuits uses these materials in various forms.

The materials are characterised by a dielectric constant, dielectric loss and dielectric strength. The two properties the dielectric constant and dielectric loss are strongly frequency dependent. Ferroelectrics are ceramic dielectric materials which exhibit a very high dielectric constant and low dielectric loss. The dielectric loss manifests in the form of heat. The dielectric heating can be employed in cases where heating by other means is not possible such as food processing.

3.02 DIELECTRIC CONSTANT

Dielectrics can be used for insulating purposes as they prevent the flow of current through them and therefore called insulators. Such dielectric materials when inserted between the conducting plates of a capacitor, it modified the electric field between the plates and increases capacitance of capacitor by a dimension less factor ϵ_r which is called the dielectric constant of the material. The dielectric constant varies from one material to another. The effect of a dielectric in a capacitor can be shown in following way:

Let us consider a parallel plate capacitor connected with a D.C. voltage V. The capacitor without dielectric has charge Q and a capacitance C [Fig. 3.1(*a*)]. The capacitance C is related with charge Q stored on either plate is given by

$$C = \frac{Q}{V} \qquad ...(1)$$

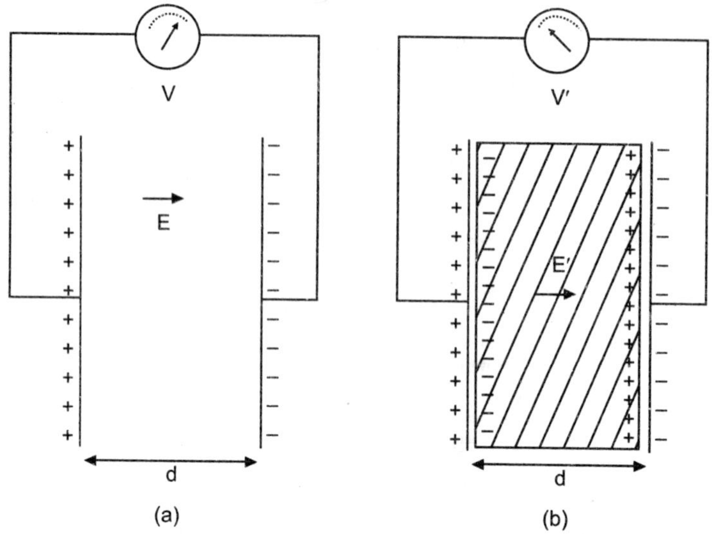

Fig. 3.1

Now the plates of capacitor be disconnected from the source voltage V. The circuit is now open and there is no path by which charge can flow and alter the charge on capacitor plates. If a dielectric is now inserted between the plates, the voltage between plates decreases from V to V′ [Fig. 3.1(*b*)]. It further increases to V if dielectric is pulled out of the capacitor. The voltage with or without dielectric are related by a factor $\in_r$

$$V = \in_r V' \qquad ...(2)$$

as $$V' < V \quad \text{hence } \in_r > 1$$

because Q on capacitor plates does not change, we conclude that capacitance must change to the value C′ which is described as

$$C' = \frac{Q}{V'} = \frac{Q}{V/\in_r} = \in_r \frac{Q}{V}$$

$$\boxed{C' = \in_r C} \qquad ...(3)$$

That is, the capacitance increases by a factor of $\in_r$ when the dielectric completely fills the region between the plates. For a parallel plate capacitor $C = \frac{\in_0 A}{d}$. A is the area of each plate and d is the separation between the plates. In case parallel plate capacitor is completely filled with a dielectric, the capacitance is

$$C' = \frac{\in_o \in_r A}{d} \qquad ...(4)$$

Table 3.1 The relative permittivity of some selected materials at 20°C

Dielectrics	ϵ_r
Vacuum air	1.0006
Rubber	2.94
Quartz	3.8
Bakelite	5
Glass	5–10
Mica	3–6
Germanium	16
Diamond	5.5
Porcellian	5.5
Polyethyline	2.3
PVC	3.5
Teflon	2.1
$BaTiO_3$	> 3000
Polystryne	2.6
Cotton	5
Ethyle Alcohol	26
Benzene	2.3
Toluene	2.4

ϵ_r is dimensionless quantity, always greater than unity for dielectrics. It is independent of size or shape of dielectric. It is the ratio of permittivity of the material ϵ with permittivity of free space ϵ_o and called relative permittivity.

$$\epsilon = \epsilon_o \epsilon_r \qquad ...(5)$$

The value of ϵ_o in SI units is 8.854×10^{-12} F/M. The value of ϵ_r of some dielectric materials are listed in Table 3.1.

3.03 DIELECTRIC POLARIZATION

Dielectric polarization is nothing but the displacement of charge particles under the action of the electric field. Let a dielectric be made up of polar molecules and placed in a uniform electric field parallel to plate capacitor. The dipoles of materials are randomly oriented in absence of an electric field. When an external electric field E due to charges on capacitor plates is applied, a torque is exerted on the dipoles causing them to partially align with the field. We can now describe the dielectric being polarized. The alignment of the molecules with the electric field depends on temperature and the magnitude of the field. In general alignment increases with decreasing temperature and with increasing external electric field E.

If the molecules of dielectric are non-polar then the electric field due to charges on plates produces some charge separation in dielectric and an induces dipole moment. These induces dipole moments tends to align with the external field and the dielectric is polarized. Thus we can polarize a dielectric with an external field regardless whether the molecules are polar or non-polar. The polarization of material P is proportional to electric field E and can be written as

$$P = \epsilon_o \chi_e E \quad ...(6)$$

where χ is the proportionality constant and known as dielectric susceptibility of material. It describe that how easy a dielectric can be influenced by an external field.

To describe the combine effect of the applied electric field E and electric polarization P we define an additional vector D, the dielectric displacement, which can be expressed as

$$D = \epsilon_o E + P \quad ...(7)$$

using equation (6)

$$D = \epsilon_o E + \epsilon_o \chi_e E$$

$$= \epsilon_o (1 + \chi_e) E$$

$$= \epsilon_o \epsilon_r E$$

$$\boxed{D = \epsilon E} \quad ...(8)$$

where

$$\epsilon_r = \frac{\epsilon}{\epsilon_0} = 1 + \chi_e$$

For vacuum, where there is no matter to polarize, the susceptibility χ_e is zero and hence

$$\chi_e = P = 0$$

$$\boxed{D = \epsilon_o E}$$

3.04 POLAR AND NON-POLAR DIELECTRIC MOLECULES

A molecule is a neutral system in which the algebraic sum of all charges is equal to zero. However, all positive charges of a molecule may be considered equivalent to one positive charge located at the centre of gravity of positive charges. Similarly all negative charges of molecule may be replaced with a single equivalent negative charge located at the centre of gravity of all negative charges. These two equivalent charges are equal in magnitude. Their point of action in space may coincide or may not coincide. If the point of action coincides, the molecule will not possess a permanent dipole moment. Such molecules are called non-polar molecules and the dielectric is known as non-polar dielectric [see Fig. 3.2(*a*)].

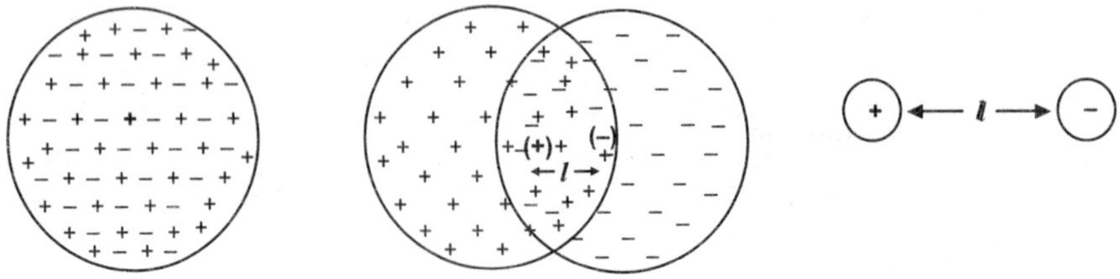

(a) (b) (c)

Fig. 3.2

If the point of action of the resultant charges of a molecule do not coincide in space, the molecule will possess a permanent dipole moment. Such molecules are called polar molecules and the dielectric is known as polar dielectric [see Fig. 3.2(*b*)]. The magnitude of the electric dipole moment of a molecule is given by

$$p = ql \qquad ...(9)$$

where q is the charge and l is the separation of the charges [see Fig. 3.2(*c*)].

Polar or non-polar molecule can be identified from their structure. The symmetric molecules are non-polar as their centre of action for both negative and positive charge coincides each other. Such non-polar molecules are He, Ne, Ar and Xe—and the molecules consisting of two identical atoms linked by a homopolar bond such as H_2, N_2, Cl_2 are non-polar.

3.4.1 Polarizability

Consider an electrically neutral slab of an isotropic dielectric in a uniform electric field E. Since no charge is free in the dielectric all are bound electrons which are induced in this external electric field. The action of electric field is to displace the positive charges in the direction of field and negative in the opposite direction of field. Each atom or molecule of dielectric then acquires an electric dipole moment parallel to the direction of field. This induced dipole moment is proportional to the field.

$$p = \alpha \text{ E} \qquad ...(10)$$

The constant of proportionality α is called atomic polarizability.

When all such dipole are pointing along the direction of field then the dielectric material becomes polarized and this effect is known as dielectric polarization and defined as dipole moment per unit volume. So the polarization of material P becomes

$$\text{P} = \text{N}p$$
$$= \alpha \text{ NE} \qquad ...(11)$$

N is the number of dipoles (atom) per unit volume.

3.4.2 Types of Polarization

When a dielectric material is subject to an electric field, the displacement of charge particles undertake. This displacement of charge particles results in the formation of electric dipole moment in atoms, ions or in molecules of material which ultimate polarize the material. There are three important types of polarization. These are:

(*i*) Electronic Polarization (*ii*) Ionic Polarization

(*iii*) Orientation Polarization.

(*i*) *Electronic Polarization.* In this type of polarization the atom is initially unpolarized in absence of external field. As the field E applied, there is displacement of electron cloud of atoms or molecules with respect to heavy fixed nuclii to a distance that is less than the dimension of atom or molecule (Fig. 3.3). The electronic polarization sets in over a very short period of time (10^{-14} to 10^{-15} seconds). It is also independent of temperature.

The polarization is described by

$$\text{P}_e = \text{N } \alpha_e \text{ E} \qquad ...(12)$$

where α_e is the electronic polarizibility and N is the number of atoms or molecules per unit volume of material.

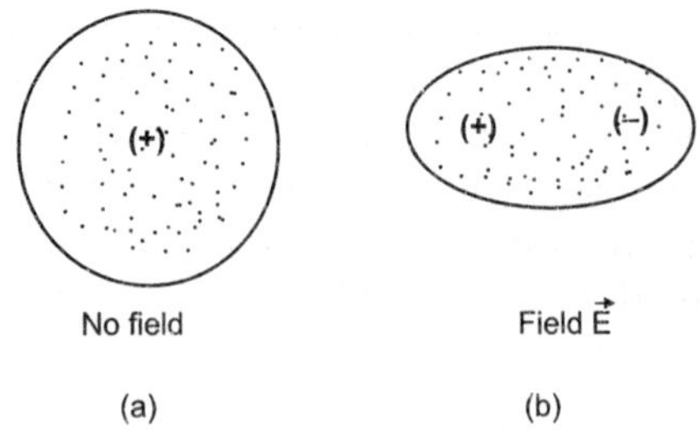

Fig. 3.3. Electronic Polarization of Molecule.

We know that dielectric constant ϵ_r is

$$\epsilon_r = 1 + \chi_e$$

$$= 1 + \frac{P_e}{\epsilon_o E} \qquad [\because P_e = \chi_e \epsilon_0 E]$$

$$= 1 + \frac{N \alpha_e E}{\epsilon_o E}$$

$$\epsilon_r = 1 + \frac{N\alpha_e}{\epsilon_o} \qquad \text{...(13)}$$

Hence dielectric constant is sole dependent on electronic polarizibility. In case of monoatomic gas

$$\alpha_e = 4\pi \epsilon_0 R^3$$

So, $$\epsilon_r = 1 + 4\pi NR^3 \qquad \text{...(14)}$$

where R is the radius of atom.

(*ii*) *Ionic Polarization.* Ionic polarization occurs in ionic crystals. Let a sodium chloride molecule composed of sodium and chlorine ions which are bounded through ionic bond. The inter-ionic distances of these ion charges q are l [Fig. 3.4(*a*)]. These molecules exhibit a permanent dipole moment equal to ql.

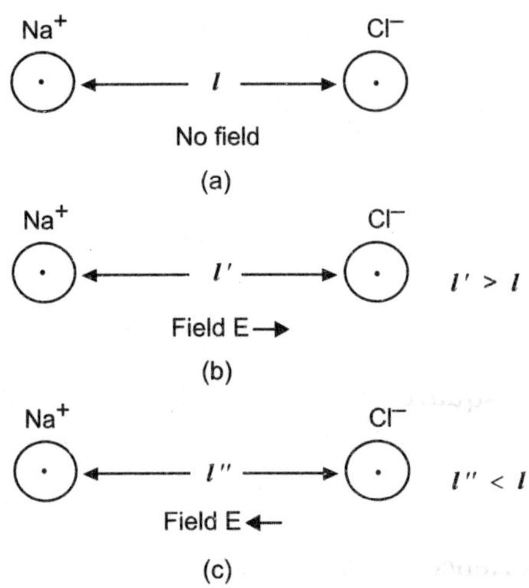

Fig. 3.4. Ionic Polarization of Molecule.

Now if this molecule is subjected to an external electric field E, the sodium and chlorine ions are further displaced in opposite direction until ionic bonding is forces to stop them. The dipole moment in this process is increases because of increasing the value of l [Fig. 3.4 (*b*)]. The same is reverse if the direction of field E is changed [Fig. 3.4(*c*)]. So the displacement and hence the polarization P_i (dipole moment per unit volume) is proportional to field E, which can be expressed as

$$P_i = N \alpha_i E \quad \text{...(15)}$$

Here α_i is the ionic polarizibility. For most of the materials, the ionic polarizibility is less than the electronic polarizibility. A typical value of α_i is

$$\alpha_i = \frac{1}{10} \alpha_e \quad \text{...(16)}$$

(*iii*) *Orientation Polarization.* Orientation polarization occurs in polar dielectrics whose molecules have permanent dipole moment. In absence of any external field, the alignment of such dipoles is random and cancelled the effect of others [Fig. 3.5(*a*)]. But when field E is applied across the material, the molecular dipoles rotate about their axis to align with field. [Fig. 3.5(*b*)]. This alignment of dipoles is affected by the thermal agitation which randomizes the dipoles alignment. The effect of thermal agitation is so effective that even in case of liquid and gases where these dipoles are more free, could not aligned completely.

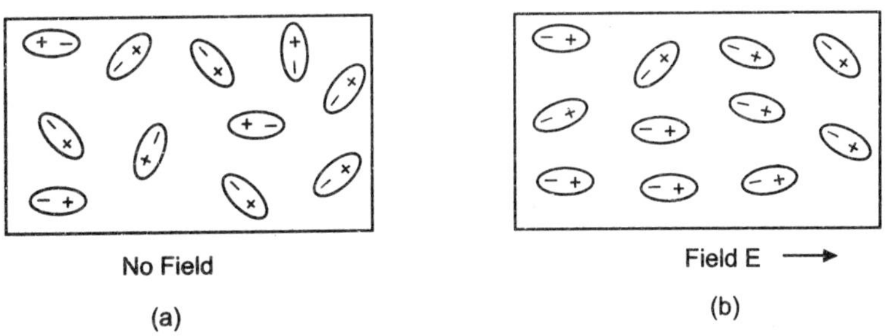

Fig. 3.5. Orientation polarization.

So the orientation polarization is temperature dependent. Since it involve rotation of molecules, it takes relatively longer time to build up than electronic or ionic polarization. The build up time for orientation polarization is order of 10^{-10} sec.

The orientation polarizibility $$\alpha_o = \frac{p^2}{3kT} \quad \text{...(17)}$$

and the orientation polarization $$P_o = \frac{Np^2}{3kT} E \quad \text{...(18)}$$

P_o is proportional to square of dipole moment p and inversely proportional to temperature T.

3.4.3 Total Polarization

When a material experiences all three types of polarization, then total polarizibility is the sum of electronic, ionic and orientation polarizibilities *i.e.*,

$$\alpha = \alpha_e + \alpha_i + \alpha_o \qquad ...(19)$$

and the total polarization P

$$P = P_e + P_i + P_o \qquad ...(20)$$

The total polarization of a polar gaseous dielectric may be expressed as

$$P = N\left[\alpha_e + \alpha_i + \frac{p^2}{3\,kT}\right]E \qquad ...(21)$$

But $P = \chi_e \in_0 E = (1 - \in_r) \in_o E$

$$\frac{P}{E} = (1 - \in_r) \in_0 \qquad ...(22)$$

Therefore,

$$(1 - \in_r) \in_o = N\left[\alpha_e + \alpha_i + \frac{p^2}{3\,kT}\right] \qquad ...(23)$$

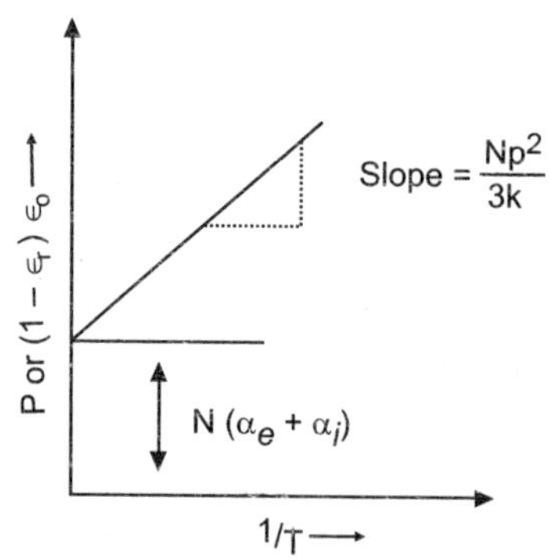

Fig. 3.6

If P or $(1 - \in_r) \in_o$ is plotted against $\frac{1}{T}$ a straight line is obtained [Fig. 3.6]. The intercept of line with y axis at $\frac{1}{T} = 0$ gives $N(\alpha_e + \alpha_i)$. Hence $\alpha_e + \alpha_i$ can be calculated. The slope of line $\frac{Np^2}{3k}$ can determine the value of dipole moment p.

It can also be observe that contribution by one or two polarization either be absent or negligible in magnitude in P. As an example P_0 is not exist in non-polar dielectric and the P_i is not found in covalent bond materials. Also P_e is negligible compared to P_0 in polar dielectrics.

3.05 EQUATION OF INTERNAL FIELDS IN LIQUIDS AND SOLIDS

The interatomic spaces in gases are very large in comparison to solid and liquids. The atoms almost touch each other in solids and liquids as the attraction force between them is more than in gaseous atoms. Therefore, when an electric field E is applied, the gaseous atoms sees only the applied field E. However, the solid and liquid atoms experiences field in different way. This field at atom site or seen by atom in liquid or solid is called internal field or local field.

As the atoms in solid and liquids are closely surrounded by the polarized atoms, their effect is to be taken into account for internal field. Therefore, the internal field E_i at the location of any atom be the sum of fields created by the neighbouring polarized atoms E′ and the external field E. So

$$E_i = E + E' \quad ...(24)$$

To evaluate the value of E′ let us consider the one dimensional array of atoms as shown in Fig. 3.7.

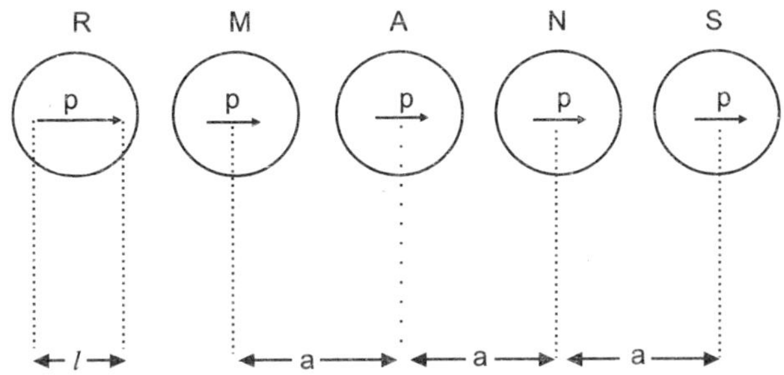

Fig. 3.7. One dimensional array of atoms with dipole moment *p*.

This figure shows an array of equispaced atomic dipoles with dipole moment $p_i(=ql)$ separated by a distance a. Let an electric field E is applied from left to right then dipole moment p_i is induced in each atom. The internal field seen by an atom A is the sum of E and the field produced by p_i of all other atoms M, N, R, S... lying on the left and right of A.

The field at A by p_i of M is given by

$$E_{AM} = \frac{2p_i}{4\pi\epsilon_o a^3} \text{ (field by a dipole at } a) \quad ...(25)$$

Similarly p_i of N is

$$E_{AN} = \frac{2p_i}{4\pi\epsilon_o a^3}$$

and the field together by M and N at A (*i.e.*, by left and right dipoles)

$$E_{AM} + E_{AN} = \frac{p_i}{\pi\epsilon_o a^3} \quad ...(26)$$

In similar way

$$E_{AR} + E_{AS} = \frac{p_i}{\pi\epsilon_o (2a)^3} \quad ...(27)$$

So the internal field at A is

$$E_i(A) = E + \frac{p_i}{\pi\epsilon_o a^3} + \frac{p_i}{\pi\epsilon_o (2a)^3} + \dots$$

i.e., $$E_i = E + \frac{p_i}{\pi \varepsilon_0 a^3} \sum_{i=1}^{\infty} \frac{1}{n^3}$$

But $$\sum_{i=1}^{\infty} \frac{1}{n^3} = 1.2$$

So, $$E_i = E + \frac{1.2\, p_i}{\pi \epsilon_o a^3} \quad ...(28)$$

From equation (28) the local field E_i exceeds the applied field. Greater the polarizibility $a_e \left(= \frac{p_i}{E}\right)$ or smaller the interatomic spacing, larger is the internal field.

In three dimensional case, the internal field would be complicated to evaluate however, the general expression is similar to equation (28) except that $\frac{1}{a^3}$ is replaced by N, the number of atoms per unit volume and $\left(\frac{1.2}{\pi}\right)$ by γ. γ is a constant and depends on the structure of crystal and known as internal field constant. Then

$$E_i = E + \frac{\gamma N p_i}{\epsilon_0} \quad ...(29)$$

or $$E_i = E + \frac{\gamma}{\epsilon_o} P \qquad [\because P = Np_i] \quad ...(30)$$

In case of crystal possessing cubic symmetry, the internal field constant $\gamma = \frac{1}{3}$ and E_i then

$$E_i = E + \frac{P}{3\epsilon_o} \quad ...(31)$$

The field given by equation (31) is called Lorentz field.

3.06 THE CLAUSIUS – MOSSOTTI EQUATION

It is possible to relate the dielectric constant of an insulator to the polarizibility of atoms. Let us consider a simple case of a solid dielectric which exhibits only electronic polarizibility α_e.

We know that $$\alpha_e = \frac{P}{NE_i} \quad ...(32)$$

where N is the number of atoms per unit volume and E_i is the local field. Further,

$$\alpha_e = \frac{P}{N\left[E + \frac{\gamma P}{\epsilon_0}\right]} \quad ...(33)$$

But $$P = \epsilon_0 (\epsilon_r - 1)E \text{ and}$$

$$\gamma = \frac{1}{3}$$

For Lorentz field then putting these value in equation (33), we get

$$\alpha_e = \frac{P}{N\left[\frac{P}{\epsilon_o(\epsilon_r - 1)} + \frac{P}{3\epsilon_o}\right]}$$

or
$$\frac{N\alpha_e}{\epsilon_o} = \frac{1}{\left[\frac{1}{\epsilon_r - 1} + \frac{1}{3}\right]} = \frac{1}{\frac{(\epsilon_r + 2)}{3(\epsilon_r - 1)}}$$

or
$$\frac{3(\epsilon_r - 1)}{\epsilon_r + 2} = \frac{N\alpha_e}{\epsilon_o}$$

$$\boxed{\frac{\epsilon_r - 1}{\epsilon_r + 2} = \frac{N\,\alpha_e}{3\,\epsilon_o}} \quad ...(34)$$

Equation (34) is called Clausius–Mossotti equation which is valid for non-polar solids having a cubic crystal structure.

3.07 RELATION BETWEEN DIELECTRIC CONSTANT AND THE REFRACTIVE INDEX

There can be a definite relation between dielectric constant and refractive index of medium. When an electromagnetic wave propagate through medium, the magnetic and electric field vectors varies with time. These vectors are perpendicular to each other as well as to the direction of propagation. The speed of wave in any medium whose permittivity is ϵ and magnetic permeability is μ, can be defined by

$$C = \frac{1}{\sqrt{\mu\epsilon}} \quad ...(35)$$

The refractive index of a medium is determined by the ratio of the speed of wave in vacuum to the speed in medium *i.e.,*

$$n = \frac{\text{Speed in Vacuum}}{\text{Speed in medium}}$$

$$= \frac{1/\sqrt{\epsilon_0\mu_o}}{1/\sqrt{\epsilon\,\mu}}$$

$$n = \sqrt{\frac{\epsilon}{\epsilon_o}} \qquad [\because \mu = \mu_0]$$

$$n = \sqrt{\epsilon_r} \quad ...(36)$$

Using this relation equation (34) can be written as

$$\boxed{\frac{n^2 - 1}{n^2 + 2} = \frac{N\alpha}{3\epsilon_o}} \qquad ...(37)$$

Equation (37) is known as Lorentz-Lorentz equation.

3.08 FERROELECTRICITY

There are certain dielectric materials, which below certain temperature, possess spontaneous polarization even in absence of any external applied electric field. Such materials are called ferroelectric materials. In ferroelectric materials there are large number of polarised regions with an electric dipole moment in its own directions. These regions are known as domain. Analogous to ferromagnetic materials, ferroelectric domains of material are oriented in different directions keeping net polarization of substance zero. However, in presence of an external electric field these domains are aligned and polarises the material as whole. Above critical temperature T_C called curies temperature, the material looses its ferroelectric property of spontaneous polarization. The materials which exhibit this property are barium titanate ($BaTiO_3$), lead titanate ($PbTiO_3$), potassium nabate ($KNbO_3$) and potassium dihydrogen phosphate (KH_2PO_4) etc. The dielectric constant of these materials changes with temperature according to following relation

$$\epsilon = \frac{C}{T - T_C} \qquad \text{provided } T > T_C$$

where C is the constant known as curies constant and T_C the curie temperature. The above relation is known as Curie–Weiss law.

Analogous to ferromagnetic materials, ferroelectric materials also exhibit hysteresis phenomena in which curve is plotted between polarisation versus applied electric field. In absense of any external electric field the ferroelectric domains are oriented in random direction hence the substance as whole has zero polarization. When an electric field is applied across the material its polarization increases rapidly to a value above which the dependence is linear [Fig. 3.8(*a*)]. At this stage the material got saturated polarization. Now when field is reduced to zero, the polarization does not fall to zero but the material retains some polarization which is called remanance polarization P_r. This remanance polarization eliminated when electric field is applied in opposite direction. The amount of field in opposite direction which make the substance of zero polarization is called coercive field E_C. By further increasing field in opposite direction material polarizes in opposite direction (see Fig. 3.8(*a*)). The different stages of material can be seen in a close loop between P and E for a complete cycle or a hysteresis loop.

The shape of hysteresis loop is changes with increase in temperature. The height and width of loop decreases with increase in temperature. At ferroelectric curie temperature the loop merges into a straight line [Fig. 3.8(*b*)] and ferroelectric behaviour of material is disappear.

These ferroelectric materials are used in manufacturing small size capacitors with high capacitance. The property of hysteresis suggests their use as memory devices for electronic computers. Piezoelectric acoustic transducers and pyroelectric infrared detectors are the devices based on ferroelectric material.

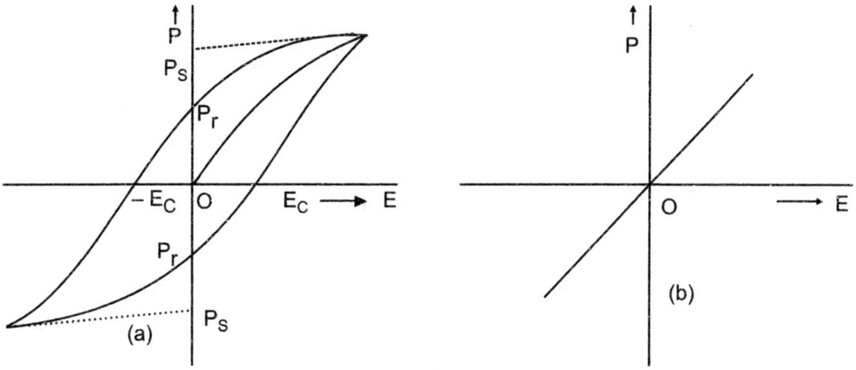

Fig. 3.8. Hysteresis Phenomena in Ferroelectric material.

3.09 PIEZOELECTRICITY

There are certain dielectric materials in which electric charges appears on their surface if the dimension of material is changed by application of stress (mechanical force). The effect is reversible that the crystal become strained (changes dimension of crystal) when electric field is applied across the material. (Fig. 3.9). This phenomenon is known as piezoelectricity and the material which possess this property are called peizoelectric materials. Quartz, rochella salt, tourmaline are the familiar piezoelectric substances. The piezoelectric strain in the crystals are very small corresponding to large value of applied electric field. For example, in quartz a field of 1000 V/cm produces a strain of the order 10^{-7}. In reverse, small strain can produce large electric field.

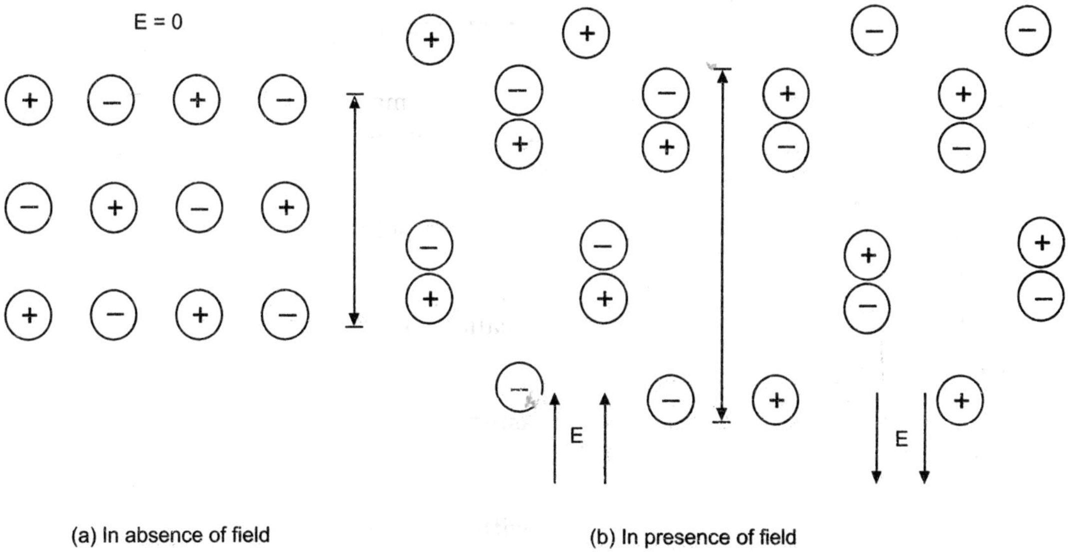

Fig. 3.9. Shifting of ions in presence of electric field.

Normally in any crystal the distribution of ionic charges about their lattice sites is symmetrical and the internal electric field is zero. But this symmetrical distribution of charges is disturb or charges are displaced when the crystal is stressed or a mechanical force is applied across it. Therefore a net polarization results in such crystal and we observe the peizoelectric effect. However, on other side in non-peizoelectric crystal the distribution of charges maintain its symmetry even after the displacement. Such non-piezoelectric crystal exhibit no net polarization. It can be understand that piezoelectricity is related to crystal symmetry.

These materials can be used as crystal oscillators, delay lines, ultrasound applications, tranducers, piezoelectric transformers etc.

3.10 FREQUENCY DEPENDENCE OF DIELECTRIC CONSTANT

When a dielectric material is subjected to an alternating electric field, its polarization is required to follow field reversal. The total polarization depends on the ability of dipoles to orient themselves with change of applied electric field. The relative permittivity ϵ_r of dielectric material varies with the frequency of alternating electric field. The field frequency dependence of dielectric constant of material for polar dielectric is shown through Fig. 3.10.

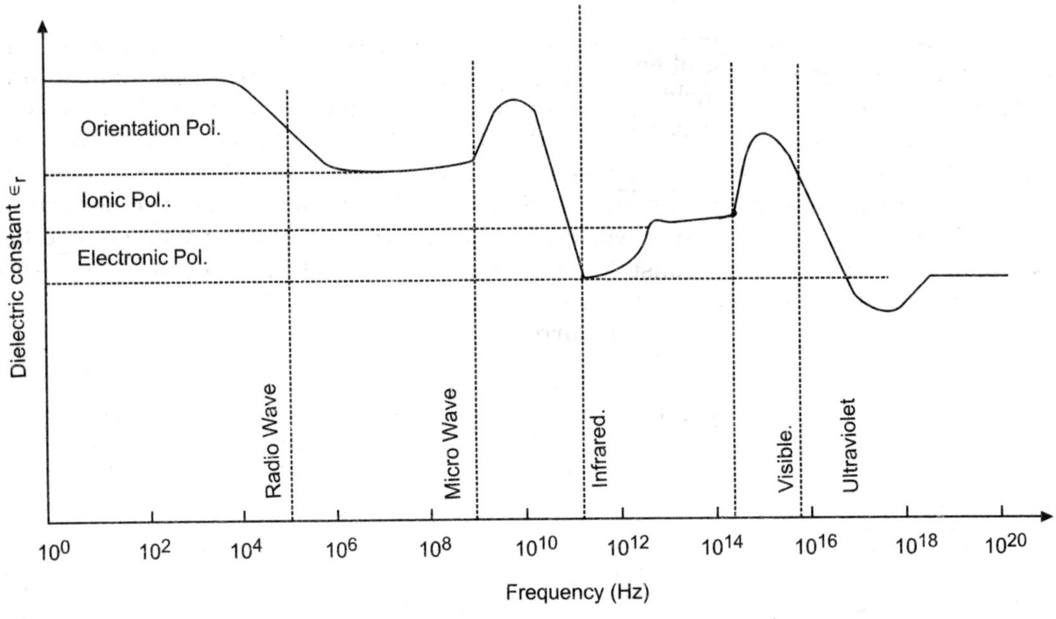

Fig. 3.10. Frequency dependence of dielectric constant ϵ_r.

Since we know that total polarization is contribution of electronic, ionic and orientation polarizations which are individually frequency dependent. When an alternate electric field is applied, the electric dipoles are oriented themselves each time with change of electric field. The average time taken by the dipole to reorient in the direction of field is known as relaxation time and the reciprocal of relaxation time is called relaxation frequency. The relaxation frequencies of different types of polarization mechanism are different. If the frequency of applied field is more enough than the relaxation frequency of particular polarization mechanism *i.e.*, electric dipole could not got sufficient time to prepare themselves for the polarization, then that polarization mechanism is stopped.

In audio frequency region, which is at lower side of frequency range, all three types of polarization are possible. In radio and microwave region, the permanent dipoles fails to follow the field reversal, as the dipoles do not have sufficient time to change themselve with the field and hence there is no orientation polarization. The permittivity of medium is also decreases. In further high range frequency of infrared region, the ionic polarization fails to follow the phase reversal and hence only electronic polarization contribute to the total polarization. In visible region the electrons cloud follow the field reversal so the material exhibits electronic polarizibility α_e. The dielectric constant for this region of frequency will be equal to the square of refractive index n of the dielectric.

$$\epsilon_r = n^2 \qquad ...(38)$$

The electrons cloud fails to follow the field attraction in ultraviolet region and hence the contribution of electronic polarization is zero so as the total polarization of material. The ϵ_r for this UV region and above approach unity.

$$\epsilon_r = 1 \qquad ...(39)$$

3.11 DIELECTRIC LOSS

When an electric field is applied across a dielectric material, the dipoles of material are oriented in a certain direction. Due to this orientation, heat energy is produced in the material which is lost at latter. Thus loss of energy as heat in the dielectric is called dielectric loss. The power loss in material for AC field is more than a DC field. An ideal dielectric does not loss any energy. However, in real sense it always causes some electrical energy loss.

3.11.1 Loss Angle and Loss Tangent

Let a parallel plate capacitor with plate area A and separation d is filled with a dielectric material of dielectric constant ϵ_r. The capacitor with capacitance C is connected with an alternating voltage $V = V_0 e^{jwt}$. The current flows through the capacitor consists of two components, the displacement current I_d and conduction current I_c.

$$I = I_d + I_c \qquad ...(40)$$

The displacement current is the rate of change of electric field between the capacitor plates *i.e.*,

$$I_d = \epsilon_0 \epsilon_r \frac{d}{dt} A \frac{V}{d}$$

$$= \frac{d}{dt}\left(\frac{\epsilon_0 \epsilon_r A}{d} V\right)$$

$$= jw \epsilon_0 \epsilon_r \frac{A}{d} V \qquad ...(41)$$

$$= jw\, CV \qquad [\because C = \epsilon_0 \epsilon_r \frac{A}{d}]$$

The conduction current is the current passes through dielectric which is zero in ideal case.

$$I_C = XV \qquad ...(42)$$

where X is conductance

Hence $$I = jw\, CV + XV \qquad ...(43)$$

In case of ideal dielectric, conductance X is zero and therefore $I_C = 0$. Hence dielectric would not absorb electrical energy. Therefore the total current through capacitor would be equal to displacement current I_d which is ahead of the voltage V by a phase of 90°. (see Fig. 3.11). The magnitude of I_d would be

$$|I_d| = wCV \quad \text{...(44)}$$

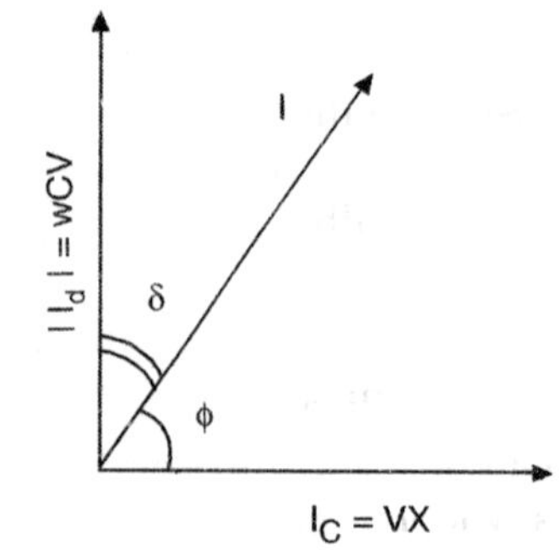

Fig. 3.11

In case of lossy dielectric the conduction current also contribute with displacement current for being total. The phase angle ϕ between I and V is now slightly less than 90°. The angle $\delta = (90 - \phi)$ is called loss angle and given by

$$\tan \delta = \left|\frac{I_C}{I_d}\right|$$

$$= \frac{VX}{wCV} = \frac{X}{wC} \quad \text{...(45)}$$

$\tan \delta$ is known as loss tangent. It represents the loss of electrical energy as heat and therefore also called the dissipation factor.

The real power loss in dielectric is given by

$$P_R = V|I_c| = V|I_d| \tan \delta \quad \text{...(46)}$$

$$P_R = wCV^2 \tan \delta$$

$$= \epsilon_o \epsilon_r \, w \frac{A}{d} V^2 \tan \delta$$

$$= \epsilon_0 \epsilon_r \, (2\pi f)\, (Ad)\, E^2 \tan \delta \qquad \left[\because E = \frac{V}{d}\right]$$

$$= 5.565 \times 10^{-11} \; v \, . \, E^2 f \epsilon_r \tan \delta \quad \text{...(47)}$$

where v is volume of dielectric, E is electric field and f is frequency of field. From equation (47) the power loss P_R in a dielectric is related to dissipation factor, dielectric constant, frequency and value of electric field and volume of dielectric material.

3.12 APPLICATION OF DIELECTRIC

Dielectric material is basically an insulator and hence it has a vide scale of applicability. These are used in number of electrical and electronics devices as per need of the property of material. Some material have large electronic and ionic polarizibilities and therefore used for making of large capacitance capacitors. There are many dielectric material like glass, mica, ceramic, asbestos resin, rubber etc. which have different uses. Some of them are described briefly.

Glass

Glass is basically a silica (SiO_2) which is best insulator and act as dielectric. Its dielectric constant is very high in the range of 4 to 10 and tangent loss is about .002. It is widely used in making electrical bulbs, electronic valves, switches and in capacitors.

Mica

Mica is an organic compound of silicates of aluminium, soda potash and magnesia. It is crystaline in nature and can be easily split into very thin flat sheets. It also has good mechanical strength. It is widely used in making electrical devices and machine such as switchgear, armature winding, electrical iron, and hot plates etc.

Ceramics

Ceramics is a non metalic inorganic compound and has dielectric properties. These are widely used as insulator in switchgear, plug holder, vacuum type ceramic metal seal etc. This material also used in capacitors which can be operated at high temperature.

Asbestos

Asbestos is a good dielectric and used as insulator in the form of paper, tape, cloth and board. It is also used to make panel board, insulating tubes and cylinders in constructions of air cooled transformers.

Resin

It is widely used in radio, TV, power and submarine cables.

Rubber

Rubber is very common insulating material which has a very wide application. However, it is restricted in use by temperature. It is mainly used in making electrical wires, cables, tapes coating etc.

In addition to above, there are many more dielectric as liquid and gases and have a wide applicability.

SOLVED EXAMPLES

1. *A parallel plate capacitor is charged through a battery of 50 volt. The capacitor is now disconnected with battery and a dielectric slab is inserted in the gap between the plate. If potential difference is reduced to half that of original, find the dielectric constant of slab.*

Sol. The capacitance of a parallel plate capacitor in absence of slab is

$$C = \frac{Q}{V}$$

where Q is the charge on each plate and V = 50 volt

When slab of dielectric constant $\in_r$ is inserted into the capacitor the reduced potential is V′ = V/2 then capacitance

$$C' = \frac{Q}{V'} = \frac{2Q}{V}$$

We know that $C' = \in_r C$ hence

$$\in_r = \frac{C'}{C} = \frac{2Q}{V} / \frac{Q}{V} = \mathbf{2}$$

2. *A parallel plate capacitor of area 6.5 cm*^{2} *and a plate separation d of 4 mm has a charge of 200 p.c. Calculate potential difference between the plates if a dielectric of* $\in_r = 3.5$ *is introduced between the plates.*

Sol. Area of plate of capacitor = $6.50 \times 10^{-4}\ m^2$

Distance between plates $d = 4 \times 10^{-3}\ m$

The capacitance of a capacitor given as

$$C = \frac{\in_0 \in_r A}{d} = \frac{Q}{V}$$

So $$V = \frac{Q \cdot d}{\in_0 \in_r A}$$

$$V = \frac{200 \times 10^{-12} \times 4 \times 10^{-3}}{\left(8.86 \times 10^{-12}\right) \times 3.5 \times 6.5 \times 10^{-4}} = \mathbf{39.8\ volt}$$

3. *If a NaCl crystal is subjected to an electric field of 1000 V/m and the resulting polarization is 4.3 × 10*$^{-8}$ *C/m*^{2}*. Calculate the relative permittivity of NaCl.*

Sol. Polarization of material is related with electric field as

$$P = \in_0 (\in_r - 1) E \text{ then}$$

$$\in_r = 1 + \frac{P}{\in_0 E}$$

$$= 1 + \frac{4.3 \times 10^{-8}}{8.86 \times 10^{-12} \times 1000} = \mathbf{5.86}$$

4. *Calculate the electric polarizability of an argon atom whose* $\in_r = 1.0024$ *at* NTP *and* N = 2.7 × 10^{25} *atoms/m*^{3}*.*

Sol. As we know that

$$P = \in_0 (\in_r - 1)E \text{ also}$$

$$P = N\,\alpha_e\,E$$

So $$N\,\alpha_e\,E = \epsilon_o\,(\epsilon_r - 1)E$$

The polarizability $$\alpha_e = \frac{\epsilon_o\left(\epsilon_r - 1\right)}{N}$$

$$= \frac{8.86 \times 10^{-12} \times (1.0024 - 1)}{2.7 \times 10^{25}}$$

$$= \mathbf{7.9 \times 10^{-40}\ F.m^2}$$

5. *The number of atoms in volume of one cubic meter of hydrogen gas is 9.8 × 10^{26}. The radius of hydrogen atom is 0.53 Å. Calculate the polarizability and relative permittivity.*

Sol. The atomic polarizability is described by following relation

$$\alpha_e = 4\,\pi\epsilon_o\,R^3$$

$$= 4 \times 3.14 \times 8.85 \times 10^{-12} \times (0.53 \times 10^{-10})^3$$

$$= 1.66 \times 10^{-41}\ F.m^2$$

The relative permittivity,

$$\epsilon_r = 1 + 4\pi\,NR^3$$

$$= 1 + 4 \times 3.14 \times 9.8 \times 10^{26} \times (0.53 \times 10^{-10})^3$$

$$= \mathbf{1.0018}$$

6. *The atomic weight and density of sulphur are 32 and 2.08 gm/cm^3 respectively. The electronic polarizability of the atom is 3.28 × 10^{-40} F.m^2. If a sulphur solid has cubic symmetry, what will be its relative permittivity?*

Sol. From the relation of Claussius-Mossotti equation

$$\frac{\epsilon_r - 1}{\epsilon_r + 2} = \frac{N\alpha_e}{3\epsilon_o}$$

As $$N = \frac{\text{Avogadro No.} \times \text{density}}{\text{Atomic weight}}$$

$$= \frac{6.023 \times 10^{26} \times 2.08 \times 10^3}{32}$$

Then $$\frac{\epsilon_r - 1}{\epsilon_r + 2} = \frac{6.023 \times 10^{26} \times 2.08 \times 10^3}{32} \times \frac{3.28 \times 10^{-40}}{3 \times 8.86 \times 10^{-12}}$$

$$= 0.483$$

$$\epsilon_r - 1 = 0.483\,(\epsilon_r + 2)$$

$$\epsilon_r\,(1 - 0.483) = 1 + 2 \times 0.483$$

$$\epsilon_r = \mathbf{3.8}$$

MULTIPLE CHOICE QUESTIONS

1. When a capacitor is inserted with a dielectric material, its capacitance.

(*a*) Increases (*b*) Decreases

(*c*) Remain same (*d*) None of the above

2. Pure dielectric medium is a

(*a*) Conductor (*b*) Insulator

(*c*) Semi-conductor (*d*) None of these

3. When a monoatomic gas atom is placed in a uniform electric field E, the displacement of nucleus is proportional to

(*a*) E^2 (*b*) E

(*c*) E^3 (*d*) Independent of E

4. A capacitor is inserted by a dielectric of constant ϵ_r, its potential.

(*a*) Decreases (*b*) Increases

(*c*) Remain same (*d*) None of these

5. The flux density is related to the field as

(*a*) $D = \epsilon + E$ (*b*) $D = \epsilon - E$

(*c*) $D = \frac{\epsilon}{E}$ (*d*) $D = \epsilon E$

6. If two equal and opposite charges q are separated by a distance d, this makes a dipole whose dipole moment will be

(*a*) $2qd$ (*b*) qd

(*c*) $\frac{qd}{2}$ (*d*) qd^2

7. The relation between dipole moment, polarizability and electric field is

(*a*) $p = \frac{\alpha}{E}$ (*b*) $E = p\alpha$

(*c*) $p = \alpha E$ (*d*) $p = \alpha E^2$

8. The orientation polarizability per molecule in a polyatomic gas is proportional to

(*a*) T (*b*) T^2

(*c*) $\frac{1}{T}$ (*d*) $\frac{1}{T^2}$

9. The orientational polarizability per molecule in a polyatomic gas is given by

(*a*) $\frac{p^2}{3kT}$ (*b*) $\frac{p}{3kT}$

(*c*) $\frac{p^3}{3kT}$ (*d*) $\frac{p^2}{3kT^2}$

10. The dipole moment per unit volume of a solid is the sum of all the individual dipole moments and is called
(*a*) Polarization of solid (*b*) Permittivity of solid
(*c*) Electrostatic moment (*d*) Polarizability of solid

11. For a given dielectric, as the temperature increases, the ionic polarizability
(*a*) Increases (*b*) Decreases
(*c*) Remain unchange (*d*) First increases then decreases

12. The Clausius-Mossotti relation is
(*a*) $\frac{\epsilon_r - 1}{\epsilon_r + 1} = \frac{N\alpha}{3\epsilon_o}$ (*b*) $\frac{\epsilon_r + 1}{\epsilon_r - 2} = \frac{N\alpha}{3\epsilon_o}$
(*c*) $\frac{\epsilon_r + 1}{\epsilon_r - 1} = \frac{3N\alpha}{\epsilon_o}$ (*d*) $\frac{\epsilon_r - 1}{\epsilon_r + 2} = \frac{N\alpha}{3\epsilon_o}$

13. For a given dielectric, the electronic polarizability α_e
(*a*) Increases with temperature (*b*) Decreases with temperature
(*c*) Independent of temperature (*d*) None of the above

14. In a dielectric, the polarization is
(*a*) Linear function of applied field (*b*) Square function of applied field
(*c*) Exponental function of applied field
(*d*) None of the above

15. At high frequency generally, which polarization is not accountable
(*a*) Electronic (*b*) Ionic
(*c*) Orientational (*d*) Electronic & ionic

16. In ferroelectric material, as the applied electric field is gradually reduced to zero, the polarization still left is known as
(*a*) Coercive polarization (*b*) Remanance polarization
(*c*) Zero polarization (d) Positive polarization

17. The factor responsible for spontaneous polarization is
(*a*) Free electrons (*b*) Atoms of material
(*c*) Permanent dipole (*d*) None of these

18. Piezoelectric effect is the production of electricity by
(*a*) Chemical effect (*b*) Varying field
(*c*) Temperature (*d*) Pressure

19. Above curie temperature, the hysteresis loop of a ferroelectric material merges into a
(*a*) Parabola (*b*) Straight line
(*c*) Point (*d*) Ellipse

20. In a dielectric, the power loss is proportional to
(*a*) w (*b*) w^2
(*c*) $\frac{1}{w}$ (*d*) $\frac{1}{w^2}$

21. Loss tangent is the electrical power loss which is often in the form of

(*a*) Sound (*b*) Spark

(*c*) Chemical (*d*) Heat

ANSWERS

1. (*a*)	**2.** (*b*)	**3.** (*b*)	**4.** (*a*)	**5.** (*d*)	**6.** (*b*)	**7.** (*c*)
8. (*c*)	**9.** (*a*)	**10.** (*a*)	**11.** (*c*)	**12.** (*d*)	**13.** (*c*)	**14.** (*a*)
15. (*c*)	**16.** (*b*)	**17.** (*c*)	**18.** (*b*)	**19.** (*b*)	**20.** (*a*)	**21.** (*d*).

LONG QUESTIONS

1. Explain dielectric polarization. Find the relation between P, E and D.

2. What is meant by polarization of a material? Discuss electronic polarization.

3. What are polar and non-polar materials. Discuss the orientational polarization for polar molecule.

4. Describe polarization. Explain how electronic polarization takes place in a dielectric.

5. What is meant by ionic polarization? How does it vary with temperature?

6. What are internal fields in solid and liquid. Describe these.

7. Describe Langevin theory for orientation polarization.

8. Deduce the Clausius-Mossotti equation.

9. What do you understand by dielectric loss? Describe tangent loss.

10. What are ferroelectric materials? Discuss the hysteresis phenomenon for ferroelectric materials.

11. What do you understand by piezoelectricity? Describe the behaviour of peizoelectric materials.

12. What are the dielectrics? Write some applications of dielectrics.

13. Explain the different type of polarization mechanism in dielectrics and sketch their dependency on the frequency of the applied electric field.

14. What is meant by local field in a dielectric and how is it calculated for cubic structure?

15. Explain electronic polarization in atoms and obtain an expression for electronic polarizability in terms of the radius of atom.

4

Magnetic Properties of Materials

4.01 INTRODUCTION

Some knowledge of ferro-magnetism, like static electricity, existed at the time of the ancient Greeks. Thales of Miletus is said to have written on both subjects in about 6000 B.C. The mariner's compass was invented by 1190, since the first reference to it in the literature of Western Europe dates from that year. The practical beginnings of the study of magnetism, however, date from 1269 with the writing of the *Epistola* by Peter Peregrinus. In 1600, William Gilbert published *De Magnete,* a compendium of the then known facts about magnetism as well as additional facts which he obtained as the result of his own studies. (This is the work in which Gilbert also summarised the existing knowledge of static electricity). The information in *De Magnete* is essentially that which any school-going student would now be expected to have concerning magnetism. Let us now discuss a brief summary of this knowledge.

4.02 BAR MAGNET AND ITS PROPERTIES

A piece of iron can be treated in such a way as to produce a magnet. This could be done in Gilbert's time by stroking it with a lodestone—a natural magnet consisting of a piece of magnetite, an iron ore. The magnetic properties of a magnet can also be destroyed in several ways, one of which is by heating it. When a magnet is suspended so that it is free to rotate about a vertical axis, it will line up along a geographical meridian or nearly so. This fact is the basis for the magnetic compass. A magnet will attract iron objects and objects containing iron alloys. Ordinarily, the ends of the magnet are the centres of this attraction. The two centres of attraction in any magnet are called its poles. After a magnet is suspended so that it is free to rotate about a vertical axis and it has lined up along a meridian, the pole in the northern end of the magnet is called a north pole and the pole in the southern end is called a south pole. Forces are exerted between the poles of two adjacent magnets such that unlike poles attract each other while like poles repel each other.

The region around a magnet in which it exerts forces on other magnets and on objects made of iron is called a magnetic field.

To map the shapes of magnetic fields in a qualitative fashion, we can scatter iron filings in the region. The iron filings tend to line up and form chains. The orientation of these chains indicates the direction along which magnets line up. (The iron filings in-fact become little magnets by induction). Just as we can visualise electrostatic fields with the aid of electric lines of force, we can also visualise magnetic fields by magnetic lines of force. A magnetic line of force at any point in the field is the line along which a small suspended magnet (a compass needle) will orient itself at that point. The direction of the field at this point is arbitrarily chosen to be the direction of the force on the north pole of the test magnet.

The magnetic field in the vicinity of an isolated bar magnet is as shown in Fig. 4.1.

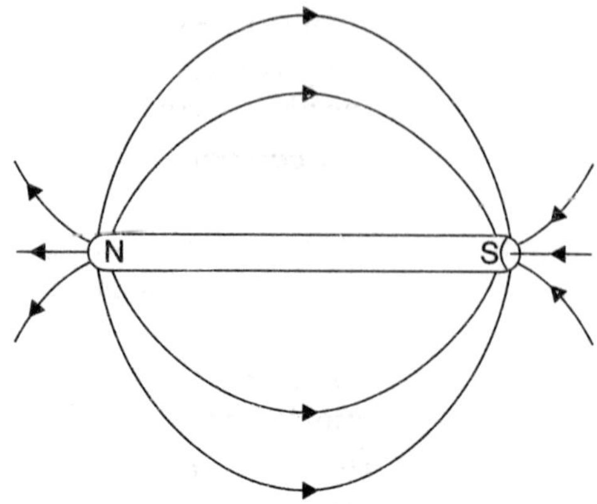

Fig. 4.1. Magnetic field of an isolated bar magnet.

Let us now turn our attention to making magnets. In this connection, the properties of soft iron and hard steel are quite different from each other. These two materials give contrasting responses to attempts to make magnets out of them. Let us now discuss the case of a steel rod. Inside a steel rod, there are many tiny "atomic magnets" (like so many compass needles). Initially, all these are *oriented in random directions* (Fig. 4.2). We can make a magnet out of a steel rod by placing it in the north-south direction and repeatedly hammering it. The repeated hammering tends to align all the "atomic magnets" in the direction of the earth's field (Fig. 4.3). The magnetism so acquired by the steel rod can be retained by the steel rod. That end of the magnetised steel rod which points towards or seeks the earth's geographical north pole is called the north (N) pole. The other end which seeks the geographical south pole is called the south (S) pole.

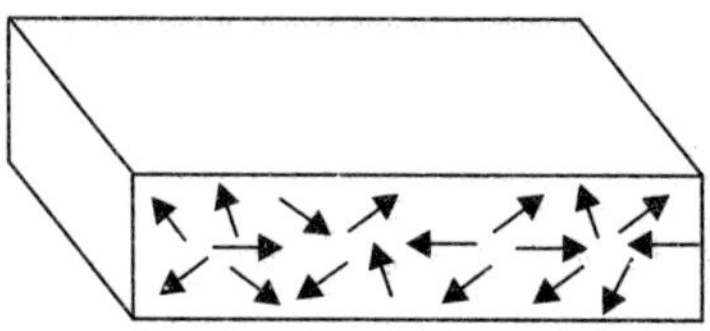
Fig. 4.2

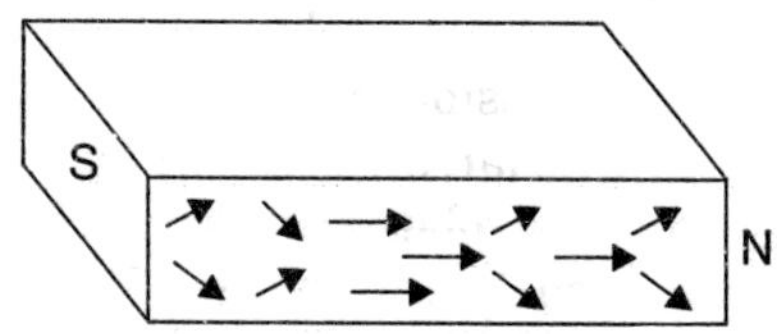

Fig. 4.3

It was observed that if we had two long thin magnets lying parallel and near to each other as shown in Fig. 4.4, then there exists a force of repulsion between them. However, if the magnets are placed as shown in Fig. 4.5, then there exists a force of attraction between them. Thus, while the like poles repel each other, the unlike poles attract each other. This is similar to the behaviour of electric charges.

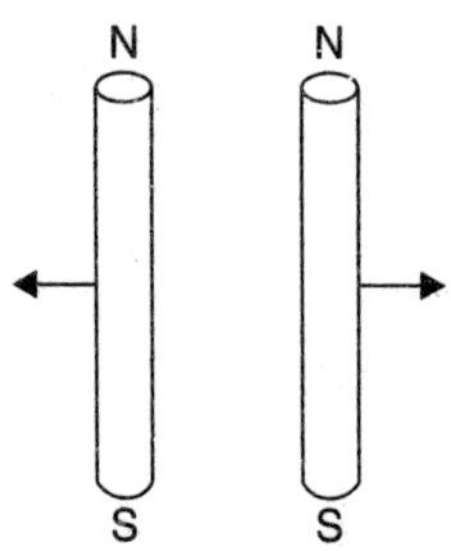

Fig. 4.4. Force of repulsion. between long thin magnets

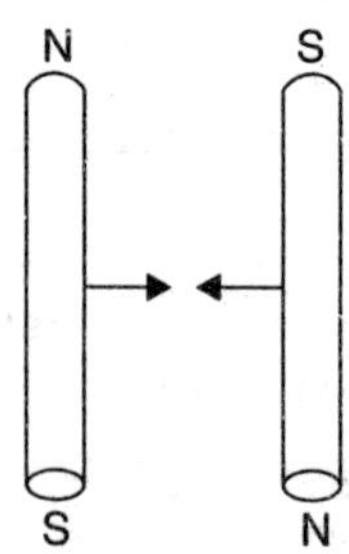

Fig. 4.5. Force of attraction between long thin magnets.

Note. Since the south pole of a magnet points towards geographical south pole of earth and the unlike poles attract each other therefore we can say that the north pole of the imaginary bar magnet in the earth is located somewhere near the geographical south pole of the earth.

The main **properties** of a bar magnet can be summed up as follows :

1. Directive property. When a magnet is suspended freely, it aligns itself with one end pointing towards north of earth and other towards south of earth.

2. Force between poles. Unlike poles attract and like poles repel (Fig. 4.6). The force of attraction or repulsion acts along the line joining the two poles and is directly proportional to the product of the pole strengths and inversely proportional to the square of the distance between them. This law was discovered by Charles Augustin de Coulomb in the year 1785.

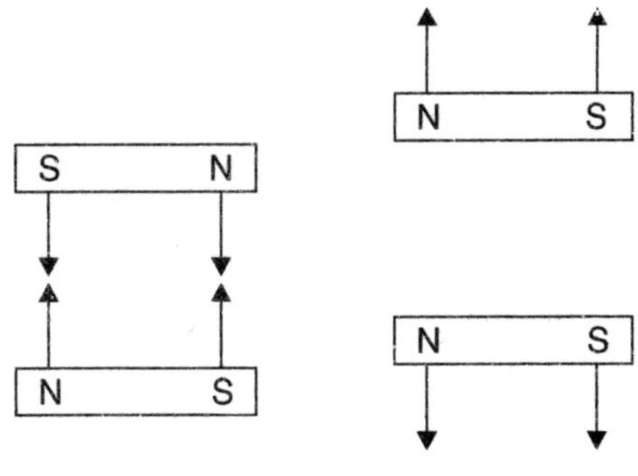

(*a*) Unlike poles attract (*b*) Like poles repel

Fig. 4.6

3. Attractive property. A magnet attracts magnetic materials such as iron, steel, cobalt, nickel etc. This attraction is maximum at the poles. *It may be noted that the poles are situated near the ends of the magnet and not exactly at the ends.*

4. Isolated magnetic pole does not exist i.e., magnetic poles always exist in pairs. The similarity in the behaviour of electric charges and poles seems to suggest the possibility of 'magnetic charges'—N-type and S-type. But, so far, every attempt at finding single magnetic charges or to isolate the poles of a magnet from one another have failed. If we break a long bar magnet into two (or even more) pieces, then we shall observe that each piece is complete magnet in itself (Fig. 4.7).

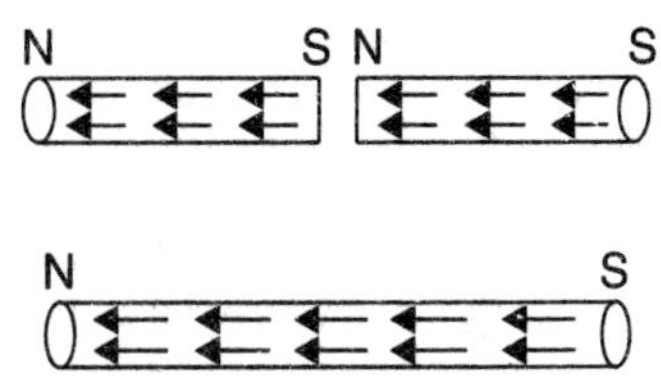

Fig. 4.7. Breaking of a magnet.

Since there is no way to isolate pole therefore magnets have to be necessarily regarded as dipoles. This brings out an essential difference between electricity and magnetism. While in electricity, the fundamental or simplest object is point charge, for magnetism the simplest object is a magnetic dipole.

5. A magnet can induce magnetism in substances like soft iron, cobalt, nickel and various ferrous alloys. When an iron, nickel or cobalt rod is placed near a bar magnet, the bar magnet induces magnetism in the rod.

4.03 REPULSION—SUREST TEST OF MAGNETISATION

An iron rod is attracted towards a magnet. The opposite poles also attract each other. So, attraction is not a sure test of magnetisation. On the other hand, if there is repulsion between a magnet and a given rod, we can be sure that the given rod is magnetised.

4.04 ATOMIC THEORY OF MAGNETISM

Both the orbital and spin motions of electrons gives rise to tiny circular currents. These tiny current loops behave like small magnets. These small magnets are called elementary or atomic magnets. When the material is not magnetised, these magnets form closed chains thereby annulling each other's effect (Fig. 4.8).

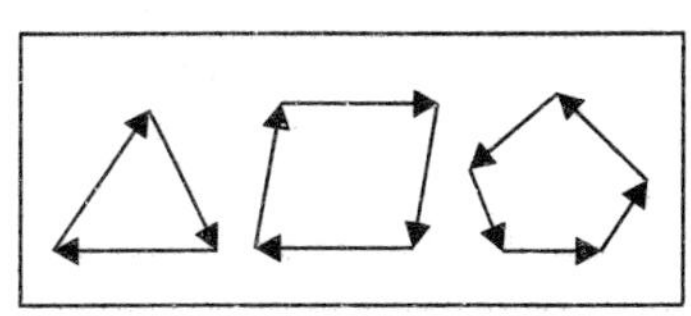

Fig. 4.8. Closed chains of elementary magnets.

When the material is magnetised, the elementary magnets are aligned in the same direction (Fig. 4.9).

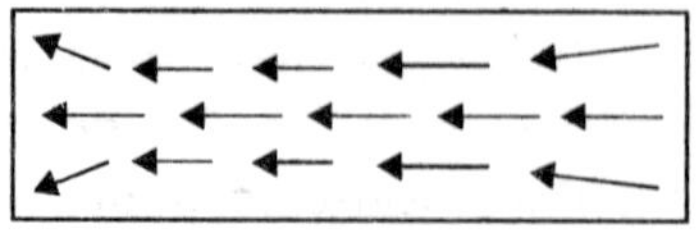

Fig. 4.9. Elementary magnets aligned in the same direction.

The atomic theory of magnetism explains the following facts :

(i) Single poles cannot exist. Poles always exist in pairs. ***(ii)*** The magnetic poles are of equal strength. ***(iii)*** When a magnet is heated, the thermal energy of the elementary magnets increases. These again form closed chains and magnetism is lost.

4.05 REPRESENTATION OF UNIFORM MAGNETIC FIELD

A magnetic field is said to be uniform over a region if its magnetic field induction $\vec{B}$ has the same magnitude and direction at all points in the region.

Example. Earth's magnetic field may be regarded as uniform over a small region.

If $\vec{B}$ varies in magnitude or in direction or both, then the magnetic field is non-uniform.

A uniform magnetic field in the plane of paper is represented by a set of equal, equidistant and parallel lines pointing in the same direction as shown in Fig. 4.10 (*a*).

A uniform magnetic field perpendicular to the plane of paper and directed outwards is represented by a group of equidistant tiny dots as shown in Fig. 4.10 (*b*).

A uniform magnetic field perpendicular to the plane of paper and directed inwards is represented by a group of equidistant crosses as shown in Fig. 4.10 (*c*).

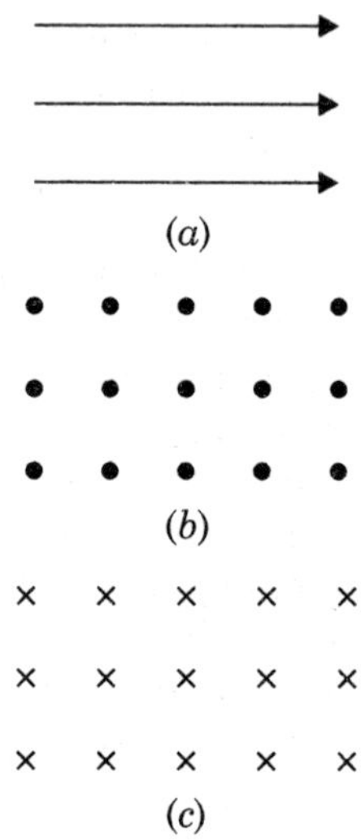

Fig. 4.10. Representation of uniform magnetic field.

4.06 PROPERTIES OF MAGNETIC LINES OF FORCE

Let us consider a sheet of paper on which soft iron filings are sprinkled. If a bar magnet is placed below this sheet and the sheet is tapped, then the iron filings shall arrange themselves approximately in the positions indicated by dotted lines as shown in Fig. 4.11. This gives a picture of the magnetic field lines produced by a bar magnet.

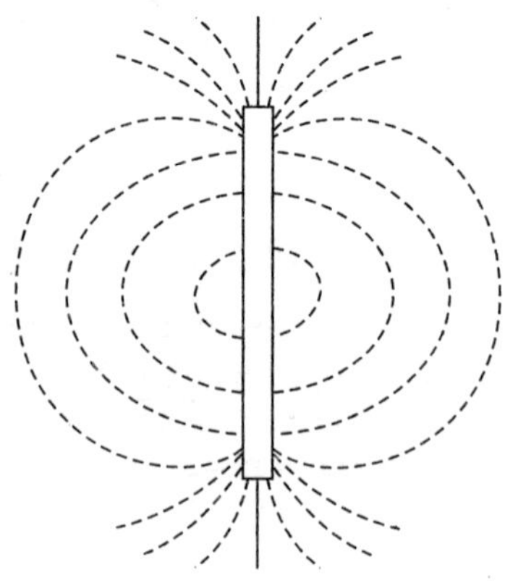

Fig. 4.11

The formation of the "line" pattern by the iron filings or by the directional indications of the compass needle is due to the action of the magnetic energy in the lines of force. The magnetic energy magnetises each filing and urges it (or the compass needle which is already a magnet) to take positions along the lines of force. In the process of magnetising a filing (or some other magnetisable object), the energy in the magnetic field induces magnetism in the filing. The end of the filing entered by the lines of force becomes the South pole of the magnetised bit of iron, the end where the lines of force leave the filing is the North pole. Thus, the magnetic lines of force pass lengthwise through each magnet they create.

Being positioned along the lines of force, each tiny magnet in the field is threaded by the greatest number of flux lines. *This is a rule of magnetic behaviour*. A magnetisable object located in a magnetic field will position itself so that it will be threaded by the greatest number of lines of force. Given sufficient energy in the magnetic field, the lines of force exert turning, pulling or pushing force on other magnets on which they act, so that the magnets will be threaded by the greatest number of flux lines.

Two magnets placed near each other have separate magnetic fields which occupy an area common to both. One might gather from this that the lines of force cross or intersect. But this does not happen ; *lines of force never cross*. When like poles face each other, the lines of force of each field remain separate entities, but each field becomes distorted. When unlike poles face each other, the lines of force of each field interact with the other and produce a strong resultant field.

Wherever the lines of force of the two fields are parallel and have the same direction, a strengthened field is created. This follows from the condition that a greater number of lines of force acting in the same direction are present per unit area than if only one magnet and one magnetic field were present. Where the lines of force of the two fields act in opposite directions, the energy of one field tends to offset the energy in the other field. The result is a weakened field in that area. This condition is equivalent to fewer lines of force traversing the area.

Still another important behaviour of magnetic lines of force is that each magnetic line behaves as though it were a stretched rubber band which, if given the opportunity, would contract. This explains why unlike poles attract each other. The lines of magnetic force passing through the magnet behave as though they shrink, thus pulling the magnets towards each other. This is the simplified version of the condition whereby one portion of a resultant of two fields is strengthened and another is weakened. The movable sources of the field move from the stronger to the weaker portion of the field. Strengthened and weakened magnetic fields in an area are illustrated by more or fewer lines of force.

4.07 BAR MAGNET AS AN EQUIVALENT SOLENOID

Fig. 4.12 shows the magnetic lines of force due to a solenoid carrying current. For points like P close to a single turn, the wire behaves like a long straight wire. The magnetic field lines

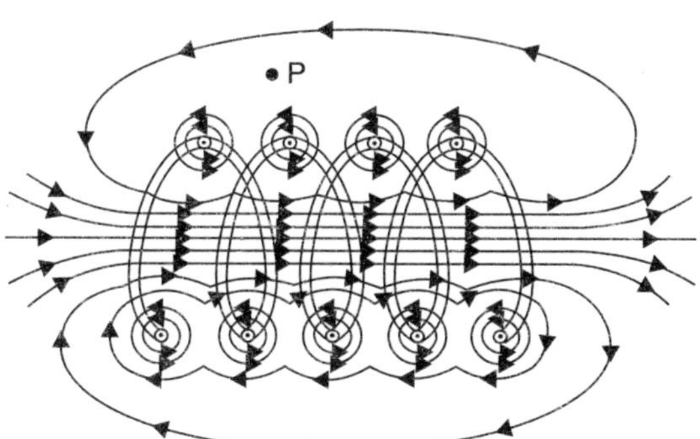

Fig. 4.12

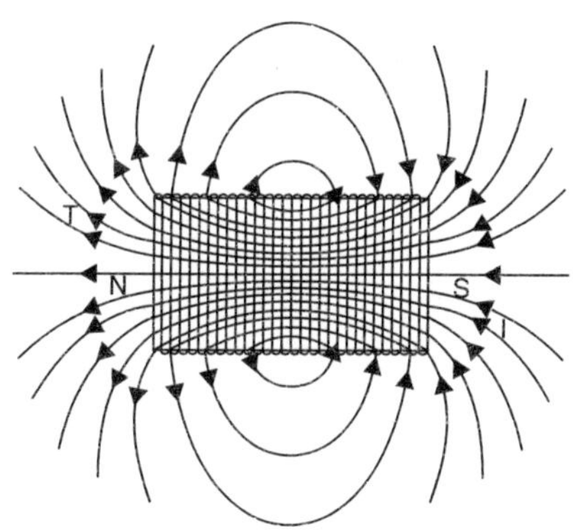

Fig. 4.13

due to a single turn are nearly concentric circles. The magnetic field of a current-carrying solenoid is the vector sum of the fields set up by all the turns of the solenoid. Fig. 4.12 shows the magnetic field lines for a loosely wound solenoid. Fig. 4.13 shows the magnetic field lines of a closely wound solenoid of finite length. The right end of the solenoid from where the lines of force emerge out of the solenoid behaves like a N pole. On the other hand, the left end (where field lines enter the solenoid) behaves like a south pole. So, we may conclude that the current-carrying solenoid behaves like a bar magnet.

4.08 COMPARISON OF MAGNETIC AND ELECTRIC CIRCUITS

Lines of magnetic flux do not end at magnetic poles, but are continuous or closed on themselves like the lines of electron flow in an electric circuit. In fact, a magnetic circuit is, in many ways, similar to an electric circuit. To produce an electric current, a voltage called the electromotive force (emf) is required. Similarly, to produce a magnetic flux, a force called *magnetomotive force* (mmf) is required. In an electric circuit, for a given amount of voltage or electromotive force, the current is determined by the circuit resistance. Similarly, in a magnetic circuit, for a given amount of magnetomotive force, the flux density depends upon the amount of opposition or *reluctance*. Thus, in the magnetic circuit, as in the electric circuit, we can say that the result produced is directly proportional to the force that produces it and inversely proportional to the opposition encountered.

There are two important differences in the relationship of the electric circuit to that of magnetic circuit. The first difference is that in the electric circuit, the resistance is a constant value and can be determined by measuring the ratio of voltage to current. In a magnetic circuit, however, the reluctance is not a constant but depends upon the flux or field strength. The second difference is that in electric circuits, current actually flows (electron flow) from one point to another. In the magnetic circuit, there is no actual flow of flux but merely an indication of the intensity and direction of the magnetic field.

The magnetising force set up due to current flowing in a coil or solenoid is known as the magnetomotive force (abbreviated mmf). The practical unit of magnetomotive force is the ampere-turn. The magnetomotive force is proportional to the current (in ampere) in the circuit and to the number of turns of the coil. The magnetic flux (ϕ) comprises the total number of lines of force in the magnetic circuit.

4.09 MAGNETOMOTIVE FORCE (mmf), RELUCTANCE AND PERMEANCE

Consider an iron ring of cross-sectional area A and permeability μ. Let l be its mean circumferential length. Suppose a current of i ampere passes through the closely-fitted toroidal winding of N turns. Using Ampere's circuital law,

$$\oint \vec{H} \cdot \vec{dl} = Ni$$

or

$$\oint H dl = Ni$$

Now,

$$\phi_B = BA = \mu HA \quad \text{or} \quad H = \frac{\phi_B}{\mu A}$$

$\therefore$

$$\oint \frac{\phi_B}{\mu A} dl = Ni$$

or

$$\phi_B \oint \frac{dl}{\mu A} = Ni$$

This is called magnetic circuit equation. The line integral of magnetising field H around a magnetic circuit is called *magnetomotive force (mmf)

$$mmf = \oint \text{H}dl = \text{N}i$$

$$= \phi_B \oint \frac{dl}{\mu \text{A}}$$

or

$$\phi_B = \frac{mmf}{\oint \frac{dl}{\mu \text{A}}}$$

The flux ϕ_B of a magnetic circuit is equivalent to electric current of an electric circuit. The mmf of a magnetic circuit is equivalent to emf of an electric circuit. By the same analogy, the magnetic resistance known as reluctance of the magnetic circuit is given by

$$\text{R} = \oint \frac{dl}{\mu \text{A}}$$

mmf = flux × reluctance

If A is constant over length l of the circuit, then

$$\text{R} = \frac{l}{\mu \text{A}}$$

If magnetic circuits of different permeabilities $\mu_1, \mu_2, \mu_3, \ldots$ of lengths $l_1, l_2, l_3, \ldots\ldots$ and cross-sectional areas $\text{A}_1, \text{A}_2, \text{A}_3, \ldots\ldots$ respectively are joined in series, then the total reluctance of the circuit is given by

$$\text{R} = \frac{l_1}{\mu_1 \text{A}_1} + \frac{l_2}{\mu_2 \text{A}_2} + \frac{l_3}{\mu_3 \text{A}_3} + \ldots\ldots$$

If the reluctances are joined in parallel, then the equivalent reluctance is given by

$$\frac{1}{\text{R}} = \frac{1}{\text{R}_1} + \frac{1}{\text{R}_2} + \ldots\ldots$$

The reciprocal of reluctance is called **permeance**. It is analogous to conductance in an electric circuit.

4.10 MAGNETIC CIRCUIT OF A ROWLAND RING

Consider a Rowland ring of mean radius r and cross-sectional area A. Suppose it is surrounded by a toroidal winding of N turns carrying current of i ampere.

Reluctance of the ring is given by

$$\text{R} = \oint \frac{dl}{\mu \text{A}} = \frac{1}{\mu \text{A}} \oint dl$$

or

$$\text{R} = \frac{1}{\mu \text{A}} \times 2\pi r = \frac{2\pi r}{\mu \text{A}}$$

If n be the number of turns per unit length,

then $$mmf = \text{N}i = 2\pi rni$$

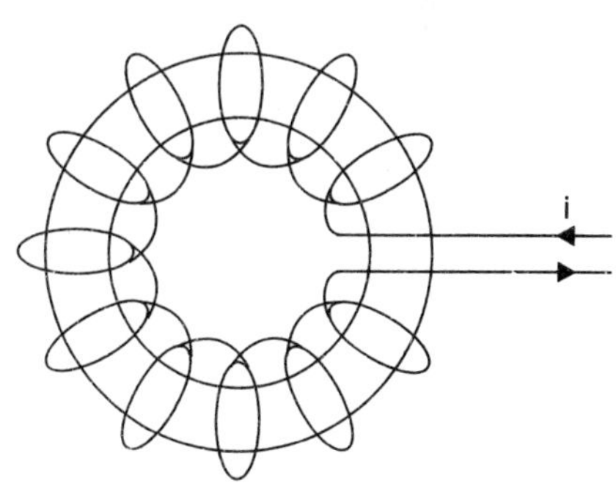

Fig. 4.14

*Note that the line integral of $\vec{E}$ around a closed path is called electromotive force (emf).

Now, $$\text{magnetic flux} = \frac{mmf}{\text{reluctance}}$$

$$\phi_B = \frac{2\pi rni}{\dfrac{2\pi r}{\mu A}} \quad \text{or} \quad \phi_B = \mu ni\,A$$

or $$BA = \mu niA \quad \text{or} \quad B = \mu ni$$

or $$\mu H = \mu ni \quad \text{or} \quad H = ni$$

4.11 MAGNETIC CIRCUIT WITH AN AIR GAP

Consider a ferromagnetic circuit in the form of a ring having an air gap of length l_a and cross-section A_a. The ring has N turns of wire carrying current i as shown.

Let, l_i = length of magnetic circuit (excluding air gap)

A = cross-sectional area.

If R be the total reluctance of the magnetic circuit and air gap, then

$$R = \frac{l_a}{\mu_0 A_a} + \frac{l_i}{\mu A}$$

Here, μ_0 is the permeability of air gap and μ is the permeability of the ferromagnetic material.

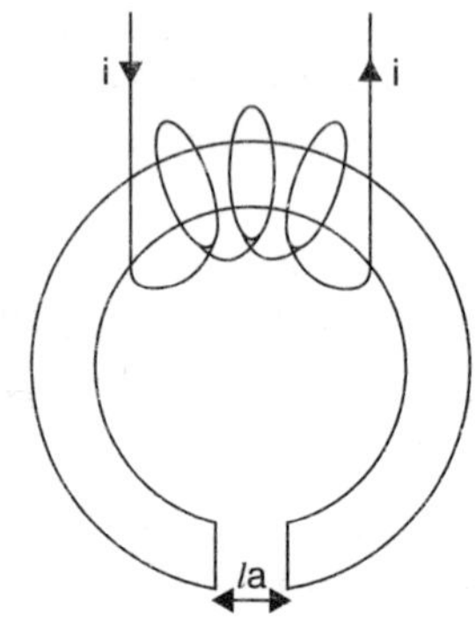

Fig. 4.15

4.12 MAGNETIC INTENSITY H

A given amount of magnetomotive force may be distributed over a magnetic circuit of considerable length, or it may be concentrated in a circuit of shorter length. In the latter case, the magnetic effect is more intense than in the former case. These differences can be described by the term 'magnetic intensity'. This is denoted by H and is defined as magnetomotive force per unit of length.

When the flux density is measured in weber m^{-2}, then the appropriate unit for magnetic intensity H is ampere-turns per linear metre.

When the flux density is in gauss, H is expressed in oersted.

4.13 RELATION BETWEEN B AND H

The relation between B and H is as follows :

$$\boxed{B = \mu H}$$

If B is in weber m^{-2}, then H is expressed as ampere-turn per metre. The permeability is expressed in weber per ampere-metre. The value of the permeability μ_0 of empty space has the value $4\pi \times 10^{-7}$ Wb/ampere-m.

If B is expressed in gauss, H in oersted then μ is expressed in gauss per oersted. The numerical value for the permeability of empty space in this system is $\mu_0 = 1$.

4.14 INTENSITY OF MAGNETISATION (OR MAGNETISATION)

When a ferromagnetic material is placed in a magnetic field, it acquires a magnetic dipole moment $\vec{M}$ in the direction as shown in Fig. 4.16. This magnetic dipole moment per unit volume of the substance is known as intensity of magnetisation. It is given by

$$I = \frac{M}{V}$$

where V is the volume of the ferromagnetic specimen. If m is the pole strength developed in the specimen and $2l$ is the magnetic length of the specimen, then

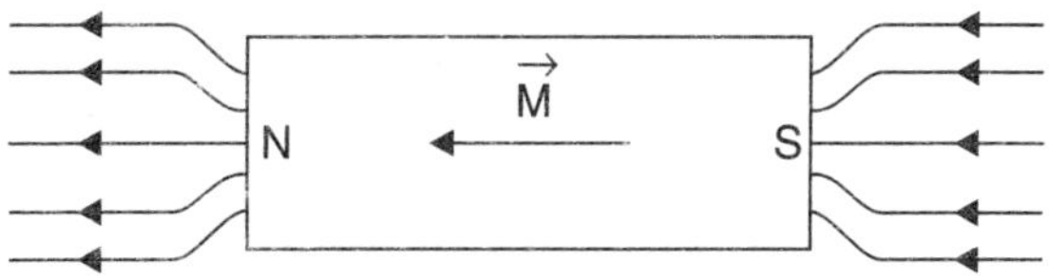

Fig. 4.16. Magnetisation

$$I = \frac{m \times 2l}{A \times 2l}$$

where A is the cross-sectional area of the specimen.

$$\therefore \qquad I = \frac{m}{A}$$

So, intensity of magnetisation is the pole strength per unit area developed in a ferromagnetic substance when it is placed in a uniform magnetic field.

4.15 MAGNETIC SUSCEPTIBILITY

The term "magnetic susceptibility" of a substance refers to the ability of the substance to acquire magnetisation when placed in a magnetic field. *It is a measure of how easily a material is magnetised in a magnetising field.* It has been observed experimentally that the intensity of magnetisation is proportional to the strength of the magnetising field (the magnetic field in which the substance is placed).

$$\therefore \qquad \vec{I} \propto \vec{H} \quad \text{or} \quad \vec{I} = x_m \vec{H}$$

where x_m is a "constant" known as the susceptibility of the substance.

If $\quad H = 1, \quad$ then $\quad x_m = I.$

Magnetic susceptibility is the intensity of magnetisation developed in the substance when placed in a magnetic field of unit strength.

Both $\vec{I}$ and $\vec{H}$ have the same units. So, x_m is a pure number.

In vacuum, there is no magnetisation. So, $\vec{I} = 0$. Thus, $x_m = 0$.

We can classify magnetic materials on the basis of magnetic susceptibility.

x_m has a large positive value for ferromagnetic and ferrimagnetic materials. For these materials, $\vec{I}$ and $\vec{H}$ are not exactly proportional.

x_m is small and positive for anti-ferrimagnetic substances.

For paramagnetic substances, x_m has a small positive value. The presence of paramagnetic material increases magnetic induction.

For diamagnetic materials, the magnetic susceptibility is negative.

The net magnetic induction in a magnetised material is given by

$$\vec{B} = \mu_0(\vec{H} + \vec{I})$$

where $\vec{I}$ represents magnetisation and $\vec{H}$ is the strength of the applied magnetic field.

But $\qquad \vec{I} = x_m \vec{H}$

$$\therefore \qquad \vec{B} = \mu_0(\vec{H} + x_m\vec{H})$$

or $$\vec{B} = \mu_0(1 + x_m)\vec{H}$$

or $$\mu\vec{H} = \mu_0(1 + x_m)\vec{H}$$

or $$\mu = \mu_0(1 + x_m)$$

Comparing with $\mu = \mu_0\mu_r$, we get

$$\mu_r = 1 + x_m$$

4.16 MAGNETIC INDUCTION

According to Gauss' law in magnetism, 4π magnetic lines of force are associated with a unit magnetic pole. Thus, if a pole strength m is induced in a specimen, $4\pi m$ lines of magnetic force are produced. These lines (shown dotted in Fig. 4.17) are in addition to the lines of force due to the external magnetic field. Thus, we can conclude that the magnetic field at any point inside a material is the sum of the external magnetic field and the field produced due to the magnetisation of the material. **The total number of lines of force threading the specimen per unit area is called magnetic induction.** It is also known as magnetic flux density and is denoted by B.

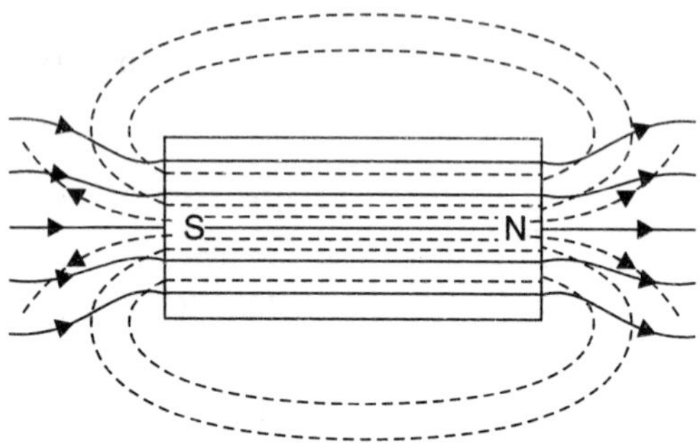

Fig. 4.17. Magnetic induction.

$$\therefore \qquad B = H + \frac{4\pi m}{A}$$

But $$\frac{m}{A} = I$$

$$\therefore \qquad B = H + 4\pi I.$$

4.17 MAGNETIC PERMEABILITY

The term 'magnetic permeability' deals with the **degree of concentration of lines of force in a specimen.** ***It is the ratio of the number of lines of force passing per unit area (held perpendicular to the lines of force) in the specimen to the lines of force that would exist there due to the field in the absence of specimen.***

The magnetising field is expressed by H. Its unit is ampere-turn/metre. The magnetic flux density within the sample is denoted by B. Its unit is tesla or Wbm^{-2}. The absolute permeability of the material medium is given by :

$$\mu = \frac{B}{H}$$

For air, the value of μ is 1.0000004. For vacuum, $\mu = 1$. For iron, the value of μ may exceed 1000.

Relative permeability is the ratio of magnetic induction within the sample to the magnetising field when expressed in the same units.

If B is magnetic induction within the sample in tesla and B_0 the applied external magnetic induction in free space in tesla, then

$$\mu_r = \frac{B}{B_0}$$

Relative permeability is also defined as the ratio of permeability of medium to the permeability of free space.

$$\mu_r = \frac{\mu}{\mu_0}$$

Clearly, μ_r has no units.

4.18 CURRENT LOOP AS A MAGNETIC DIPOLE AND ITS MAGNETIC DIPOLE MOMENT

The current loop behaves as a magnetic dipole (Fig. 4.18). It has been observed experimentally that the magnetic dipole moment of a current loop is directly proportional to

(*a*) the current I in the loop.

(*b*) face area 'A' of the loop.

$\therefore$ $M \propto IA$ or $M = KIA$

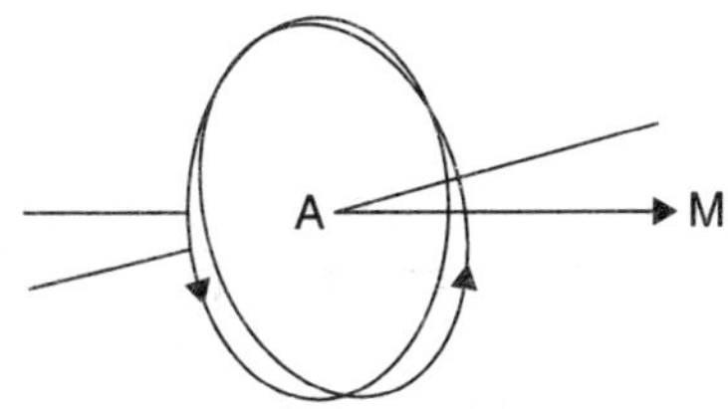

Fig. 4.18. Current loop.

where the constant of proportionality K depends upon the system of units.

If we choose unit dipole moment as that of a small one-turn loop of unit area through which unit current is flowing, then K will also be unity.

In that case, $\boxed{M = IA}$

The direction of dipole moment is perpendicular to the face area and along the axis of the loop. According to **right hand rule**, it is directed from south to north *i.e.*, outwards from north.

In vector form, $\vec{M} = I\vec{A} = IA\hat{n}$

Here $\hat{n}$ is a unit vector perpendicular to the plane of the loop.

The SI unit of dipole moment is **A m². It is the magnetic moment of one turn loop of an area one square metre when a current of one ampere is flowing through it.**

It is interesting to arrive at the relation M = IA in the following way.

Torque on a current loop in a magnetic field,

$$\tau = BIA \sin\theta \qquad ...(1)$$

where θ is the angle between the direction of $\vec{B}$ and the vector perpendicular to the plane of the loop.

Torque experienced by a magnetic dipole in a magnetic field,

$$\tau = MB \sin\theta \qquad ...(2)$$

From (2) and (1), we get $M = IA$

4.19 MAGNETIC DIPOLE MOMENT OF A REVOLVING ELECTRON (ATOM AS A MAGNETIC DIPOLE)

Consider an atom in which an electron is revolving in a circular orbit of radius *r* around a positively charged nucleus (Fig. 4.19). Suppose it is revolving in the counter-clockwise direction. The motion of this electron constitutes an electric current I in the clockwise direction.

Now, $$I = \frac{e}{T}$$

where e is the magnitude of the charge on an electron and T is the period of revolution of the electron. If ω is the angular velocity of the revolving electron, then

$$I = \frac{e}{2\pi/\omega} = \frac{e\omega}{2\pi}$$

Magnetic moment of current loop, M = IA

where A is the face area of the orbit.

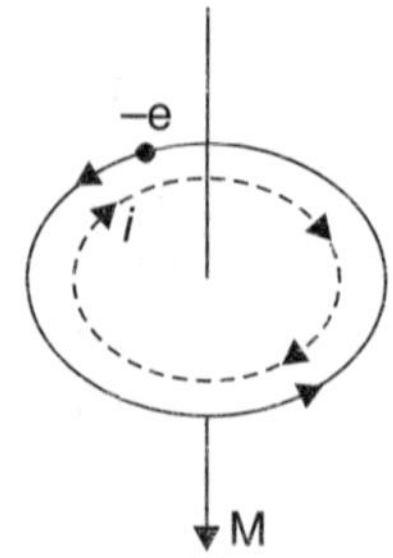

Fig. 4.19. Atom as a magnetic dipole.

Now, $$M = I \times \pi r^2 = \frac{e\omega}{2\pi} \times \pi r^2$$

or $$M = \frac{e\omega r^2}{2} = \frac{e}{2m}(mr^2\,\omega), \quad ...(1)$$

where m is the mass of the electron.

Again, $$M = \frac{e}{2m} L$$

where L is the angular momentum of the electron.

In vector form, $$\vec{M} = -\frac{e}{2m}\vec{L} \quad ...(2)$$

The negative sign in equation (2) indicates that the direction of the magnetic moment is opposite to the direction of the angular momentum.

Since the electron is supposed to be moving in counter-clockwise direction, therefore, the conventional current flows in the clockwise direction. So, the upper face of the 'current loop' acts as a south pole while the lower face acts as the north pole. Since the magnetic dipole moment vector is always directed from south pole to north pole, therefore, we can say that the magnetic dipole moment in the present case is perpendicular to the plane of the loop and directed downwards. On the other hand, the direction of the angular momentum is perpendicular to the plane of the loop and directed upwards.

According to *Bohr's theory*, an electron (in an atom) is permitted to revolve only in that orbit for which the angular momentum ($mr^2\omega$) is an integral multiple of $h/2\pi$. Here, h is Planck's constant.

$$\therefore \quad mr^2\omega = n\frac{h}{2\pi}, \text{ where } n = 1, 2, 3, ... \quad ...(3)$$

Combining equations (1) and (3), we get $M = \frac{e}{2} \times \frac{nh}{2\pi m}$ or $M = n\frac{eh}{4\pi m}$

It is clear from this equation that the value of the magnetic moment will be least when the value of n is least *i.e.*, one.

$$\therefore \quad \text{Least value of magnetic moment} = \frac{eh}{4\pi m}$$

This least value is considered as one Bohr magneton. It is denoted by μ_B or β.

Both magneton is defined as the orbital magnetic moment of an electron revolving in the innermost orbit.

$$1 \text{ Bohr magneton } \mu_B = \frac{eh}{4\pi m} = \frac{1.6 \times 10^{-19} \times 6.63 \times 10^{-34} \times 7}{4 \times 22 \times 9.1 \times 10^{-31}} \text{ A m}^2 = 9.27 \times 10^{-24} \text{ A m}^2$$

If the number of electrons is more than one in an atom, then the resultant magnetic moment of the atom is given by the *vector sum* of the magnetic moments due to all the electrons. This has been experimentally confirmed by Stern and Gerlach.

In the above treatment, the dipole moment has been calculated for the orbital motion of electron. The calculation of dipole moment for the spin motion of electron is however, slightly complicated. As in the case of orbital dipole moment, the total spin dipole moment is equal to the vector sum of the spin dipole moments of all the individual electrons. The magnetic dipole moment of the atom as a whole is the vector sum of the net magnetic moments due to orbital and spin motions of electrons. 'The magnetic dipole moment of the atom' is an important factor in determining the magnetic behaviour of a material.

MAGNETIC PROPERTIES OF MATERIALS

It was Faraday who showed that all materials can be magnetised, feebly or strongly, when they are subjected to a strong magnetising field. Materials can be divided broadly into the following three categories :

(*i*) Diamagnetic Substances (*ii*) Paramagnetic Substances
(*iii*) Ferromagnetic Substances

4.20 DIAMAGNETIC SUBSTANCES

Diamagnetic substances are those substances which are repelled by a magnet.

Examples : Antimony, bismuth, copper, lead, gold, silver, zinc, quartz, mercury, alcohol, sodium chloride, water, hydrogen, air, argon, etc.

Some of the **main properties** of diamagnetic substances are as under :

(*i*) When a diamagnetic substance is placed in a non-uniform magnetic field, it tends to move from the stronger to the weaker regions of the magnetic field. When a diamagnetic liquid contained in a watch glass is placed on two pole-pieces lying close to each other, it shows a depression in the middle and gets accumulated at the sides as shown in Fig. 4.20. This is because the magnetic field is stronger in the middle and weaker at the sides. However, when the pole pieces are sufficiently apart as shown in Fig. 4.21, the magnetic field is weaker in the middle and stronger at the sides. The liquid now accumulates at the middle and shows depression at the sides.

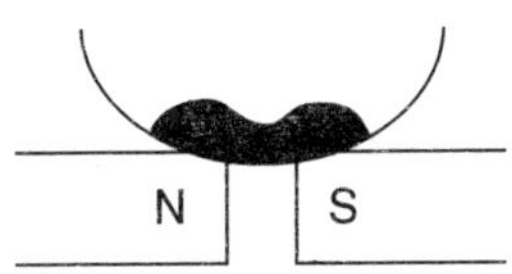

Fig. 4.20

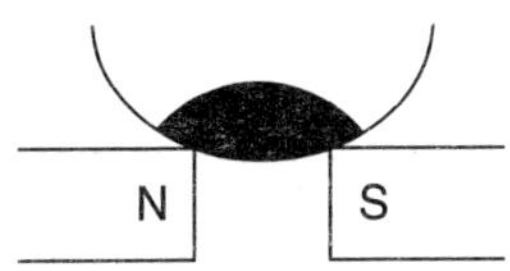

Fig. 4.21

Explanation. *Potential energy of dipole placed in uniform magnetic field is '– MB cos θ'. The value of θ is 180° for a diamagnetic substance. So, the potential energy is MB. Since every system in nature tends to have the least potential energy, therefore, the diamagnetic substance moves to a region of minimum B.*

(*ii*) When a diamagnetic rod is suspended freely in a uniform magnetic field, the rod aligns itself in a direction perpendicular to the direction of the magnetic field as shown in Fig. 4.22.

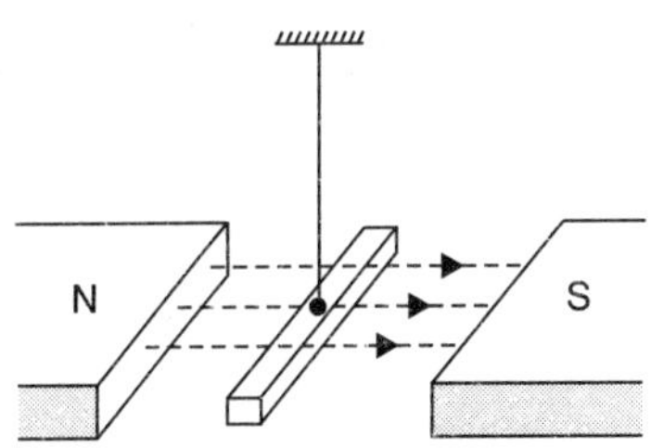

Fig. 4.22

Explanation. *Potential energy per unit volume in the case of diamagnetic substance is* $\frac{M}{V}$ B *i.e.,* IB. Clearly, the diamagnetic rod will have minimum energy if its length is perpendicular to the direction of magnetic field.

(iii) When a diamagnetic substance is placed in a magnetic field, the lines of force prefer to pass through the surrounding air as shown in Fig. 4.23. This is because poles are induced in the diamagnetic in a direction opposite to the direction of magnetic field.

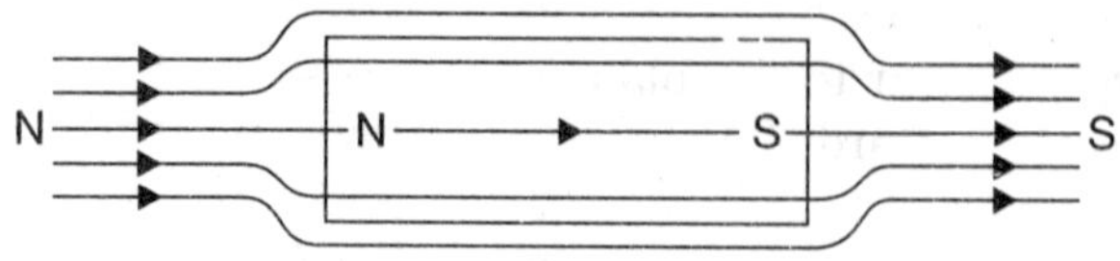

Fig. 4.23. Diamagnetic substance in a magnetic field

(iv) The permeability of a diamagnetic substance is less than one.

(v) When a diamagnetic substance is placed in a magnetic field, it develops weak magnetisation in a direction opposite to the direction of the magnetising field.

(vi) A diamagnetic substance loses its magnetisation as soon as the magnetisation field is removed.

(vii) The magnetic susceptibility of a diamagnetic substance does not depend upon the temperature.

(viii) The magnetic susceptibility of a diamagnetic substance has a small negative value.

Table 4.1. Magnetic susceptibilities of certain diamagnetic substances at room temperature

No.	*Diamagnetic substance*	χ_m	*No.*	*Diamagnetic substance*	χ_m
1	hydrogen (1 atmosphere)	-2.1×10^{-9}	5	copper	-9.6×10^{-6}
2	nitrogen (1 atmosphere)	-5.0×10^{-9}	6	lead	-1.6×10^{-5}
3	water	-9.1×10^{-6}	7	diamond	-2.2×10^{-5}
4	sodium	-2.4×10^{-6}	8	mercury	-3.2×10^{-5}

4.21 EXPLANATION OF DIAMAGNETISM

Inside an atom, an electron possesses two types of motion—spin motion about an axis which passes through the electron itself and the orbital motion around the nucleus. Thus, the electron possesses two dipole moments—magnetic dipole moment $\vec{M}_o$ due to the orbital motion of electron [Fig. 4.24 (*a*)] and magnetic dipole moment $\vec{M}_s$ due to the spin motion of electron [Fig. 4.24 (*b*)]. The net dipole moment $\vec{M}$ of the electron is the vector sum of $\vec{M}_o$ and $\vec{M}_s$.

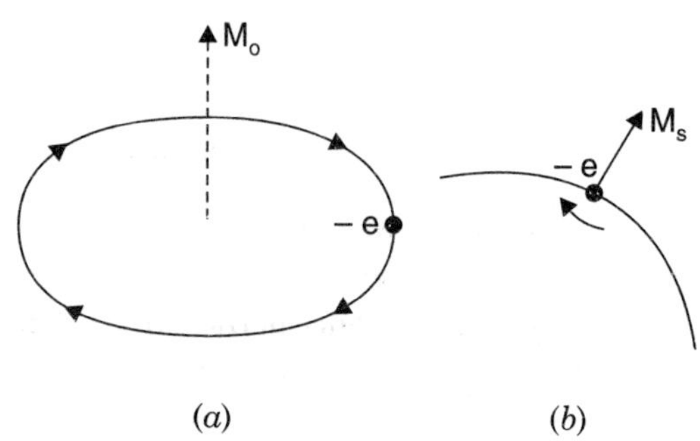

(*a*) (*b*)

Fig. 4.24

The electrons in an atom have either clockwise or anticlockwise spin. Similarly, the electrons move in their orbits in either clockwise or anticlockwise direction. Thus, in a diamagnetic substance, the atoms possess no net magnetic moment. For the sake of convenience, the motion of all the electrons in an atom can be considered to be equivalent to the motion of only two electrons, both moving with the same angular velocity ω in opposite directions. The net magnetic dipole moment of the electron having anti-clockwise orbital motion acts perpendicular to the plane of the paper and directed inwards [Fig. 4.25 (*a*)]. The net magnetic dipole moment of the electron having clockwise orbital motion acts perpendicular to the plane of the paper and directed upwards [Fig. 4.25 (*b*)]. If the atom is not subjected to any magnetic field, then the two magnetic dipole moments cancel each other's effect. Consequently, the atom as a whole has no net magnetic dipole moment.

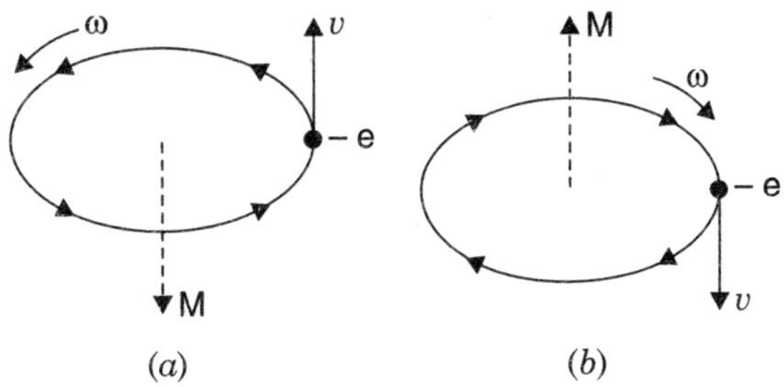

Fig. 4.25

Let us now consider an atom of a diamagnetic substance subjected to a uniform magnetic field perpendicular to the plane of rotation of the electrons and directed upwards. Let us study the effect of the applied magnetic field on the electron having anti-clockwise orbital motion. The force $\vec{F}_B$ experienced by the electron due to the applied magnetic field is given by

$$\vec{F}_B = -e(\vec{v} \times \vec{B})$$

This force acts along the radius of the loop and towards the centre of the loop. The velo-city of the electron is increased from v to $v + \Delta v$ and the angular velocity is increased from ω to $\omega + \Delta\omega$ [Fig. 4.26 (*a*)]. An increased angular velocity implies an increase in the value of current. So, the magnetic dipole moment increases from M to M + ΔM.

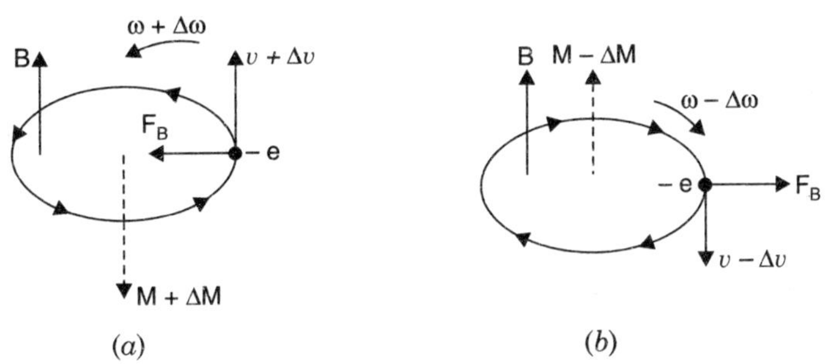

Fig. 4.26. Explanation of diamagnetism.

In the case of the electron orbiting in the clockwise direction, the effect of the applied magnetic field is to decrease the magnetic dipole moment from M to M – ΔM [Fig. 4.26 (*b*)].

Under the influence of the external magnetic field B, the net magnetic moment of the atom is (M + ΔM) – (M – ΔM), *i.e.*, 2 ΔM. Its direction is opposite to that of the external magnetic field B.

If all the atoms of a diamagnetic substance are taken into consideration, then there is a sufficient 'net magnetic moment' in a direction opposite to the direction of the applied magnetic field. This accounts for all the properties of diamagnetic substances. It is because of the interaction between the acquired opposite magnetic moment and the magnetic field that the substance is moved away from (out of) the field. It is said to be 'repelled'.

4.22 LANGEVIN'S THEORY OF DIAMAGNETISM

Consider an electron in an atom revolving counter-clockwise in a circular orbit of radius r in xy plane as shown in Fig. 4.27.

If v is the speed of electron, then the magnetic moment of the revolving electron is given by :

$$\vec{\mu}_m = -\frac{evr}{2}\hat{k}$$

If ω be the angular speed of the electron, then

$$v = r\omega$$

$$\vec{\mu}_m = -\frac{e(r\omega)r}{2}\hat{k}$$

or

$$\vec{\mu}_m = -\frac{e\omega r^2}{2}\hat{k}$$

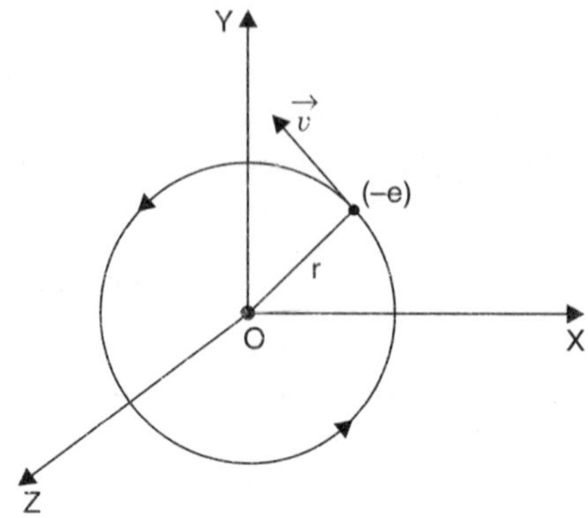

Fig. 4.27

The direction of this magnetic moment is along negative z-axis.

The centripetal force acting on the electron is given by

$$F_c = \frac{mv^2}{r}$$

or

$$F_c = mr\omega^2$$

Suppose an external magnetic field $\vec{B}$ is applied on the electron in the z-direction.

The Lorentz magnetic force experienced by the electron is given by

$$\vec{F}_m = -e(\vec{v} \times \vec{B})$$

Applying right hand rule for the cross product of vectors, we find that $\vec{F}_m$ acts in the direction of $\vec{F}_c$.

In magnitude, $F_m = Be\,v \sin\theta$

Here, $\theta = 90°$. $[\because \vec{v} \perp \vec{B}]$

$\therefore$ $F_m = B\,ev$

or $F_m = B\,er\omega$

or $F_m = er\,\omega B$

The net force acting on the electron is given by

$$F = F_c + F_m$$

$$F = mr\omega^2 + er\,\omega B.$$

Clearly, the force on the electron has increased due to the application of external magnetic field. Consequently, the electron shall move faster in its orbit. Let the new angular speed increase from ω to $\omega + \Delta\omega$. The new magnetic moment of the electron is given by :

$$\vec{\mu}_m{}' = -\frac{e(\omega + \Delta\omega)r^2}{2}\hat{k}$$

Let $\Delta\vec{\mu}_m$ be the change in magnetic moment of electron due to magnetic field.

Then, $\Delta\vec{\mu}_m = \vec{\mu}_m{}' - \vec{\mu}_m$

or $$\vec{\Delta\mu}_m = \left[-\frac{e(\omega+\Delta\omega)r^2}{2}\hat{k}\right] - \left[-\frac{e\omega r^2}{2}\hat{k}\right]$$

$$= \frac{e\omega r^2}{2}\hat{k} - \frac{e(\omega+\Delta\omega)r^2}{2}\hat{k}$$

$$= -\frac{e\Delta\omega r^2}{2}\hat{k} \qquad ...(i)$$

In the presence of external magnetic field, the centripetal force on the electron is given by

$$F_2 = mr(\omega + \Delta\omega)^2$$

Clearly, $mr\omega^2 + er\,\omega B = mr\omega^2 + mr\Delta\omega^2 + 2mr\omega\,\Delta\omega$

When $\Delta\omega$ is small, $\Delta\omega^2$ can be neglected.

$$mr\omega^2 + er\,\omega B = mr\omega^2 + 2mr\omega\Delta\omega$$

or $$er\,\omega B = 2mr\omega\,\Delta\omega$$

or $$\Delta\omega = \frac{er\,\omega B}{2mr\omega} \quad \text{or} \quad \Delta\omega = \frac{eB}{2m}$$

From = n (i), $$\vec{\Delta\mu}_m = -\frac{er^2}{2}\left(\frac{eB}{2m}\right)\hat{k}$$

or $$\vec{\Delta\mu}_m = -\frac{e^2r^2}{4m}\vec{B} \quad \text{or} \quad \vec{\Delta\mu}_m = -\frac{e^2r^2\mu_0}{4m}\vec{H} \qquad (\because\ \vec{B} = \mu_0\vec{H})$$

The negative sign indicates that the change in magnetic moment takes place in a direction opposite to $\vec{H}$.

Conclusion. The electron develops an induced magnetic moment $\vec{\Delta\mu}_m$ in a direction opposite to the direction of the external magnetic field.

Note. In the above treatment, we have considered an electron revolving anti-clockwise. However, the result holds good even for an electron revolving clockwise.

Let us now consider the case of an atom which has Z electrons and orbits of all the electrons are in the same plane.

Let $\vec{\Delta\mu}_a$ be the total magnetic moment induced in the atom.

Then $$\vec{\Delta\mu}_a = Z\vec{\Delta\mu}_m \quad \text{or} \quad \vec{\Delta\mu}_a = -\frac{Ze^2r^2\mu_0}{4m}\vec{H}$$

If there are N atoms per unit volume, then the magnetisation induced in the sample is given by

$$\vec{M} = N\,\vec{\Delta\mu}_a = -\frac{NZe^2r^2\mu_0}{4m}\vec{H}$$

This equation can be written as

$$\vec{M} = \chi_m\vec{H}$$

Here $$\chi_m = -\frac{NZe^2r^2\mu_0}{4m}\vec{H}$$

This is the magnetic susceptibility of diamagnetic substances.

From the above discussion, we arrive at the following conclusions :

(i) The magnetisation $\vec{M}$ acquired by a diamagnetic substance is in a direction opposite to $\vec{H}$.

(ii) The magnetisation $\vec{M}$ induced in a diamagnetic substance is proportional to $\vec{H}$.

(iii) The magnetic susceptibility is nearly independent of temperature.

(iv) The diamagnetic effect is a small effect.

(v) The diamagnetism arises due to the orbital motion of electrons.

An estimate of the magnetic susceptibility of diamagnetic substances

$$\chi_m = -\frac{NZ\mu_0 e^2 r^2}{4m}$$

Here, $N = 10^{29}$, $Z = 2$, $r = 5 \times 10^{-11}$ m

$$\therefore \quad \chi_m = -\frac{10^{29} \times 2 \times 4\pi \times 10^{-7}(1.6 \times 10^{-19})^2(5 \times 10^{-11})^2}{4 \times 9.1 \times 10^{-31}}$$

On simplification, $\chi_m = -2.4 \times 10^{-6}$.

4.23 PARAMAGNETIC SUBSTANCES

Paramagnetic substances are those substances which are attracted by a magnet but very feebly. Aluminium, chromium, manganese, oxygen, platinum, alkali and alkaline earth metals are examples of paramagnetic substances. Some properties of paramagnetic substances are as under :

(i) As soon as the magnetising field is removed, the paramagnetics lose their magnetisation.

(ii) A paramagnetic substance develops weak magnetisation in the direction of magnetic field.

(iii) The magnetic lines of force prefer to pass through a paramagnetic material rather than air.

(iv) When a paramagnetic rod is suspended freely in a uniform magnetic field, it aligns itself in the direction of magnetic field.

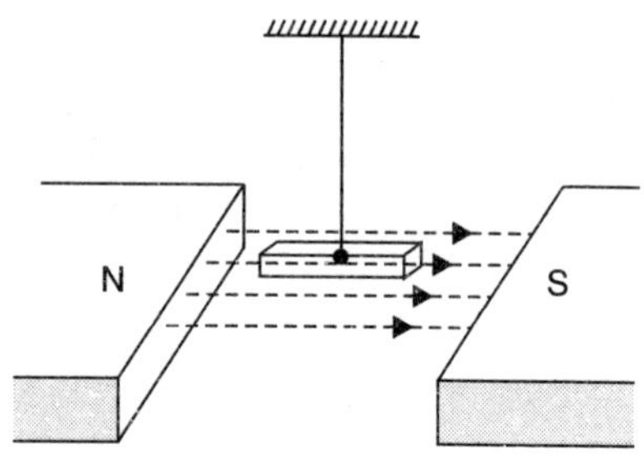

Fig. 4.28

Explanation. *We know that $U = -MB\cos\theta$. Potential energy per unit volume is $-\frac{M}{V}B\cos\theta$ i.e., $-IB\cos\theta$. In the case of paramagnetic substance, $\theta = 0°$.*

$\therefore$ *Potential energy per unit volume = $-IB$*

Clearly, this will be minimum if volume is maximum. This will be possible only if the length of the paramagnetic rod is in the direction of the magnetic field.

(v) A paramagnetic substance moves from weaker to stronger parts of the magnetic field.

Explanation. *We know that $U = -MB\cos\theta$. In the case of paramagnetic substance,*

$$\theta = 0°.$$

$$\therefore \quad U = -MB$$

Clearly, potential energy will be minimum if paramagnetic substance is in a region of maximum B.

(vi) The intensity of magnetisation I of a paramagnetic material is directly proportional to the strength H of the magnetising field and inversely proportional to the absolute temperature T of the material.

$$\therefore \qquad I \propto \frac{H}{T} \quad \text{or} \quad I = C\frac{H}{T}$$

If we plot a graph between I and $\frac{H}{T}$, it will be observed that the graph is almost linear for low values of $\frac{H}{T}$. The region over which the graph is linear is called Curie's law region. Above a certain value of $\frac{H}{T}$, the curve attains saturation. This represents a stage where a complete alignment of magnetic moments is attained.

(vii) The magnetic permeability of a paramagnetic is slightly greater than one.

Table 4.2. Magnetic susceptibilities of different paramagnetic substances at room temperature

No.	*Paramagnetic substance*	χ_m	*No.*	*Paramagnetic substance*	χ_m
1.	Oxygen (1 atmosphere)	2.1×10^{-6}	4.	Titanium	7.1×10^{-5}
2.	Aluminium	2.3×10^{-5}	5.	Platinum	3.0×10^{-4}
3.	Tungsten	6.8×10^{-3}	6.	Gadolinium chloride	2.8×10^{-3}

4.24 LANGEVIN THEORY OF PARAMAGNETISM

Langevin derived an expression for susceptibility of a paramagnetic gas, based on kinetic theory of gases. Like other paramagnetic material, paramagnetic gas also have molecules with permanent magnetic dipole moment. When the external magnetic field is applied across the paramagnetic gas, the dipole moment try to orient in the direction of field. However due to thermal agitation dipole do not aligned with direction of field but makes some angle with the field.

To find the expression for gas molecules, let us consider a paramagnetic gas whose each molecule has permanent magnetic moment μ_m which makes an angle θ with the direction of applied magnetic field B. Then potential energy of molecule in the direction of field is

$$U = -\mu_m B \cos\theta \qquad ...(1)$$

Let n be the number of atoms per unit volume of gas and dn be the molecules having energy U at an absolute temperature T, oriented at angles between θ and $d\theta$ with respect to the field as shown in figure.

Then according to classical statistical theory, dn is defined as

$$dn = C\,e^{-U/kT}\,dw \qquad ...(2)$$

Where C is proportionality constant and k is Boltzman constant. dw is the solid angle subtended between θ and θ + $d\theta$ angles and is given as

$$dw = \frac{\text{Area of the ring beween } \theta \text{ and } \theta + d\theta}{r^2}$$

$$= \frac{2\pi(r\sin\theta)(rd\theta)}{r^2} = 2\pi\sin\theta\, d\theta \qquad ...(3)$$

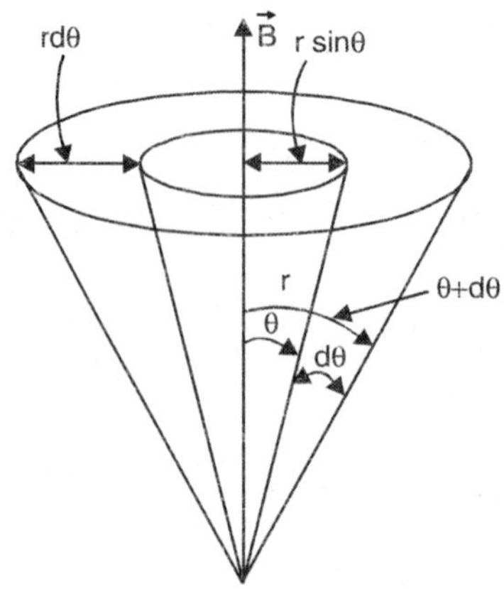

Fig. 4.29

Substituting value of U from eq. (1) and dw from (3) in eq. (2) we get

$$dn = C\, e^{\mu_m B\cos\theta/kT}(2\pi\sin\theta\, d\theta)$$

$$= 2\pi C\, e^{\mu_m B\cos\theta/kT}\sin\theta\, d\theta$$

$$= A\, e^{\mu_m B\cos\theta/kT}\sin\theta\, d\theta \quad ...(4)$$

where $A = 2\pi C$

In order to obtain the total number of molecule per unit volume *i.e.*, *n*, equation (4) should be integrated between limits of $\theta = 0$ to $\theta = \pi$.

i.e.,
$$n = A\int_0^{\pi} e^{\mu_m B\cos\theta/kT}\sin\theta d\theta \quad ...(5)$$

Let $\frac{\mu_m B}{k'T} = x$ and $\cos\theta = y$, $-\sin\theta\, d\theta = dy$

then eq. (5) becomes

$$n = -A\int_{+1}^{-1} e^{xy} dy$$

$$= -A\left[\frac{e^{xy}}{x}\right]_{+1}^{-1}$$

$$= \frac{-A}{x}\left(e^{-x} - e^{x}\right)$$

$$= \frac{2A}{x}\sinh x \qquad \left[\because \frac{e^{-x} - e^{x}}{2} = \sinh x\right] \quad ...(6)$$

The component of a magnetic dipole along field inclined at an angle θ be $\mu_m \cos\theta$. The total magnetic moment of all dn molecules will be $\mu_m \cos\theta\, dn$. Therefore the total magnetic moment of unit volume will be the intensity of magnetization

$$I = \int_0^{\pi} \mu_m \cos\theta\, dn \qquad ...(7)$$

Substituting the value of dn from eq. (4) we get

$$I = \int_0^{\pi} \mu_m \cos\theta \, A e^{\mu_m B \cos\theta / kT} \sin\theta \, d\theta \qquad ...(8)$$

$$= -\mu_m A \int_{+1}^{-1} y\, e^{xy} dy \text{ where } \frac{\mu_m B}{kT} = x \text{ and } \cos\theta = y; -\sin\theta d\theta = dy$$

$$= -\mu_m A \left[\frac{ye^{xy}}{x} - \int \left(\frac{d}{dy}(y) \int e^{xy} dy \right) dy \right]$$

$$= -\mu_m A \left[\frac{ye^{xy}}{x} - \int \left(\frac{e^{xy}}{x} \right) dy \right]$$

$$= -\mu_m A \left[\frac{ye^{xy}}{x} - \frac{e^{xy}}{x} \right]_{+1}^{-1}$$

$$= -\mu_m A \left[\left(-\frac{e^{-x}}{x} - \frac{e^{-x}}{x^2} \right) - \left(\frac{e^{x}}{x} - \frac{e^{x}}{x^2} \right) \right]$$

$$= +\frac{\mu_m A}{x} \left[\left(e^{-x} + \frac{e^{-x}}{x} \right) + \left(e^{x} - \frac{e^{x}}{x} \right) \right]$$

$$= +\frac{\mu_m A}{x} \left[\left(e^{x} + e^{-x} \right) - \frac{1}{x} \left(e^{x} - e^{-x} \right) \right]$$

$$= +2\frac{\mu_m A}{x} \left[\cos h\, x - \frac{\sin hx}{x} \right] \left[\because \frac{e^{x} + e^{-x}}{2} = \cos h\, x \right]$$

$$= n\mu_m \left[\frac{\cos hx}{\sin hx} - \frac{\sin hx}{x \sin hx} \right] \left[\because \frac{2A}{x} = \frac{n}{\sin hx} \right]$$

$$= n\mu_m \left[\cot hx - \frac{1}{x} \right] \qquad ...(9)$$

The term $\left(\cot hx - \frac{1}{x} \right)$ is known as Langevin's function and usually denoted by $L(n)$.

therefore $\quad I = n\, \mu_m\, L(x) \qquad ...(10)$

The value of x is very small even for a strong magnetic field except for absolute zero temperature. Therefore series expansion of $L(x)$ shown

$$L(x) = \left(\coth x - \frac{1}{2} \right) = \frac{x}{3}$$

so $$I = n\,\mu_m \frac{x}{3}$$

$$= \frac{n\mu_m{}^2 B}{3kT} \qquad ...(11)$$

For a paramagnetic $B = \mu_0 H$, since I is small

so $$I = \frac{n\mu_m^2 \mu_0 H}{3kT} \qquad ...(12)$$

The magnetic susceptibility of paramagnetic gas

$$\chi_m = \frac{I}{H} = \frac{n\mu_m^2 \mu_0}{3kT}$$

$$\chi_m \propto \frac{1}{T} \qquad ...(13)$$

Susceptibility for a gram mole of paramagnetic gas will be

$$\chi_m = \frac{\mu_0 N \mu_m^2}{3kT} \quad [N \rightarrow \text{Avogadro number}]$$

$$= \frac{C}{T} \quad \left[\because C = \frac{\mu_0 N \mu_m^2}{3k}\right] \qquad ...(14)$$

C is called curie constant. Eq. (14) shows that the magnetic susceptibility for paramagnetic substance is inversely proportional to absolute temperature.

4.25 PARAMAGNETISM ON THE BASIS OF ELECTRON THEORY

The atoms of paramagnetics possess permanent magnetic dipole moments. These 'moments' arise in the case of atoms which do not have their inner electron shells completely filled. The magnetic moments are due to the existence of electrons having unpaired spins. The magnetic moment of such an atom includes some contribution from the orbital motion too. The magnetic moment of atoms of paramagnetic substances lie between one and five Bohr magneton.

When a paramagnetic substance is not under the influence of any external magnetic field, then the magnetic moments are oriented in all possible directions. There is weak interaction between magnetic dipole moments of various atoms. Since the average dipole moment vector per unit volume of a paramagnetic material is zero, therefore, the material does not show any magnetisation.

When an external magnetic field is applied to a paramagnetic substance, the magnetic field tries to align the magnetic moments in the direction of the field. Complete alignment is, however, not possible on account of thermal vibrations. A better alignment can be achieved by increasing the strength of the applied magnetic field. The net result of the applied magnetic field is that the material acquires magnetic moment in the direction of the field.

4.26 FERROMAGNETIC SUBSTANCES

Ferromagnetic substances are those substances which are attracted by magnets and can also be magnetised. These behave just like paramagnetic materials but the effect is much more intense.

Iron (Fe), Cobalt (Co), Nickel (Ni), Gadolinium (Gd) and Dysprosium (Dy) are important examples of ferromagnetic substances.

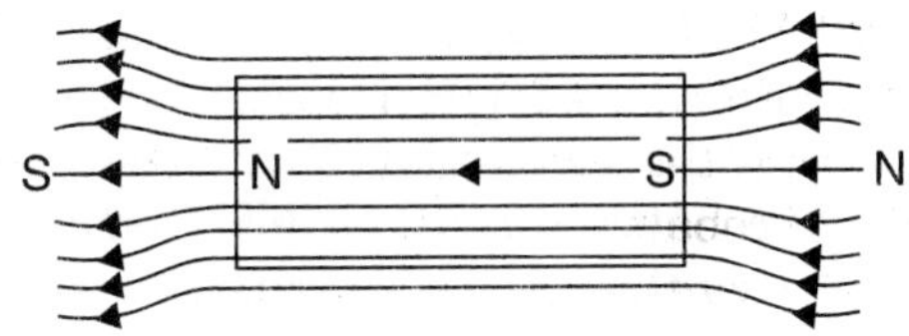

Fig. 4.30

Some properties of ferromagnetic substances are as under :

(i) When a ferromagnetic substance is placed in a magnetic field, the magnetic lines of force tend to crowd into the specimen as shown in Fig. 4.30.

(ii) When a ferromagnetic substance is placed in a non-uniform magnetic field, it moves from weaker to stronger parts of the magnetic field. Let us consider the behaviour of a ferromagnetic liquid contained in a watch glass and placed on two closely placed magnetic pole pieces. It would be observed that the ferromagnetic liquid flows from the sides (weaker magnetic field) to the middle (stronger magnetic field). Thus, there will be accumulation of ferromagnetic liquid in the middle as shown in Fig. 4.31. If the pole pieces are moved farther apart, the liquid will be observed to be depressed in the middle and accumulated at the sides as shown in Fig. 4.32. This is because the magnetic field will become weaker at the middle and stronger at the sides.

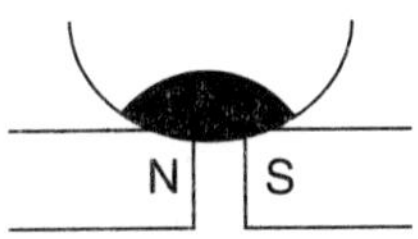

Fig. 4.31

Fig. 4.32

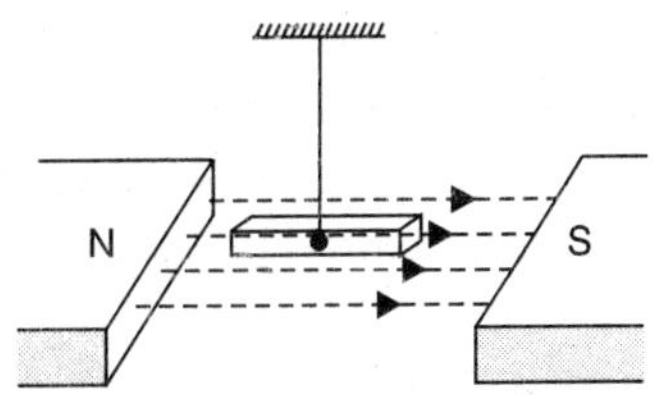

Fig. 4.33

(iii) When a ferromagnetic rod is freely suspended in a uniform magnetic field, it aligns itself parallel to the direction of the magnetic field as shown in Fig. 4.33.

(iv) When a ferromagnetic substance is placed in a magnetic field, it develops strong induced magnetism. So, the ferromagnetic substance is strongly attracted by a magnet.

(v) Ferromagnetic substances retain their magnetism even after the magnetising field is removed.

(vi) The *permeability of ferromagnetic substances is extremely large as compared to the permeability of free space. To put it differently, the magnetic induction B is much greater than the magnetising field H.

(vii) Ferromagnetics have large positive susceptibility. The susceptibility in some cases may be as high as 10^5.

(viii) When a ferromagnetic substance is placed in a magnetic field, it is strongly magnetised in the direction of magnetic field. And the magnetisation is retained even after the magnetising field is removed.

---

*The magnetic permeability of iron is nearly 1000.

(*ix*) The magnetic susceptibility of ferromagnetics decreases with the increase of temperature.

(*x*) The intensity of magnetisation has a large positive value.

(*xi*) A ferromagnetic substance develops strong magnetisation in the direction of the magnetising field.

(*xii*) The ferromagnetism decreases with rise of temperature. It is maximum at absolute zero of temperature and drops to zero at a temperature called Curie temperature. Above the Curie temperature, the ferromagnetic material becomes paramagnetic. The Curie temperatures for iron, nickel and cobalt are 770°C, 365°C and 1075°C respectively. The Curie temperature in certain cases is below the room temperature. As an example, the Curie temperature of Gadolinium is 15°C.

4.27 DOMAIN THEORY OF FERROMAGNETISM

Ferromagnetic substances are paramagnetic in nature but because of extremely strong paramagnetism shown by them, they form a class by themselves. Since iron (ferrum) is the most representative substance in this category, therefore, these are called ferromagnetic substances and the property is named as ferromagnetism.

In the atoms of elements iron, cobalt, nickel, gadolinium and dysprosium (all ferromagnetic materials), there are vacancies in the inner shells. So, the electrons in these shells are not paired off with anti-parallel spins and equal and opposite orbital magnetic moments. Thus, the atoms of these elements possess considerable magnetic moments. Moreover, an unpaired electron in one atom interacts strongly with the unpaired electron in the adjacent atom. Consequently, the magnetic moments get aligned in the same direction. This is called exchange interaction.

The atoms in all the five elements mentioned above group themselves together in tiny assemblies called domains. All the ferromagnetic materials have a characteristic 'domain structure'.

A domain is a region which is magnetised to saturation *i.e.*, in which all the atoms are arranged with their magnetic moments parallel. These domains are of the size of 10^{-4} m to 10^{-5} m. These domains are thus visible under a good microscope. The domains thus have a net magnetic moment of high value and the part played by individual atoms in a paramagnetic material is similar to the part played by domains in a ferromagnetic material. Each domain has its own magnetic moment oriented in its own direction. So, the net result of the assembly of domains is zero. But within a domain, the adjacent dipole moments are bound together by strong forces which give rise to a quantum mechanical interaction called exchange interaction.

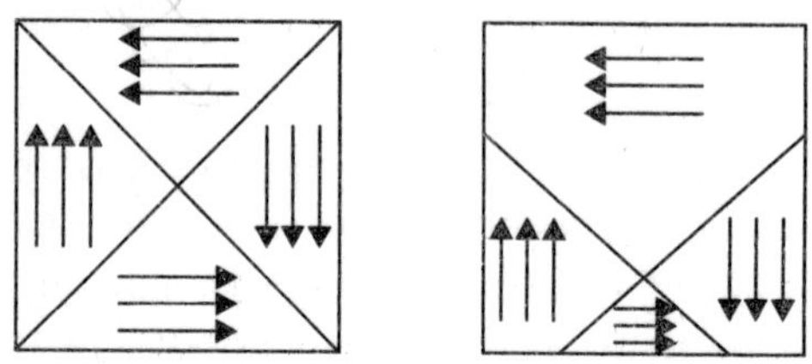

Fig. 4.34

In the absence of an external magnetic field, the magnetic moments of the domains are randomly oriented as shown in Fig. 4.34 and the material shows no magnetism. When an external field is applied, the domains having magnetic moments in the direction of magnetic field start growing in size at the cost of other domains. Thus, the number of magnetic moments pointing in the direction of the magnetic field increases. If the strength of the applied magnetic field is sufficient, then the ferromagnetic material gets strongly magnetised in the direction of the applied magnetic field.

4.28 PHENOMENON OF HYSTERESIS

Let us take a specimen of unmagnetised steel in the form of a ring and wind over it a wire as shown in Fig. 4.35. Arrangements are made to send current through the wire and also to vary the magnitude and direction of this current. When the current is gradually increased from zero to maximum in one direction, the magnetising field H also increases gradually from zero to maximum in one direction. When the current is being increased, the corresponding values of magnetic induction B in the ring are noted at different stages.

When the current is now gradually decreased from maximum to zero, the magnetising field H is also gradually decreased from maximum to zero. When the current is being decreased, the corresponding values of the magnetic induction B in the steel ring are again determined.

Next, the current is increased gradually from zero to maximum in the other direction. The magnetising field H will also gradually increase from zero to maximum in the opposite direction. While the current is being increased, the corresponding values of magnetic induction in the steel ring are determined.

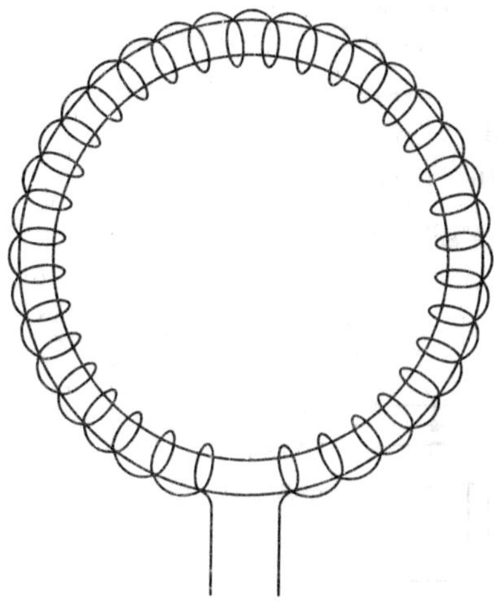

Fig. 4.35. Hysteresis.

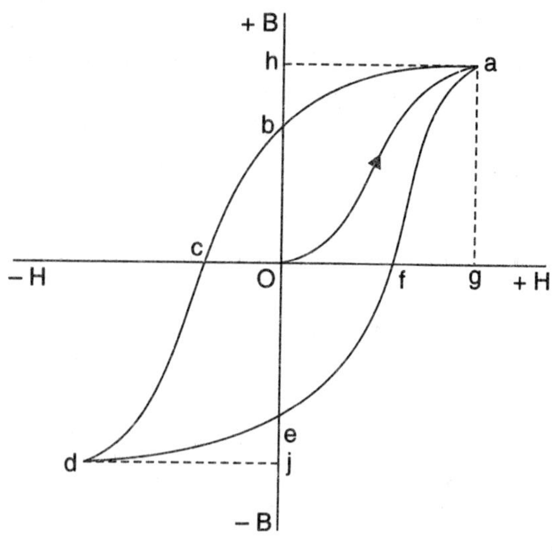

Fig. 4.36. Hysteresis loop.

The current is again reduced from maximum to zero. The magnetising field H also decreases from maximum to zero. The corresponding values of magnetic induction B in the steel ring are determined.

Finally, the current is increased again from zero to maximum in the first direction and different corresponding values of magnetic induction B are noted.

When magnetising field H and the magnetic induction B are varied from *zero to a maximum in one direction, and then back through zero to a maximum in the opposite direction and finally back again through zero to the first maximum*, a cycle of magnetisation is said to be completed.

Using the values of H and B obtained above, we can plot a graph by taking H along x-axis and B along y-axis. The curve obtained is shown in Fig. 4.36. As is clear from the Fig. 4.36, when the magnetising field H is increased from zero to the maximum value O*g*, the induction B follows the curve O*a*. However, when the magnetising field is brought to zero, the induction does not follow back the curve O*a* but a different curve *ab*. The same pattern of variations is seen for the rest of the curve. This situation can be described as follows.

"The B-H curve for decreasing H does not coincide with the B-H curve for increasing H."

In layman's language, B lags behind H. This is known as *hysteresis* which literally means "to lag behind".

Hysteresis is defined as the lagging of the magnetic induction B behind the corresponding magnetising field H.

As is clear from Fig. 4.36, even when the magnetising field H is zero, some magnetic induction (O*b* or O*e*) is still present in the specimen. It is denoted by B_r and is called *remanence* or *retentivity or residual magnetism*.

The value of the intensity of magnetisation of a material when the magnetising field is reduced to zero is called retentivity or residual magnetism of the material.

After the specimen has been magnetised to saturation (O*h* or O*j*), a reversed magnetising field (equal to O*c* or O*f*) is required to reduce the magnetic induction to zero. This is called the *coercivity* or *coercive force* H_c.

The value of the magnetising field required to reduce residual magnetism to zero is called coercivity of the material.

In the situation discussed above, the magnetic intensity H was varied between equal positive and negative values. The corresponding changes in the flux density resulted in a symmetrical curve. Other ways of changing H would result in other forms of hysteresis curves.

When a ferromagnetic material is taken through a cycle of magnetisation, heat is produced. It has been observed that in each cycle of magnetisation, the heat developed per unit volume of the material is proportional to the area enclosed by the hysteresis loop. Thus, the hysteresis loop gives a clear indication of the dissipation of energy due to hysteresis. A narrow hysteresis loop indicates less dissipation of energy. In certain cases, we have to ensure that the material being used has a narrow hysteresis loop. As an illustration, when iron is used as the core of a transformer, we select iron or iron alloy having narrow hysteresis loop. Similarly, while making permanent magnets, we select that material which has high retentivity.

Demagnetisation. As discussed above, the reversal of the magnetising field does not ensure a change in the value of flux from ϕ in one direction to an equal amount in the opposite direction (giving a change of 2ϕ) unless the effect of any stronger magnetisation has been removed from iron. The demagnetisation requires a very careful manipulation of the rather elaborate process of demagnetisation. Briefly speaking, the magnetising current is reversed a number of times. The magnitude of each current is reduced with each reversal. Thus, the specimen to be demagnetised is carried through more and more narrow hysteresis loops.

4.29 HYSTERESIS LOSS PER UNIT VOLUME PER CYCLE

Considering domain theory of ferromagnetic material, when magnetic field is applied across the material, the domains which are aligned with the direction of field grow in size. The boundries of favourable domain shifted to increase its size on the cost of unfavourable domain and the energy is absorbed during this process. When field is completely removed the shifting boundries do not recover at their original position and hence material retains some magnetization. Therefore the energy supplied during magnetization is not fully recovered and remains in the material is lost as heat. This loss of energy is the form of heat is called hysteresis loss.

For obtaining an expression for hysteresis loss per unit volume per cycle let us consider a coil of N turns wound uniformly on a ring of ferromagnetic material of cross-sectional area A. Suppose a current i in the coil produces a magnetic flux density B in the material. The total magnetic flux linked with the coil is given by

$$\phi_B = NBA$$

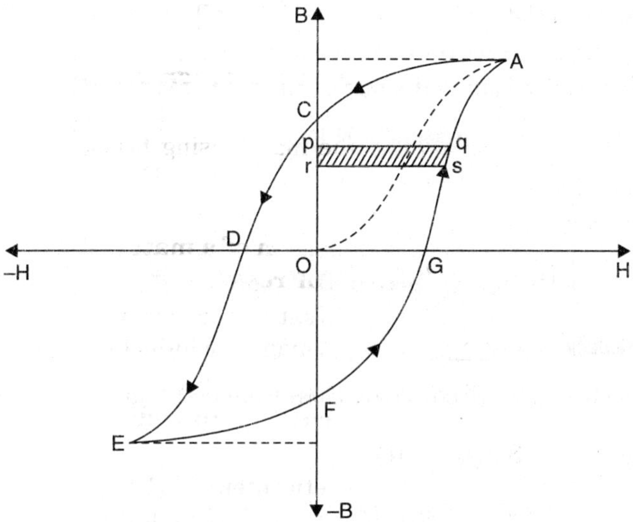

Fig. 4.37

Magnetic intensity, $H = \frac{Ni}{l}$ or $i = \frac{Hl}{N}$

Here l is the circumferential length of the ring. When the magnetisation current is increased, the flux linked with the coil changes. An induced emf e is developed in the *coil*. Using Faraday's law,

$$e = -\frac{d\phi_B}{dt}$$ [Negative sing is due to Lenz's law]

$$e = -\frac{d}{dt}(NBA)$$

or $$e = -NA\frac{dB}{dt}$$

The negative sign indicates that the emf opposes the increase in magnetisation current.

Let dW' be the work done by the current i against the induced emf e in dt second.

$$dW' = ei\,dt$$

$$= NA\frac{dB}{dt}\frac{Hl}{N}dt$$

$$= HAl\,dB$$

Here dB is the change in flux density corresponding to change in magnetisation current.

Increase in energy/volume of material, dW

$$= \frac{HAl\,dB}{Al} = HdB$$

Now $$W = \int_{B_1}^{B_2} H\,dB$$

Its SI unit is Jm^{-3}

In the B-H curve, the area of the shaded strip *pqrs* is HdB.

In taking the material from G to C through A, the energy dissipated per unit volume of material is given by area GACOG.

The net energy dissipated per unit volume of the material during a complete cycle of magnetisation is equal to the area of the B-H loop *i.e.*, the area GACDEFG.

$$\therefore \quad \oint H\,dB = \text{area of B-H loop}$$

This is valid for SI units.

If we use cgs units, then

$$\oint H\,dB = \frac{1}{4\pi} \text{ area of B-H loop.}$$

4.30 WARBURG'S LAW

This law helps us to calculate hysteresis loss from M-H loop

We know that $$\vec{B} = \mu_0(\vec{H} + \vec{M})$$

or $$d\vec{B} = \mu_0(d\vec{H} + d\vec{M})$$

Multiplying by H and integrating over the whole hysteresis loop, we get

$$\oint H\,d\vec{B} = \mu_0 \oint H\,d\vec{H} + \mu_0 \oint H\,d\vec{M}$$

But $\oint H\,dH$ represents the area enclosed by a straight line. This area is equal to zero.

$$\therefore \quad \oint H\,dH = 0$$

$$\therefore \quad \oint H\,dB = \mu_0 \oint H\,dM$$

$\therefore$ hysteresis loss = μ_0 × area of M-H loop

So, the energy loss per unit volume of the material per cycle is equal to μ_0 times the area of the M-H loop. This is Warburg's law.

4.31 APPLICATIONS OF PHENOMENON OF HYSTERESIS

The study of the hysteresis curve of different ferromagnetic materials helps us to select the material which has minimum hysteresis loss. The study of the hysteresis curve gives us a fair idea of the magnetic properties like susceptibility, retentivity, coercivity, permeability etc.

The choice of a magnetic material for the construction of permanent magnets, electromagnets, cores of transformers and magnetic shielding can be made from a study of the hysteresis curve of the specimen.

Electromagnets (temporary magnets)

These magnets are generally used for the purpose of lifting ferromagnetic substances. Soft iron is highly suitable for making electromagnets. The material of an electromagnet must have the following properties :

(i) The material should have *low retentivity*. When the magnetising field is removed, the material would be completely demagnetised.

(ii) The material must have high value of *saturation magnetisation*. This would make the electromagnet very strong.

(iii) The material must have *low coercivity*. Such a material can be easily demagnetised.

(iv) The *hysteresis loss* of the material should be small. Such a material would not be heated up during magnetisation and demagnetisation.

Permanent Magnets

These magnets are generally used in measuring instruments. The material of a permanent magnet must have the following properties :

(i) The material should have *high retentivity*. So, it would remain magnetised in the absence of the magnetising field.

(ii) The material should have *high saturation magnetisation*. This would help making the magnet strong.

(iii) The material should have *high coercivity*.

The *hysteresis loss* of the material of a permanent magnet is immaterial in the case of permanent magnets. This is because the permanent magnet need not be taken through cycles of magnetisation and demagnetisation.

*Steel is generally used for making permanent magnets. It may be clearly noted that steel does not meet all the above-mentioned requirements. As an example, it does not have high saturation magnetisation. But the strength of the magnet is increased by making it laminated. Also, steel does not have high coercivity. In recent times, alloys of high retentivity and high coercivity have been developed. Examples : Alini, **Alnico and Tinconol.

Transformer Cores

The magnetic material needs to have the following characteristics.

(*i*) Large value of permeability (*ii*) Small coercive force.

(*iii*) Low core loss

Alloy of iron and nickel, called permalloy, Mumetal (77% Ni, 16% Fe, 5% cu, 2% Cr) are suitable for making the cores of transformers and diaphragms of telephones. These are suitable for power frequencies (50–60 Hz).

Magnetic Shielding

The magnetic materials show have low coercivity and high saturation induction.

Super mendur (49% Fe, 49% Co, 2% V) and permalloy are most suitable.

---

*"Cobalt steel" is a suitable material for permanent magnets.

**Alnico [Ni 18%, AL 10%, Co 12%, Cu 6%, Fe 54%].

SOLVED EXAMPLES BASED ON MAGNETIC PROPERTIES

Example 4.01. *A bar magnet has a coercivity of 5×10^3 A m^{-1}. It is desired to demagnetise it by inserting it inside a solenoid 10 cm long and having 50 turns. What current should be sent through the solenoid ?* **[UPTU, B. Tech. I Sem. 2002]**

Solution. A magnetic intensity of 5×10^3 Am^{-1} is required to demagnetise the magnet.

Now, $$H = ni = \frac{N}{l} i$$

or $$i = \frac{Hl}{N}$$

$$= \frac{5 \times 10^3 \times 10 \times 10^{-2}}{50} \text{ A}$$

$$= \mathbf{10\ A.}$$

Example 4.02. *A iron rod of length 50 cm and cross-sectional area 4 cm^2 is in the form of a closed circular ring. If the permeability of iron is 65×10^{-4} Hm^{-1}, calculate the number of ampere-turns required to produce a flux of 4×10^{-4} Wb.*

[UPTU, B. Tech. I, Q. Bank, 2001]

Solution. Let Ni be the number of ampere-turns required to produce a magnetic flux ϕ_B by a circular ring of length l and cross-sectional area A.

$$\phi_B = BA$$

$$\phi_B = \mu n iA$$

or $$\phi_B = \mu \frac{N}{l} iA$$

or $$Ni = \frac{\phi_B l}{\mu A}$$

$$= \frac{4 \times 10^{-4} \times 0.5}{65 \times 10^{-4} \times 4 \times 10^{-4}}$$

$$= 76.9 \approx \mathbf{77.}$$

Example 4.03. *A solenoid consisting of 100 turns and carrying 10 ampere current is 0.5 m long. Calculate (a) magnetomotive force (b) total flux. Given : area of cross-section = 0.0004 m^2.*

[UPTU, B. Tech. I, Q. Bank, 2001]

Solution. (*a*) magnetomotive force, mmf = Ni

Here, N = 100 and i = 10 ampere

∴ mmf = 100 × 10 = **1000 ampere-turn.**

(*b*) $Hl = Ni$

$$H = \frac{Ni}{l} = \frac{1000}{0.5}$$

= 2000 ampere turn/m

$$\therefore \quad B = \mu_0 H = 4\pi \times 10^{-7} \times 2000$$

$$= 4 \times 3.14 \times 10^{-7} \times 2000 \text{ Wb m}^{-2}$$

$$= 2.512 \times 10^{-3} \text{ Wb m}^{-2}$$

$$\phi_B = BA = 2.512 \times 10^{-3} \times 0.0004 \text{ Wb}$$

$$= \mathbf{1.0048 \times 10^{-6}\ Wb.}$$

Example 4.04. *The magnetic susceptibility of a medium is 948×10^{-11}. Calculate the permeability and relative permeability.*

[UPTU, B. Tech. I, Q. Bank, 2001]

Solution. $\mu_r = 1 + x_m$

$$\mu_r = 1 + 948 \times 10^{-11}$$

Clearly, relative permeability is *only slightly greater than one.*

Again, $$\mu_r = \frac{\mu}{\mu_0} \text{ or } \mu = \mu_r \mu_0$$

or $$\mu = 1 + 4\pi \times 10^{-7} = \mathbf{4\pi \times 10^{-7}\ H\ m^{-1}.}$$

Example 4.05. *An electron in the ground state of hydrogen atom is revolving in anticlockwise direction in a circular orbit. The atom is placed in uniform magnetic induction $\vec{B}$ such that the normal to the electron orbit makes an angle of 30° with the magnetic field. Calculate the torque experienced by the orbiting electron.*

Solution. $\tau = MB \sin \theta$

$$= \frac{eh}{4\pi m} B \times \sin 30° = \mathbf{\frac{ehB}{8\pi m}}.$$

Example 4.06. *A Rowland ring of mean radius 15 cm has 3500 turns of wire wound on a ferromagnetic core of relative permeability 800. What is the magnetic field (B) in the core for a magnetising current of 1.2 A ?*

Solution. Magnetic field,

$$B = \frac{\mu NI}{l} \quad \text{But } \mu = \mu_0 \mu_r \text{ and } l = 2\pi r$$

$$\therefore \quad B = \frac{\mu_0 \mu_r NI}{2\pi r}$$

$$= \frac{4\pi \times 10^{-7} \times 800 \times 3500 \times 1.2}{2\pi \times 0.15} \text{ tesla}$$

$$= \mathbf{4.48\ tesla.}$$

Example 4.07. *A sample of paramagnetic salt contains 2.0 × 10^{24} atomic dipoles each of dipole moment 1.5 × 10^{-23} J T^{-1}. The sample is placed under a homogeneous magnetic field of 0.84 T and cooled to a temperature of 4.2 K. The degree of magnetic saturation achieved is equal to 15%. What is the total dipole moment of the sample for a magnetic field of 0.98 T and a temperature of 2.8 K ? (assume Curie's law).*

Solution. Initial dipole moment,

$$M_i = \frac{15}{100} \times 1.5 \times 10^{-23} \times 2.0 \times 10^{24} \text{ J T}^{-1}$$

$$= 4.5 \text{ J T}^{-1}$$

According to Curie's law,

$$M = \text{constant} \times \frac{H}{T}$$

$$\therefore \quad \frac{M_f}{M_i} = \frac{H_f}{H_i} \times \frac{T_i}{T_f}$$

or

$$M_f = M_i \times \frac{H_f}{H_i} \times \frac{T_i}{T_f}$$

$$= 4.5 \times \frac{0.98}{0.84} \times \frac{4.2}{2.8} \text{ J T}^{-1}$$

$$= \mathbf{7.875\ J\ T^{-1}.}$$

Example 4.08. *The core of a toroid having 3000 turns has inner and outer radii of 11 cm and 12 cm respectively. The magnetic field in the core for a current of 0.70 A is 2.5 T. What is the relative permeability of the core ? Take $\pi = 3.14$.*

Solution. The magnetic field in the empty space enclosed by the windings of the toroid is given by

$$B = \mu_0 n I \quad \text{...(1)}$$

where n is the number of turns per unit length and I is the current.

If the space is filled by a core of permeability μ, then eqn. (1) is rewritten as under.

$$B = \mu n I$$

Now, B = 2.5 tesla, I = 0.70 ampere, mean radius = 11.5 cm = 11.5 × 10^{-2} m

Number of turns per unit length,

$$n = \frac{3000}{2\pi \times 11.5 \times 10^{-2}} \text{ m}^{-1}$$

Here, we have ignored the variation of B across the cross-section of the toroid and taken the radius of the toroid to be the mean of inner and outer radii.

$$\therefore \quad \mu = \frac{B}{nI} = \frac{2.5 \times 2\pi \times 11.5 \times 10^{-2}}{3000 \times 0.7}$$

$$= 8.6 \times 10^{-4} \text{ Tm A}^{-1}$$

Relative permeability of core,

$$\mu_r = \frac{\mu}{\mu_0} = \mathbf{684.7.}$$

Example 4.09. *A paramagnetic gas consists of atoms each with a dipole moment of 1.5 × 10^{-23} J T^{-1}. The temperature of the gas is 27°C and its number density (number of atoms per unit volume) is 2.0 × 10^{26} m^{-3}. What is the maximum possible magnetisation of the sample when it is placed in an external magnetic field ? Do you think the sample can achieve close to maximum magnetisation for a magnetic field of 3.0 T ?*

Solution. Magnetisation I is the total dipole moment per unit volume. Maximum magnetisation is achieved when all the dipoles are aligned parallel to the direction of the external field B. In that case, the total dipole moment along B is Nm and $I_{max} = \frac{Nm}{V}$.

For $\frac{N}{V} = 2.0 \times 10^{26} \text{ m}^{-3}$,

and $m = 1.5 \times 10^{-23} \text{ J T}^{-1}$

$$I_{max} = 3.0 \times 10^3 \text{ A m}^{-1}$$

$$(1 \text{ J T}^{-1} \equiv 1 \text{ A}^2 \text{ m})$$

The actual magnetisation achieved by the sample is less than this due to the thermal disorder (as a result of incessant collisions between the atoms of the gas) which tends to disrupt the alignment of the atomic dipoles with the external field. A detailed analysis is beyond our scope here. But some idea can be obtained by comparing the thermal energy of an atom $\left(= \frac{3}{2} kT\right)$ with the magnitude of potential energy of an atomic dipole when aligned with the field (= MB).

$$\frac{3}{2} kT = \frac{3}{2} \times 1.38 \times 10^{-23} \times 300$$

$= 6.21 \times 10^{-21}$ J ;

MB $= 1.5 \times 10^{-23} \times 3.0 = 4.5 \times 10^{-23}$ J

$$\therefore \quad \frac{\frac{3}{2}kT}{MB} = 138$$

Thus the randomising effect (represented roughly by $\frac{3}{2}$ kT) is much greater than the alignment effect (represented roughly by MB). Consequently, even though this simple reasoning does not tell us the actual value of the sample's magnetisation, we can be sure that the sample is far from maximum magnetisation under the given conditions.

EXERCISES

1. An iron ring of mean radius 10 cm and area of cross-section 5 sq. cm has an air gap 1 cm wide. The ring is wound uniformly with coil of 900 turns. A flux of 10^{-4} Wb is required in the air gap. Find the strength of current required to produce the flux. Given $\mu_{iron} = 65 \times 10^{-4}$ H/m. *(UPTU, B. Tech. I, Q. Bank, 2001)*
2. Estimate the number of ampere-turns necessary to produce a flux of 0.001 Wb around an iron ring of 6 cm^2 cross-section and 20 cm mean diameter, having an air gap of 2 mm. The permeability of iron is 1200 μ_0. *(UPTU, B. Tech. I, Q. Bank. 2001)*
3. An iron ring of 10 cm radius and 5 sq. cm cross-section has an air gap 2 mm in width. A coil is wound uniformly over the ring ; a flux of 10^{-3} Wb is produced in the gap. If a current of 1.82 amp passes through the coil, find the total number of turns in the coil if the permeability of the core is 65×10^{-4} H/m. *(UPTU, B. Tech. I, Q. Bank, 2001)*
4. An iron ring has cross-sectional area 10 cm^2 and air gap of width 2 mm and mean length 60 m including the air gap. Find the number of ampere-turn required to produce the flux density 1 Wb m^2. Given $\mu_{iron} = 10^{-3}$ H/m. *(UPTU, B. Tech. I, Q. Bank, 2001)*
5. A toroidal solenoid has 3000 turns and a mean radius of 0.10 m. It has a soft iron core or relative permeability 2000. Find the magnetic field in the core when a current of 1 ampere is passed through the solenoid. (**Ans.** 12. T)
6. A magnetising field of 1600 A m^{-1} produces a magnetic flux, of 2.4×10^{-5} Wb in a bar of iron of cross-section 0.2×10^{-4} m^2. Calculate the permeability and susceptibility of the bar. (**Ans.** 7.5×10^{-4} T A^{-1} m, 595.6)

SHORT QUESTIONS

1. Fig. 4.38 shows two paramagnetic spheres located near the south pole of a bar magnet. Are (*a*) the magnetic forces on the spheres and (*b*) the magnetic dipole moments of the spheres directed toward or away from the bar magnet ? (*c*) Is the magnetic force on sphere 1 greater than, less than or equal to that on sphere 2 ?

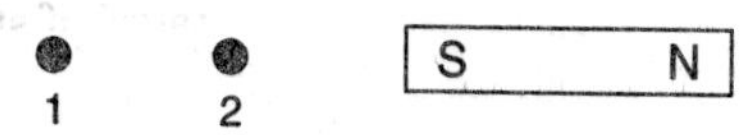

Fig. 4.38

2. Fig. 4.39 shows two diamagnetic spheres located near the south pole of a bar magnet. Are (*a*) the magnetic forces on the spheres and (*b*) the magnetic dipole moments of the spheres directed toward or away from the bar magnet ? (*c*) Is the magnetic force on sphere 1 greater than, less than or equal to that on sphere 2 ?

Fig. 4.39

3. What do you understand by the term "non-magnetic material" ?
4. Why paramagnetic and diamagnetic substances do not show hysteresis ?
5. The maximum value of the permeability of μ-metal (77% Ni, 16% Fe, 5% Cu, 2% Cr) is 0.126 Tm A^{-1}. What is the maximum relative permeability and susceptibility ?

6. How is relative permeability μ of a magnetic material related to the magnetic susceptibility χ_m ?
7. Name one property of a magnetic material used for making permanent magnet.
8. Why is it possible to make permanent magnets out of steel or alnico ?
9. What is the source of magnetism ?
10. How does the intensity of magnetisation of a paramagnetic material vary with increasing applied magnetic field ?
11. Define magnetic permeability.
12. How a ferromagnetic bar behaves when suspended in a uniform magnetic field ?
13. What is the SI unit of magnetic moment of a dipole ?
14. Besides iron, write down the names of two other ferromagnetic substances.
15. How does the intensity of magnetisation of a paramagnetic material vary with increasing applied magnetic field.
16. Give an example of a substance having negative permeability.
17. Mention two properties of ferrites.
18. Identify a substance which has negative magnetic susceptibility.
19. Which field, $\vec{B}$ or $\vec{H}$, is stronger in (*a*) ferromagnetic, (*b*) paramagnetic, (*c*) diamagnetic ?
20. An iron nail is attracted by a magnet. What is the source of kinetic energy ?
21. Why do magnetic field lines prefer to pass through ferromagnetic substances than through air ?
22. Why do magnetic lines of force prefer to pass through iron than air ?
23. Name the physical quantity which is measured in Wb A^{-1}.
24. What happens when a diamagnetic substance is placed in a varying magnetic field ?
25. What type of magnetic material is used in making permanent magnets ?
26. Give one example of each of the para, dia and ferromagnetic substances.
27. Is the product of magnetic susceptibility and absolute temperature constant for a paramagnetic substance ?
28. Give three reasons for making permanent magnets with alnico.
29. Which of the two is larger—magnetisation at saturation for a ferromagnetic substance or magnetisation at saturation for a paramagnetic substance ?
30. Name any two paramagnetic substances.
31. What happens when a diamagnetic substance is placed in a varying magnetic field ?
32. What kind of ferromagnetic material is used for coating magnetic tapes in a cassette player, or for building 'memory stores' in a modern computer ?
33. What are the advantages of studying the hysteresis loop of a substance ?
34. Compare the magnetic properties of steel and soft iron.
35. Why is diamagnetism not affected by changes of temperature ?
36. Comment on the state of magnetisation of a substance whose atoms contain odd number of electrons.
37. Why paramagnetic and diamagnetic substances cannot be used to make permanent magnets ?
38. Magnetic field lines are always nearly normal to the surface of a ferromagnet at every point. (This fact is analogous to the static electric field lines being normal to the surface of a conductor at every point). Why ?
39. Would the maximum possible magnetisation of a paramagnetic sample be of the same order of magnitude as the magnetisation of a ferromagnet ?
40. A certain region of space is to be shielded from magnetic fields. Suggest a method.
41. Does an iron bar magnet retain its magnetism when melted ? Give reason for your answer.
42. Distinguish between dia and paramagnetic substances. Give one example of each.
43. Why transformer cores are made of soft iron ?

44. What is a diamagnetic substance ?

45. What is a paramagnetic substance ?

46. How can paramagnetic and diamagnetic materials be distinguished by studying their behaviour in a magnetic field ?

47. Why soft iron is used to make electromagnets ?

48. An iron bar magnet is heated to 1000°C and then cooled in a magnetic field free space. Will it retain magnetism ?

49. Two substances A and B have their relative permeabilities slightly greater and less than unity respectively. What do you conclude about A and B ?

50. Define the terms retentivity and coercivity.

51. The hysteresis loop of a soft iron piece has a much smaller area than that of a carbon steel piece. If the material is to go through repeated cycles of magnetisation, which piece will dissipate greater heat energy ?

52. Why does a paramagnetic sample display greater magnetisation (for the same magnetising field) when cooled ?

53. Why is diamagnetism almost independent of temperature ?

ANSWERS

1. (*a*) toward (*b*) toward (*c*) less.

(*a*) Paramagnetic substances are feebly attracted towards the magnet.

(*b*) Paramagnetic substances develop magnetism in the direction of the magnetising field.

(*c*) Since magnetic forces obey inverse square law of distance, therefore, the force on sphere 1 is less.

2. (*a*) away (*b*) away (*c*) less.

(*a*) Diamagnetic substances move from stronger to weaker parts of the magnetic field, So, the magnetic forces on the spheres would be directed away from the magnet.

(*b*) Diamagnetic substances develop magnetism in a direction opposite to the direction of the magnetising field.

(*c*) Sphere 1 is farther from the magnet than sphere 2. Since magnetic forces obey inverse square law of distance, therefore, force on sphere 1 is less.

3. For the sake of convenience, a material whose permeability is close to that of vacuum is called non-magnetic. Strictly speaking, there is no magnetic material. The materials are ferromagnetic, para-magnetic or diamagnetic.

4. For a material to show hysteresis, the existence of domains is necessary. Since paramagnetic and diamagnetic substances do not have domain structure therefore they do not show hysteresis.

5. Relative permeability, $\mu_r = \dfrac{\mu}{\mu_0}$

or $\mu_r = \dfrac{0.126}{4\pi \times 10^{-7}} = 10^5$

Magnetic susceptibility, $\chi_m = \mu_r - 1 \approx 10^5$.

6. $\mu_r = 1 + \chi_m$.

7. High coercivity.

8. This is because steel or alnico has a large coercivity.

9. Electrons revolving in atoms constituting current loops. These current loops give rise to magnetism.

10. It increases directly with increase in applied magnetic field.

11. It is the ratio of magnetic induction and magnetic intensity.

12. It aligns in the direction of the field.

13. A m^2 or JT^{-1}.

14. Cobalt, Nickel.

15. Intensity of magnetisation of a paramagnetic material is directly to the applied magnetic field.

16. There is no such substance.

17. (*i*) high permeability (*ii*) low eddy currents.

18. Diamagnetic substance. Magnetic susceptibility is positive for both para and ferro-substances.

19. (*a*) B >> H, (*b*) B > H, (*c*) B < H.

20. It is the magnetic field energy which is partly converted into kinetic energy.

21. This is because the relative magnetic permeability of ferromagnetic substances is greater than one.

22. Permeability of soft iron is greater than that of air.

23. The ratio of the magnetic induction and the magnetic moment is measured in Wb A^{-1}.

24. It moves from stronger to weaker part of the magnetic field.

25. Ferromagnetic.

26. Chromium, water and nickel.

27. Yes.

28. (*i*) Large area of the hysteresis loop (*ii*) high coercivity (*iii*) high retentivity.

29. Magnetisation at saturation for a.

30. 1. Aluminium, 2. Manganese.

31. The diamagnetic substance shall move from stronger part of the magnetic field to the weaker part of the magnetic field.

32. Ceramics (specially treated barium iron oxides) also called ferrites.

33. The hysteresis loop of a materials is an important factor which guides us in choosing a suitable ferromagnetic substance for a particular use. As an example, the material used for a permanent magnet should have a large hysteresis loop. The material used for electromagnet or transformer core should have narrow hysteresis loop.

34.

S.No.	*Steel*	*Soft iron*
1.	Area of hysteresis loop is larger. So, loss of energy per unit volume is large.	Area of hysteresis loop is smaller. So, loss of energy per unit volume is small.
2.	Coercivity is large.	Coercivity is small.

35. The magnetic dipole moment induced in a diamagnetic substance is opposite to the magnetising field. This has nothing to do with the thermal motion of the atoms. So, diamagnetism is not affected by changes of temperature.

36. An atom which has odd number of electrons will have at least one electron left unpaired. So, the atom has permanent magnetic dipole moment. Thus, the atom has paramagnetic character.

37. Paramagnetic and diamagnetic substances get magnetised only when subjected to an external magnetic field. These substances immediately return to the unmagnetised state when the external magnetic field is removed.

38. Proof of this important fact [of much practical use] is based on boundary conditions of magnetic fields [B and H] at the interface of two media [when one of the media has $\mu >> 1$, the field lines meet this medium nearly normally].

39. Yes, apart from minor difference in the strength of the individual atomic dipoles of two different materials, paramagnetic sample with saturated magnetisation will have the same order of magnetisation. But of course, saturation requires impractically high magnetising fields.

40. Surround the region by soft iron rings. Magnetic field lines will be drawn into the rings, and enclosed space will be free of magnetic field. But this shielding is only approximate, unlike the perfect electric shielding of a cavity in a conductor placed in an external electric field.

41. Iron melts at a temperature which is higher than the Curie temperature of iron. So, the iron bar magnet cannot retain its magnetism when melted.

42. A diamagnetic is feebly magnetised in a direction to the direction of the magnetising field. A paramagnetic is magnetised feebly in the direction of the magnetising field.

    Cooper is an example of diamagnetic substance. Aluminium is an example of paramagnetic substance.

43. Transformer cores have to repeatedly go through cycles of magnetisation. So, they have to be made from a material of low hysteresis. Soft iron meets this requirement.

44. A substance, which when placed in a magnetic field gets feebly magnetised in a direction opposite to that of magnetising field, is called diamagnetic substance.

45. A substance, which when placed in magnetic field gets feebly magnetised in the direction of the magnetising field, is called a paramagnetic substance.

46. When placed in a non-uniform magnetic field paramagnetic substance moves from weaker to stronger parts of the magnetic field. On the other hand, the diamagnetic substance shall move from stronger to weaker parts of the magnetic field.

47. The hysteresis loop for soft iron is narrow. So, loss of energy per unit volume in one cycle of magnetisation is small. Thus, soft iron is used for making electromagnets.

48. The Curie temperature for iron is 770°C. The given temperature is much higher than Curie temperature. So, iron will behave as a paramagnetic substance. When iron is cooled, it will not retain its magnetism.

49. $\chi_m = \mu_r - 1$

    Relative permeability of A is slightly greater than 1. So, χ_m is small and positive. So, substance is paramagnetic.

    Relative permeability of B is slightly less 1.

    So, χ_m is small and negative. Clearly, substance is diamagnetic.

50. The value of the intensity of magnetisation of a material when the magnetising field is reduced to zero is called retentivity or residual magnetism of the material.

    The value of the magnetising field required to reduce residual magnetism to zero is called coercivity of the material.

51. Carbon steel piece, because heat lost per cycle is proportional to the area of the hysteresis loop.

52. The tendency to disrupt the alignment of dipoles [with the magnetising field] arising from random thermal motion is reduced at lower temperatures.

53. The induced dipole moment in a diamagnetic sample is always opposite to the magnetising field, no matter what the internal motion of the atoms is. [Note that the atoms of a diamagnetic substance do not have intrinsic magnetic moment. They acquire magnetic moment only when external magnetic field is applied.]

MULTIPLE CHOICE QUESTIONS

1. Which of the following statements is incorrect with reference to ferrites ?

   (*a*) They have low electrical conductivity as compared to ferromagnetic materials.

   (*b*) They have high permeability.

   (*c*) When heated above a certain temperature, they become paramagnetic.

   (*d*) Their chemical formula is M Fe_2O_4 where M stands for monovalent metals.

2. The energy loss per unit volume of the material per cycle is equal to μ_0 times the area of

   (*a*) B-H loop (*b*) M-H loop (*c*) I-H loop (*d*) χ-H loop.

3. The reciprocal of reluctance is known as

   (*a*) permeance (*b*) resistance

   (*c*) inductance (*d*) magnetic field induction.

4. The dimensional formula of magnetic susceptibility is :
(a) $[M^0L^0T^0]$ (b) $[LTA^{-1}]$ (c) [AT] (d) $[A^3 T^{-1}]$.

5. The magnetic moment of atomic neon is :
(a) zero (b) μ_B (c) $\frac{1}{2}\mu_B$ (d) $2\mu_B$.

6. The relative permeability of platinum is greater than unity and susceptibility has a small positive value. It must be :
(a) paramagnetic (b) ferromagnetic (c) diamagnetic (d) non-magnetic.

7. The SI unit of intensity of magnetisation is :
(a) Am (b) $A\ m^{-1}$ (c) $Wb\ m^{-1}$.

8. Domain formation is a necessary formation of
(a) ferromagnetics (b) diamagnetics (c) paramagnetics (d) all of these.

9. A hydrogen molecule is diamagnetic. A hydrogen atom is a
(a) paramagnetic (b) ferromagnetic (c) diamagnetic (d) ferrimagnetic.

10. Which of the following is ferromagnetic ?
(a) hydrogen (b) platinum (c) water (d) gadolinium.

11. A sensitive magnetic instrument can be shielded very effectively from outside fields by placing it inside a box of
(a) teak wood (b) plastic material
(c) a metal of high conductivity (d) soft iron of high permeability.

12. A small piece of an unmagnetised substance gets repelled when it is brought near a powerful magnet. Then the substance can be
(a) paramagnetic (b) ferromagnetic (c) non-magnetic (d) diamagnetic.

13. Gauss is the unit of
(a) magnetic flux density (b) magnetic flux
(c) magnetic induction.

14. Magnetic susceptibility x is defined as
(a) I/H (b) H/I (c) $I \times H$ (d) I^2/H^2.

15. Which of the following is paramagnetic ?
(a) Bismuth (b) Antimony (c) Chromium (d) Water.

16. Which of the following is suitable for the core of the electromagnets ?
(a) Air (b) Soft-iron (c) Steel (d) Cu-Ni alloy.

17. A steel wire of length l has a magnetic moment M. It is then bent into a semi-circular arc. The new magnetic moment is
(a) M (b) M/l (c) $2M/\pi$ (d) $M \times l$.

18. The magnetism of the magnet is due to
(a) the spin motion of electron (b) earth
(c) presence of big magnet inside the earth (d) cosmic rays.

19. In the case of a bar magnet, lines of magnetic induction
(a) start from the north pole and end at the south pole
(b) run continuously through the bar and outside
(c) emerge in circular paths from the middle of the bar
(d) are produced only at north pole like rays of light from a bulb.

20. If the magnetic flux is expressed in weber, then magnetic induction can be expressed as :
(a) $weber/m^2$ (b) weber/m (c) weber-m (d) $weber\text{-}m^2$.

21. Magnetic lines of force
(*a*) cannot intersect at all
(*b*) intersect at infinity
(*c*) intersect within the magnet
(*d*) intersect at the neutral point.

22. Susceptibility is positive for
(*a*) paramagnetic substances
(*b*) ferromagnetic substances
(*c*) non-magnetic substances
(*d*) diamagnetic substances.

23. The core of an electromagnet is made of soft iron because soft iron has
(*a*) small susceptibility and small retentivity
(*b*) large susceptibility and small retentivity
(*c*) large density and large retentivity
(*d*) small density and large retentivity.

24. Property possessed by ferromagnetic substance only is :
(*a*) attracting magnetic substances
(*b*) hysteresis
(*c*) susceptibility independent of temperature
(*d*) directional property.

25. The incorrect statement regarding the lines of force of the magnetic field $\vec{B}$ is :
(*a*) Magnetic intensity is a measure of lines of force passing through unit area held normal to it.
(*b*) Magnetic lines of force form a closed curve.
(*c*) Inside a magnet, its magnetic lines of force move from the north pole of a magnet towards its south pole.
(*d*) Due to a magnet, magnetic lines of force never cut each other.

26. A magnetic needle is kept in a non uniform magnetic field. It experiences
(*a*) a force and a torque
(*b*) a force but not a torque
(*c*) a torque but not a force
(*d*) neither a force nor a torque.

27. The relative permeability is represented by μ_r and the susceptibility by χ for a magnetic substance. Then for a paramagnetic substance
(*a*) $\mu_r > 1, \chi < 0$
(*b*) $\mu_r > 1, \chi > 0$
(*c*) $\mu_r < 1, \chi < 0$
(*d*) $\mu_r < 1, \chi > 0$.

28. Let $[\varepsilon_0]$ denote the dimensional formula of permittivity of the vacuum and $[\mu_0]$ that of the permeability of the vacuum. If M = mass, L = length, T = time and I = electric current :
(*a*) $[\varepsilon_0] = [M^{-1}L^{-3}T^2I]$
(*b*) $[\varepsilon_0] = [M^{-1}L^{-3}T^4I^2]$
(*c*) $[\mu_0] = [MLT^{-2}I^{-2}]$
(*d*) $[\mu_0] = [ML^2T^{-1}I]$.

29. The susceptibility of a ferromagnetic material is K at 27°C. At what temperature will its susceptibility be 0.5 K ?
(*a*) 600°C
(*b*) 237°C
(*c*) 54°C
(*d*) 327°C.

30. The ratio of the magnetic induction to the intensity of the magnetising field is called :
(*a*) susceptibility
(*b*) retentivity
(*c*) relative permeability
(*d*) absolute permeability.

31. χ_1 and χ_2 are the susceptibilities of a diamagnetic substance at temperatures T_1K and T_2K respectively. Then :
(*a*) $\chi_1 T_2 = \chi_2 T_1$
(*b*) $\chi_1 \sqrt{T_1} = \chi_2 \sqrt{T_2}$
(*c*) $\chi_1 = \chi_2$
(*d*) $\chi_1 T_1 = \chi_2 T_2$.

32. Above Curie temperature :
(*a*) a diamagnetic substance becomes paramagnetic.
(*b*) a paramagnetic substance becomes antiferromagnetic.
(*c*) a paramagnetic substance becomes ferromagnetic.
(*d*) a ferromagnetic substance becomes paramagnetic.

33. Susceptibility of a magnetic substance is found to depend on temperature and the strength of the magnetising field. The material is a :
(*a*) Superconductor
(*b*) Paramagnet
(*c*) Ferromagnet
(*d*) Diamagnet.

34. Property possessed by ferromagnetic substance only is
(*a*) directional property
(*b*) attracting magnetic substances
(*c*) hysteresis
(*d*) susceptibility independent of temperature.

35. The magnetic field lines due to a bar magnet are correctly shown in

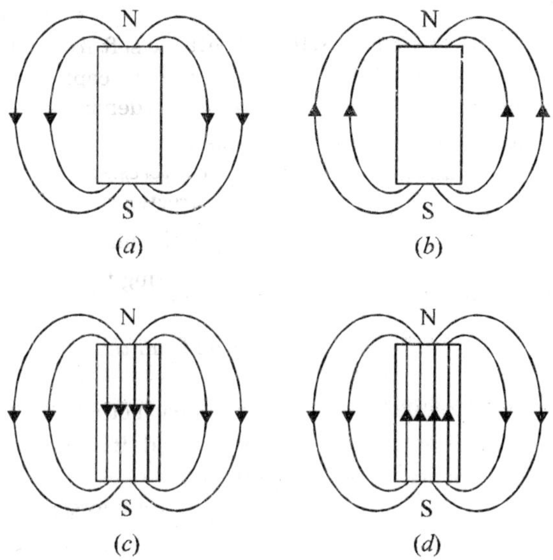

Fig. 4.40

36. Relative permeability of iron is 5500. Its magnetic susceptibility is
(*a*) 5500×10^7
(*b*) 5499
(*c*) 5501
(*d*) 5500×10^{-7}.

37. An example for diamagnetic substance is
(*a*) Nickel
(*b*) Aluminium
(*c*) Copper
(*d*) Iron.

38. The material of a permanent magnet has
(*a*) high retentivity and low coercivity
(*b*) low retentivity and high coercivity
(*c*) high retentivity and high coercivity
(*d*) low retentivity and low coercivity.

39. The area enclosed by the hysteresis curve is a measure of
(*a*) the residual magnetism of the material
(*b*) the retentivity of the material
(*c*) the magnetisation of the material
(*d*) the energy loss per cycle.

40. The susceptibility of a diamagnetic substance
(*a*) increases with temperature
(*b*) decreases with temperature
(*c*) does not vary with temperature
(*d*) first decreases and then increases with an increase in temperature.

41. Which one of the following substances is not magnetic ?
(*a*) Brass
(*b*) Cobalt
(*c*) Nickel
(*d*) Iron.

42. In cassette tapes, the sound is recorded in the form of
(*a*) variable electrical resistance of the tape
(*b*) sound wave on the tape
(*c*) electrical energy
(*d*) magnetic field on the tape.

43. In a normal iron piece,
(*a*) there are no magnetic dipoles
(*b*) magnetic dipoles are arranged in specific direction
(*c*) magnetic dipoles are arranged in random direction
(*d*) none of the above.

44. The hysteresis curve of a magnetic material is generally studied for
(*a*) diamagnetic materials (*b*) ferromagnetic materials
(*c*) paramagnetic materials (*d*) dielectric materials.

45. The absence of a magnetic monopole is explained by
(*a*) Coulomb's law (*b*) Lorentz law (*c*) Gauss's law (*d*) Fleming's law.

46. An example of diamagnetic substance is
(*a*) Aluminium (*b*) Copper (*c*) Antimony (*d*) Cobalt.

47. Which of the following instruments is used to measure magnetic field ?
(*a*) A thermometer (*b*) A pyrometer (*c*) A flux meter (*d*) A hygrometer.

48. Which of the following is ferromagnetic ?
(*a*) Aluminium (*b*) Quartz (*c*) Nickel (*d*) Bismuth.

49. For protecting a sensitive equipment from external magnetic field, it should be :
(*a*) placed inside an iron can
(*b*) wrapped with insulated wire through which a current is passed
(*c*) surrounded with fine copper gauge
(*d*) placed inside an aluminium can.

50. The area of the B-H hysteresis loop is an indication of :
(*a*) the retentivity of the material (*b*) the energy dissipated per cycle
(*c*) the permeability of the medium (*d*) the susceptibility of the substance.

ANSWERS

1. (*d*)	**2.** (*b*)	**3.** (*a*)	**4.** (*a*)	**5.** (*a*)	**6.** (*a*)	**7.** (*b*)	**8.** (*a*)
9. (*a*)	**10.** (*d*)	**11.** (*d*)	**12.** (*d*)	**13.** (*a*)	**14.** (*a*)	**15.** (*c*)	**16.** (*b*)
17. (*c*)	**18.** (*a*)	**19.** (*b*)	**20.** (*a*)	**21.** (*a*)	**22.** (*a*), (*b*)	**23.** (*b*)	**24.** (*b*)
25. (*c*)	**26.** (*a*)	**27.** (*b*)	**28.** (*b*), (*c*)	**29.** (*d*)	**30.** (*d*)	**31.** (*c*)	**32.** (*d*)
33. (*c*)	**34.** (*c*)	**35.** (*d*)	**36.** (*b*)	**37.** (*c*)	**38.** (*c*)	**39.** (*d*)	**40.** (*c*)
41. (*a*)	**42.** (*d*)	**43.** (*c*)	**44.** (*b*)	**45.** (*c*)	**46.** (*b*)	**47.** (*c*)	**48.** (*c*)
49. (*a*)	**50.** (*b*)						

LONG ANSWER QUESTIONS

1. What do you understand by a magnetic circuit ? Compare and contrast it with an electric circuit. Is it possible to have an open magnetic circuit ? *(UPTU, B. Tech. II Sem. 2001)*
2. Define $\vec{B}$, $\vec{M}$ and $\vec{H}$. Establish the relation $\vec{B} = \mu_0\vec{H} + \mu_0\vec{M}$. *(UPTU, B. Tech. I, Q. Bank, 2001)*
3. Explain the magnetic circuit, magnetomotive force and reactance. How does a magnetic circuit differ from and electric circuit. *(UPTU, B. Tech. I, Q. Bank, 2001)*
4. Distinguish between dia, para and ferromagnetic substances. Give an account of Langevin's theory of diamagnetism and show that diamagnetic susceptibility is independent of temperature.
5. Give a short account of dia-, para and ferromagnetism. *(UPTU, B. Tech. I, Q. Bank, 2001)*
6. Distinguish between paramagnetic, diamagnetic and ferromagnetic substances and illustrate their behaviour by simple experiments. How can you explain the behaviour of these substances ?
7. Distinguish dia-, para- and ferromagnetic materials. Give the Langevin's theory dia-magnetism. *(UPTU, B. Tech. I, Q. Bank, 2001)*
8. Give the physical significance of permeability and susceptibility.

9. Define permeability and magnetic susceptibility. Show that $\mu = \mu_0(1 + \chi_m)$.

10. What is magnetic circuit ? Compare and contrast the magnetic circuit with electric circuit. Derive an expression for magnetic flux in a circuit with an air gap. *(Model Q. Paper, B. Tech. I, 2001)*

11. Distinguish between diamagnetic, paramagnetic, ferromagnetic, anti-ferromagnetic and ferrimagnetic substances by giving examples.

12. With the help of domain theory explain the magnetisation in the ferromagnetic substances.

13. What is Weiss theory for ferromagnetic materials ? Establish the relation $\chi = C/(T - \theta)$. *(UPTU, B. Tech. I, Q. Bank, 2001)*

14. Discuss domain theory of ferro-magnetism ? *(UPTU, B. Tech. I, Q. Bank, 2001)*

15. How the susceptibility of a ferromagnetic substance varies with temperature explain with the help of a diagram. Define Curie-Weiss law.

16. What are paramagnetic a diamagnetic materials ? Give examples. Discuss Langevin's theory for a paramagnetic gas and obtain an expression for the paramagnetic susceptibility of a gas.

17. What are ferromagnetic domains ? How are their existence explained ? Distinguish between ferromagnetism, anti-ferromagnetism and ferrimagnetism.

18. What is Curie-Weiss law ? *(UPTU, B. Tech. I, Q. Bank, 2001)*

19. Draw a curve showing the variation of susceptibility with temperature for paramagnetic substances. Define Curie law.

20. Explain formation of domains in a ferromagnetic material and show how the hysteresis curve is explained on the basis of the domain theory.

21. What is anti-ferromagnetism ? Explain the difference between ferro-magnetism and anti-ferromagnetism. *(UPTU, B. Tech. I, Q. Bank, 2001)*

22. What is magnetisation curve ? Explain residual magnetism, coercive force and hysteresis. *(UPTU, B. Tech. I, Q. Bank, 2001)*

23. Describe the importance of hysteresis curve. How would you use the hysteresis curve for selecting the material for the construction of a permanent magnet ? *(UPTU, B. Tech. I, Q. Bank, 2001)*

24. How would you use the hysteresis curves to select material for the construction of (*i*) permanent magnets, (*ii*) electromagnets and (*iii*) transformer cores.

25. Illustrate the nature of the hysteresis loop of a sample of steel and that of a soft-iron piece. Indicate from these how the two materials differ in their magnetic behaviours.

26. What do hysteresis and cycle of magnetisation mean ? Prove that the energy dissipated per cc of the magnetisation is μ_0 times the area of M-H (or I-H) curve. *(UPTU, B. Tech. I, Q. Bank, 2001)*

27. What is meant by hysteresis ? How will you plot the B-H curve experimentally ?

28. What is meant by Hysteresis ? Explain hysteresis loss. How would you use the hysteresis curves to select material for the constructive of permanent magnets ? *(UPTU, B. Tech. I Sem. 2001)*

29. What are the factors responsible for hysteresis loss ? Prove that the area of the B-H curve is equal to the hysteresis loss per unit volume of the specimen in one cycle.

30. Define and explain the terms : Retentivity, Coercivity, Hysteresis and Hysteresis loop. Explain how a magnetised piece of iron can be completely demagnetised. Discuss the importance of hysteresis curve.

5

Ultrasonics

It is well known that sound waves are produced due to the vibration of one or other kind of particles in the medium. Sound waves are longitudinal waves and its frequency ranging from 20 to 20,000 Hz. This range of frequencies is known as audible range because human ear can hear it. The sound waves having frequency above this range are called ultrasonic waves. Although our ear are not sensible to these low wavelength waves, but some animals like dogs, mouses, cats, bats etc. can both produce and hear them. The sound wave having frequencies less than 20 Hz are called infrasonic waves.

5.01 GENERATION OF ULTRASONIC WAVES

Ultrasonic waves cannot be generated with our usual method of diaphragm loudspeaker because diaphragm can not vibrate at such high frequencies. Hence, there are number of other methods depending upon the output frequency range. The most important and common methods for producing ultrasonic waves are (*i*) Piezoelectric generator and (*ii*) Magnetostriction generator.

(*i*) *Piezoelectric Generator.* Piezoelectric generator is based on Piezoelectric effect. This method was devised by Langevin in 1917 for producing ultrasonics. In piezoelectric effect, when a crystal such as quartz, rochella salt or tourmaline is subjected to mechanical stress (strech or compress) along mechanical axis (say x) then potential difference is produced along perpendicular electrical axis (say y). Converse of this effect is also true *i.e.,*when potential difference along electrical axis is applied, the crystal is set into elastic vibration along the mechanical axis. When the frequency of applied voltage coincides with the natural frequency of crystal, resonance occurs resulting in a large amplitude of vibration of the crystal.

The experimental arrangement of piezoelectric generator is shown in Fig. 5.1. Q is a slice of a crystal placed between a parallel plate capacitor as dielectric. The plates are connected to the primary of a transformer which is coupled inductively to a oscillatory circuit of a valve. If the natural frequency of oscillations of the valve circuit is equal to the frequency of crystal mode to which we intact it to excite. The resonance will take place and the crystal will undergo linear expansion and contraction (mechanical vibrations) exactly at the same frequency and the ultrasonic waves are generated. The ultrasonic waves upto 500 K Hz with a moderate size crystal can be produced by this method. However, a tourmaline crystal can produce frequency upto 15×10^7 Hz.

The velocity of quartz crystal along x-direction is given by

$$v = \sqrt{\gamma/\rho}$$

For quartz the Young's modulus $\gamma = 8.0 \times 10^{10}$ N/m^2 and $\rho = 2654$ kg/m^2, hence $v = 5500$ m/s.

If t be the thickness of quartz slice in meters then

$$v = n\lambda = n\ (2t) \qquad n \text{ be the frequency}$$

$$n = \frac{v}{2t} = \frac{27.25}{t} \text{ Hz}$$

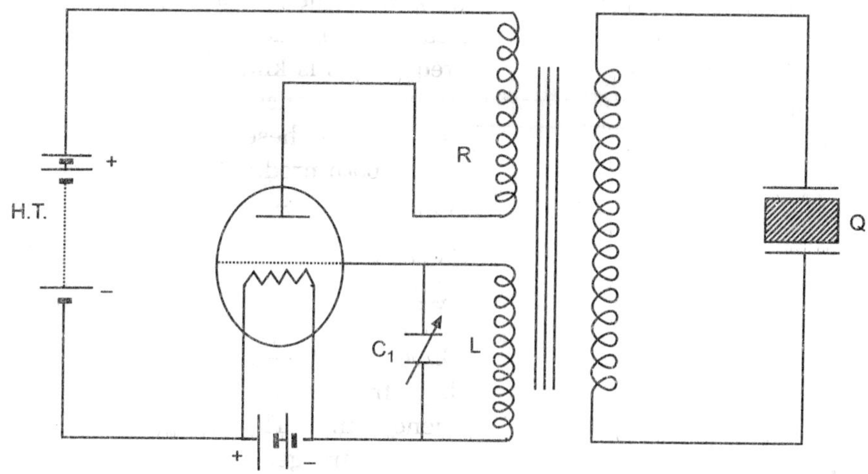

Fig. 5.1

(*ii*) *Magnetostriction Generator*. This method is based on the phenomenon of magnetostriction, which was discovered by Joule in 1874. When a rod of ferromagnetic material such as iron or nickel is subjected to a magnetic field parallel to its length, the length of rod changes. This change in length of rod is independent of direction of field but depends on the magnitude of field and nature of rod. If the applied magnetic field is alternating, the rod extends and contract in length alternately. Since the change in length is independent of the direction of applied field, the frequency of vibration of rod is twice the frequency of field. Ordinary the amplitude of vibration of rod is small. But, when the frequency of magnetic field is the same as the natural frequency of the rod, the resonance occurs and the amplitude of vibration is considerably increased. Thus sound waves now emitted from the ends of rod. If the applied frequency is the order of ultrasonic frequency, the rod produces ultrasonic waves.

As shown in Fig. 5.2 a nickel or iron rod carries a coil L at its centre and connected to DC voltage for magnetising initially. Two more coil L_1 and L_2 are wound over the rod and connected to grid and plate circuit respectively of a oscillatory circuit. The frequency of oscillator can be controlled by the capacitor C of the plate circuit. When current in L_2 changes, it changes the magnetization of rod which results to change the length of rod. Thus vibration in length occurs which causes a variation in magnetic flux through coil L_1 and developed new *emf* in the coil L_1. This *emf* acts on the grid and produces an amplified current change in plate circuit *i.e.*, in L_2. In this way, the plate current build up to a large amplitude with a frequency determined by the frequency of longitudinal vibration of the rod. When the frequency of circuit becomes equal to the frequency of vibration of rod, the sound waves of maximum amplitude are generated as ultrasonic waves.

For fundamental mod of vibration, the natural frequency of rod of length l given by

$$n = \frac{1}{2l}\sqrt{\frac{\gamma}{\rho}}$$

where γ is the Young's modulus and ρ is density of rod material.

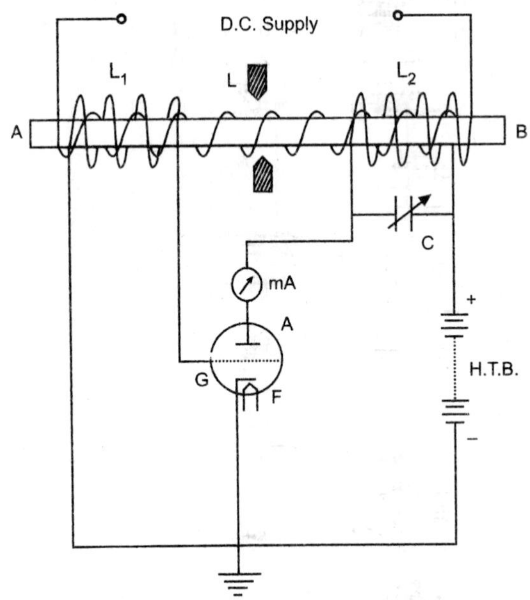

Fig. 5.2

5.02 DETECTION OF ULTRASONICS

Ultrasonic waves are high frequency waves above audible range and cannot be detected by human ear directly. However, some animal specially the bat can do so. These waves can be detected by following methods:

(*i*) *Piezoelectric detector.* When a piezoelectric quartz crystal is subjected to ultrasonic waves on its one of the pair of faces, the varying electric charge of opposite nature are developed on the other pair of faces perpendicular to previous one. The charges developed of course are very weak which can be amplified by some means and detected.

(*ii*) *Kundt's tube method.* Kundt's tube method can be used to detect ultrasonic waves when the wavelength of waves are comparatively large. When ultrasonic waves propagated through tube, Lycopodium powder sprinked in the tube collected in the form of heaps at the nodal points and blown off at antinodal points. The distance between two heaps is equal to $\lambda/2$ which may be difficult to measure if wavelengths are small or frequency of wave is very large.

(*iii*) *Sensitive flame method.* When a narrow sensitive flame moves along the medium where ultrasonic waves are present, the flame remain stationary at antinodes and flicker at nodes due to change in pressure.

(*iv*) *Thermal detection method.* This is most modern method for detection of ultrasonic waves. A fine platinum wire is moved in the medium of ultrasonic waves. Due to compression and rarefaction of wave the temperature of medium changes. The change in temperature is observed at nodes while constant at antinodes. Hence the resistance of platinum wire is changes accordingly. This change in resistance can be determined using a sensitive bridge arrangement. It will be in balance mode when platinum wire is at antinode.

5.03 PROPERTIES OF ULTRASONIC WAVES

(*i*) These waves are very high energetic being high frequency.

(*ii*) The speed of propagation of these wave increases with increase in frequency.

(*iii*) Due to small wavelength, diffraction of wave is negligible therefore wave can travel over a long distance without much appreciable loss of energy.

(*iv*) When the ultrasonic waves are travel in liquid, stationary wave pattern is formed due to reflection of wave from other end. Hence density of liquid varies along the propagation direction which can act as diffraction grating to diffract the light.

5.04 APPLICATIONS OF ULTRASONIC WAVES

Ultrasonic wave have a wide application in various field in which some are described here:

(*i*) *Detection cracks and faults in metal.* Ultrasonic waves are used to detect flaws in metal. The cracks or faults in metal changes the refractive index, which reflect the wave when subjected to it. The waves are also reflected from other end of metal. These reflected waves are pickup by receiver which shows different peaks when observed in CRO.

(*ii*) *Detection of submarine and other object in sea.* A sharp ultrasonic wave is send in all direction in sea water. The reflected wave from any direction shows the presence of any object in the sea.

(*iii*) *Detection depth of sea.* As the ultrasonic waves can be sent over a long distance without much loss of energy hence can be used to detect the depth of sea. The time internal of sending and receiving reflected wave is recorded. As the velocity of wave is known, the depth of sea can be determined.

(iv) *Application in industry*. Ultrasonic wave are used for cleaning in industry. These remove water from paper during manufacturing process, remove dust and soot from chimney, in washing, in dry cleaning cloths. These are also used in plastic welding. By applying small pressure and ultrasonic vibration to two similar surface produces sufficient thermal energy to bound the surface together. These used in drilling process in metals.

(v) *Application in medical science*. In medical science ultrasonic waves are used for diagonostic purpose. These wave can distinguish between soft tissues and liquid and between crystal and tumours. These can be used for blood less surgery and painless dental cutting and sterilization purposes. Abnormal growth of tumours in brain can also be detected by ultrasonic waves which may not possible with X-rays.

(vi) *Application in signalling*. Ultrasonic wave can be concentrated into a sharp beam due to smaller wavelength and hence can be used for signalling in particular direction.

(vii) *In ultrasonic mixing*. A colloid solution or emulsion of two non-missible liquid like oil and water can be formed by simultaneous subjecting to ultrasonic waves.

(viii) *In thermal effect*. If a liquid is subjected to ultrasonic waves its temperature will increase due to absorption of waves.

SOLVED EXAMPLES

1. *A quartz crystal of thickness of 0.001 is vibrating at resonance. Calculate the fundamental frequency. Given : γ for quartz = 7.9×10^{10} N/m^2 and ρ = 2650 Kg/mt^3*

Sol. We know that velocity of wave v is

$$v = \sqrt{\frac{\gamma}{\rho}} = \sqrt{\frac{7.9 \times 10^{10}}{2650}} = 5461 \text{ m/s}$$

For fundamental frequency of vibration, thickness of quartz should be $\lambda/2$. Hence

$$\frac{\lambda}{2} = t$$

or

$$\lambda = 2t$$

Further

$$v = n\lambda$$

or

$$n = \frac{v}{\lambda} = \frac{v}{2t} = \frac{5461}{2 \times .001} = 2.730 \times 10^6 \text{ Hz}$$

$$\boldsymbol{n = 2730 \text{ KHz}}$$

2. *A bat emits ultrasonic waves of frequency 200 kHz in air. If this sound meets a water surface, then calculate the wavelength of (i) reflected sound (ii) transmitted sound. Speed of sound in air is 340 m/s and in water is 1486 m/s.*

Sol. From the frequency and speed relation

$$\nu = n\lambda \quad \text{or} \quad \lambda = \frac{\nu}{n}$$

(*i*) Wavelength of reflected sound (in air) $= \dfrac{340}{200 \times 10^3} = 1.7 \times 10^{-3}$ m

(*ii*) Wavelength of transmitted sound (in water) $= \dfrac{1486}{200 \times 10^3} = 7.43 \times 10^{-3}$ m

So the wavelength of reflected sound is 1.7×10^{-3} m.

The wavelength of transmitted sound is 7.43×10^{-3} m.

MULTIPLE CHOICE QUESTIONS

1. Sound waves having following frequencies are above the audible range.

(*a*) 50 Hz (*b*) 500 Hz

(*c*) 5000 Hz (*d*) 50000 Hz

2. Which method is used to produce ultrasonic waves

(*a*) Doppler effect (*b*) Hall effect

(*c*) Piezo-electric effect (*d*) Peltier effect

3. The wavelength of ultrasonic waves in air is of the order of

(*a*) 1 cm (*b*) 5×10^{-8} cm

(*c*) 5×10^{5} cm (*d*) 500 cm

4. Ultrasonic waves are

(*a*) Transverse (*b*) Longitudinal

(*c*) Electromagnetic (*d*) Stationary

5. Ultrasonic waves can be detected by

(*a*) A telephone (*b*) Quincle's method

(*c*) Kundt's method (*d*) Hebb's method

6. Ultrasonic, infrasonic ad audiowave travel in medium with speed ν_u, ν_i and ν_a respectively then.

(*a*) $\nu_a = \nu_u = \nu_i$ (*b*) $v_u > v_a > v_i$

(*c*) $v_u < v_a < v_i$ (*d*) $v_a < v_u$ and $v_u = v_i$

7. Bats flying in dark can avoid obstacles because of

(*a*) Infrasonic (*b*) Ultrasonic

(*c*) Supersonic (*d*) Audible range

8. The wavelength of wave in a medium is related to frequency by

(*a*) $v = f\lambda$ (*b*) $v = f/\lambda$

(*c*) $v = \dfrac{\lambda}{f}$ (*d*) $v = f\lambda^2$

9. In magnetostriction method ultrasonic waves produces by
   (*a*) Vibration of rod (*b*) Changing of magnetic field
   (*c*) From LC circuit (*d*) Filament of valve

10. In piezoelectric effect the frequency of ultrasonic is related with Young's modulus by
   (*a*) $f = \frac{1}{2t}\sqrt{\frac{\gamma}{\rho}}$ (*b*) $f = 2t\sqrt{\frac{\gamma}{\rho}}$
   (*c*) $f = 2t\sqrt{\frac{\rho}{\gamma}}$ (*d*) $\frac{1}{2t}\sqrt{\frac{\rho}{\gamma}}$

ANSWERS

1. (*d*)	**2.** (*c*)	**3.** (*a*)	**4.** (*b*)	**5.** (*c*)
6. (*a*)	**7.** (*b*)	**8.** (*a*)	**9.** (*a*)	**10.** (*a*)

LONG QUESTIONS

1. What are the ultrasonic waves? How they are produced in piezoelectric effect?
2. Describe one method for generation of ultrasonic waves.
3. What is piezoelectric effect? How it used to produce ultrasonic waves?
4. Describe the phenomenon of magnetostriction. How it produces ultrasonic waves?
5. What are the different methods for detection of ultrasonic waves? Describe in brief.
6. State the applications of ultrasonic waves.
7. Write short notes on the following:
   (*a*) Piezoelectric effect
   (*b*) Magnetostriction
   (*c*) Ultrasonic waves and their uses
   (*d*) Ultrasonic waves and their detection

UNIT III

ELECTROMAGNETICS

This unit contains the following chapter:

6. Maxwell's Equations and Electromagnetic Waves

6

Maxwell's Equations and Electromagnetic Waves

6.01 SCALAR TRIPLE PRODUCT

The dot product of one vector with the cross product of other two vectors is called scalar triple product $\vec{A} . (\vec{B} \times \vec{C})$ is a scalar product. It is a scalar quantity.

Now,
$$\vec{B} \times \vec{C} = \begin{vmatrix} \hat{i} & \hat{j} & \hat{k} \\ B_x & B_y & B_z \\ C_x & C_y & C_z \end{vmatrix}$$
$$= \hat{i}(B_yC_z - B_zC_y) + \hat{j}(B_zC_x - B_xC_z) + \hat{k}(B_xC_y - B_yC_x)$$

Again,
$$\vec{A} . (\vec{B} \times \vec{C})$$
$$= (A_x\hat{i} + A_y\hat{j} + A_z\hat{k}) . [\hat{i}(B_yC_z - B_zC_y) + \hat{j}(B_zC_x - B_xC_z) + \hat{k}(B_xC_y - B_yC_x)]$$
$$= A_x(B_yC_z - B_zC_y) + A_y(B_zC_x - B_xC_z) + A_z(B_xC_y - B_yC_x)$$

In the form of determinant,
$$\vec{A} . (\vec{B} \times \vec{C}) = \begin{vmatrix} A_x & A_y & A_z \\ B_x & B_y & B_z \\ C_x & C_y & C_z \end{vmatrix}$$

This gives the expression for scalar triple product in terms of components.

In a scalar triple product, the dot and the cross can be interchanged.

$$\therefore \quad \vec{A} . (\vec{B} \times \vec{C}) = (\vec{A} \times \vec{B}) . \vec{C}$$

In the representation of scalar triple product, there is no need to employ any bracket to show the cross product. So, $\vec{A} . (\vec{B} \times \vec{C})$ may simply be written as $\vec{A} . \vec{B} \times \vec{C}$.

It is meaningless to write $\vec{A} . (\vec{B} \times \vec{C})$ as $(\vec{A} . \vec{B}) \times \vec{C}$. This is because the cross product of a scalar with a vector is not defined.

Additional Information

The scalar triple product gives the volume of the parallelopiped. If three vectors are coplanar, then $\vec{A} . \vec{B} \times \vec{C} = 0$. This is because the volume of the parallelopiped formed by coplanar vectors is zero.

6.02 VECTOR TRIPLE PRODUCT

Vector triple product is the cross product of one vector with the cross product of other two vectors. The vector triple product of $\vec{A}$, $\vec{B}$ and $\vec{C}$ is $\vec{A} \times (\vec{B} \times \vec{C})$. It is a vector quantity. Now,

$\vec{B} \times \vec{C} = \hat{i}\,(B_yC_z - B_zC_y) + \hat{j}(B_zC_x - B_xC_z) + \hat{k}\,(B_xC_y - B_yC_x)$

Again,
$$\vec{A} \times (\vec{B} \times \vec{C}) = \begin{vmatrix} \hat{i} & \hat{j} & \hat{k} \\ A_x & A_y & A_z \\ B_yC_z - B_zC_y & B_zC_x - B_xC_z & B_xC_y - B_yC_x \end{vmatrix}$$

On simplification, it can be proved that

$$\vec{A} \times (\vec{B} \times \vec{C}) = \vec{B}\,(\vec{A} \,.\, \vec{C}) - \vec{C}\,(\vec{A} \,.\, \vec{B})$$

This is BAC – CAB rule. It is an important vector identity.

Vector triple product is not associative. So, $\vec{A} \times (\vec{B} \times \vec{C})$ and $(\vec{A} \times \vec{B}) \times \vec{C}$ are different vectors. Note that parentheses in vector triple product needs to be retained.

6.03 SCALAR QUADRUPLE PRODUCT OF VECTORS

The dot product of a vector with triple vector product is called scalar quadruple product of vectors. For four vectors $\vec{A}$, $\vec{B}$, $\vec{C}$ and $\vec{D}$, the scalar quadruple product of vectors is $\vec{A} \,.\, [\vec{B} \times (\vec{C} \times \vec{D})]$.

6.04 VECTOR QUADRUPLE PRODUCT OF VECTORS

The cross product involving four vectors is called vector quadruple product. It is expressed as $\vec{A} \times [\vec{B} \times (\vec{C} \times \vec{D})]$.

6.05 SCALAR AND VECTOR FIELDS

Scalar and vector fields are two commonly used terms in electricity and magnetism.

A scalar field is a region of space at every point of which a scalar quantity assumes a definite value. In other words, it is a region of space within which a scalar quantity can be defined as a scalar point function.

Sometimes, a scalar quantity which is continuous and has a unique value at every point is called scalar field. So, density, temperature, electric potential are examples of scalar fields.

Mathematically, $\phi = \phi(\vec{r})$ or $\phi = \phi(x, y, z)$

A vector field is a region of space at every point of which a vector quantity assumes a definite value. In other words, a vector field is a region of space within which a vector quantity can be defined as a vector point function.

Mathematically, $\vec{A} = \vec{A}\,(\vec{r})$ or $\vec{A} = \vec{A}\,(x, y, z)$

6.06 DERIVATIVE OF A VECTOR

Let us consider a vector $\vec{A}$ which is a function of scalar variable t.

$$\vec{A} = \vec{A}\,(t)$$

For each value of t, there is a corresponding value of $\vec{A}$. When t changes to $(t + \Delta t)$, $\vec{A}$ takes the value $\vec{A}(t + \Delta t)$

$$\Delta\vec{A} = \vec{A}(t + \Delta t) - \vec{A}(t)$$

$$\frac{\Delta\vec{A}}{\Delta t} = \frac{\vec{A}(t + \Delta t) - \vec{A}(t)}{\Delta t}$$

Taking limit as $\Delta t \to 0$, we get

$$\underset{\Delta t \to 0}{\text{Lt.}} \frac{\Delta\vec{A}}{\Delta t} = \underset{\Delta t \to 0}{\text{Lt.}} \frac{\vec{A}(t + \Delta t) - \vec{A}(t)}{\Delta t}$$

or

$$\frac{d\vec{A}}{dt} = \underset{\Delta t \to 0}{\text{Lt.}} \frac{\vec{A}(t + \Delta t) - \vec{A}(t)}{\Delta t}$$

$\frac{d\vec{A}}{dt}$ is derivative of $\vec{A}$ with respect to t. It is a vector quantity.

6.07 DERIVATIVES OF PRODUCTS

(*i*) $$\frac{d}{dt}(m\vec{A}) = m\frac{d\vec{A}}{dt} + \frac{dm}{dt}\vec{A}$$

where m is a scalar and $\vec{A}$ is a vector.

Note that the ordinary calculus rule of differentiating a product of two functions is applicable to product of a scalar and a vector.

(*ii*) $$\frac{d}{dt}(\vec{A}\cdot\vec{B}) = \vec{A}\cdot\frac{d\vec{B}}{dt} + \frac{d\vec{A}}{dt}\cdot\vec{B}$$

(*iii*) $$\frac{d}{dt}(\vec{A}\times\vec{B}) = \vec{A}\times\frac{d\vec{B}}{dt} + \frac{d\vec{A}}{dt}\times\vec{B}$$

The ordinary rules of differential calculus are applicable so long as the order of the two vectors is not changed.

6.08 PARTIAL DERIVATIVES OF A SCALAR FIELD

Consider a scalar field V such that

$$V = V(x, y, z)$$

When a quantity is a function of more than one variable, we can define partial derivative of that quantity with respect to any one of the independent variables.

The partial derivative of V with respect to x is defined as :

$$\frac{\partial V}{\partial x} = \underset{\Delta x \to 0}{\text{Lt.}} \frac{V(x + \Delta x, y, z) - V(x, y, z)}{\Delta x},$$

if the limit exists.

$\frac{\partial V}{\partial x}$ gives the rate of change of V with respect to distance along x-axis at the point (x, y, z).

Now, $$\frac{\partial V}{\partial y} = \underset{\Delta y \to 0}{\text{Lt.}} \frac{V(x, y+\Delta y, z) - V(x, y, z)}{\Delta y},$$

if the limit exists.

$\frac{\partial V}{\partial y}$ gives the rate of change of V with respect to distance along y-axis at the point (x, y, z).

Again, $$\frac{\partial V}{\partial z} = \underset{\Delta z \to 0}{\text{Lt.}} \frac{V(x, y, z+\Delta z) - V(x, y, z)}{\Delta z}, \text{ if the limit exists.}$$

$\frac{\partial V}{\partial z}$ gives the rate of change of V with respect to distance along z-axis at the point (x, y, z).

6.09 THE DEL OPERATOR

The del operator, written as $\vec{\nabla}$, is a vector differential operator. It plays an important role in physics. It is defined by :

$$\vec{\nabla} = \frac{\partial}{\partial x}\hat{i} + \frac{\partial}{\partial y}\hat{j} + \frac{\partial}{\partial z}\hat{k}$$

The operator $\vec{\nabla}$ is also known as nabla. It is employed to differentiate a scalar field. It is also used to operate on a scalar field.

The del operation involves partial derivatives with respect to x, y, z.

6.10 LAPLACIAN OPERATOR

The operator $\vec{\nabla} . \vec{\nabla} = \nabla^2$ is known as Laplacian operator.

$$\vec{\nabla} . \vec{\nabla} = \nabla^2 = \left[\frac{\partial}{\partial x}\hat{i} + \frac{\partial}{\partial y}\hat{j} + \frac{\partial}{\partial z}\hat{k}\right] . \left[\frac{\partial}{\partial x}\hat{i} + \frac{\partial}{\partial y}\hat{j} + \frac{\partial}{\partial z}\hat{k}\right]$$

$$= \frac{\partial}{\partial x}\frac{\partial}{\partial x} + \frac{\partial}{\partial y}\frac{\partial}{\partial y} + \frac{\partial}{\partial z}\frac{\partial}{\partial z} = \frac{\partial^2}{\partial x^2} + \frac{\partial^2}{\partial y^2} + \frac{\partial^2}{\partial z^2}$$

In spherical polar co-ordinates (r, θ, ϕ), ∇^2 is given as

$$\nabla^2 = \frac{1}{r^2}\frac{\partial}{\partial r}\left(r^2 \frac{\partial}{\partial r}\right) + \frac{1}{r^2 \sin\theta}\frac{\partial}{\partial \theta}\left(\sin\theta \frac{\partial}{\partial \theta}\right) + \frac{1}{r^2 \sin^2\theta}\frac{\partial^2}{\partial \phi^2}$$

In cylindrical polar co-ordinates (r, θ, z),

$$\nabla^2 = \frac{1}{r}\frac{\partial}{\partial r}\left(r \frac{\partial}{\partial r}\right) + \frac{1}{r^2}\frac{\partial^2}{\partial \theta^2} + \frac{\partial^2}{\partial z^2}$$

6.11 GRADIENT OF A SCALAR FIELD

Let V and V + dV be the values of scalar field V at points P and Q respectively. Let dV be the change in V in going from P to Q.

$$dV = \frac{\partial V}{\partial x}dx + \frac{\partial V}{\partial y}dy + \frac{\partial V}{\partial z}dz$$

or $$dV = \left(\hat{i}\frac{\partial V}{\partial x} + \hat{j}\frac{\partial V}{\partial y} + \hat{k}\frac{\partial V}{\partial z}\right) . (dx\,\hat{i} + dy\,\hat{j} + dz\,\hat{k})$$

or $$dV = \vec{\nabla} V . \vec{dr}$$

$\vec{\nabla} V$ is called gradient of V. It can be written as grad. V. It is a vector which is a function of (x, y, z).

The **magnitude** of grad. V at a point gives the maximum rate of change of V at that point. The **direction** along which the rate of change of V is maximum is the direction of $\vec{\nabla} V$.

Consider an equation $V(x, y, z) =$ constant.

It represents a three dimensional surface S.

$\vec{\nabla} V$ gives a vector normal to the surface at point (x, y, z) on this surface.

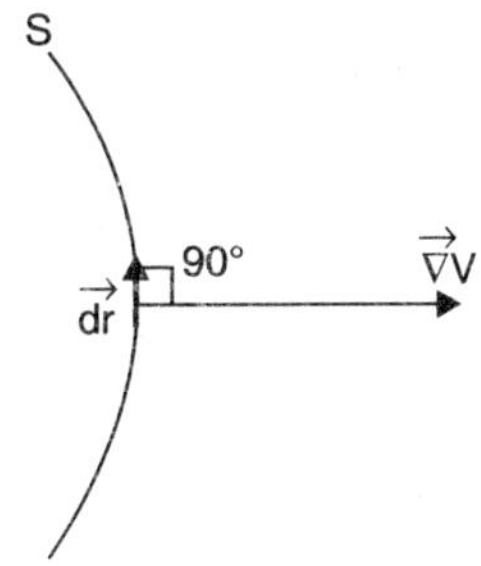

Fig. 6.1

6.12 DIVERGENCE OF A VECTOR FIELD

Consider a vector field $\vec{A}$, the magnitude and direction of which is a function of the position co-ordinates at a point. Then the scalar product of the differential operator $\vec{\nabla}$ and the vector $\vec{A}$ is a scalar function of position co-ordinates x, y, z. It is known as the divergence of the vector $\vec{A}$.

$$\therefore \qquad \text{div. } \vec{A} = \vec{\nabla} . \vec{A}$$

Here, $$\vec{\nabla} = \hat{i}\frac{\partial}{\partial x} + \hat{j}\frac{\partial}{\partial y} + \hat{k}\frac{\partial}{\partial z}$$

and $$\vec{A} = \hat{i} A_x + \hat{j} A_y + \hat{k} A_z$$

$$\therefore \qquad \vec{\nabla} . \vec{A} = \left(\hat{i}\frac{\partial}{\partial x} + \hat{j}\frac{\partial}{\partial y} + \hat{k}\frac{\partial}{\partial z}\right) . (\hat{i} A_x + \hat{j} A_y + \hat{k} A_z)$$

$$= \frac{\partial A_x}{\partial x} + \frac{\partial A_y}{\partial y} + \frac{\partial A_z}{\partial z}$$

Let us now discuss the *physical meaning* of the divergence of a vector field.

The divergence of a vector is the limiting value of the net outward flow of some physical quantity like electric flux or fluid through the surface area of a unit volume as the volume tends to approach zero.

Consider a closed volume V having surface area S in a vector field. Let $\vec{A}$ be the vector field at a point P on the surface. Let $\vec{dS}$ be a vector representing a small area surrounding the point P. The direction of $\vec{dS}$ is that of the outward drawn normal to the surface.

Flux of the field out of area $\vec{dS} = A\, dS \cos\theta = \vec{A} . \vec{dS}$

Here θ is the angle between the directions of field vector and outward drawn normal to the surface at P.

$\therefore$ Total flux of the field out of the whole surface

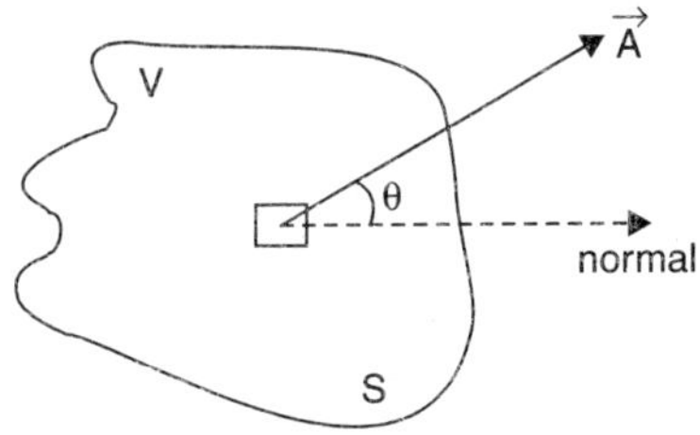

Fig. 6.2

$$= \iint_S \vec{A} \cdot d\vec{S}$$

If the volume V enclosed by the surface is vanishingly small, then the outward flux per unit volume gives the divergence of $\vec{A}$ (div. $\vec{A}$)

$$\therefore \qquad \text{div. } \vec{A} = \lim_{V \to 0} \frac{1}{V} \iint_S \vec{A} \cdot d\vec{S}$$

Examples :

1. Consider a liquid in motion. Suppose $\vec{A}$ is used to represent the velocity $\vec{V}$ of a liquid in motion. Then div. $\vec{V}$ at a point shall represent the volume of the liquid flowing out per second per unit volume enclosed by a small closed surface surrounding it.

If div. $\vec{V}$ at a point is positive, it indicates that liquid is flowing away from the point. This means that the point is a source of the liquid. If div. $\vec{V}$ at a point is negative, it indicates that liquid is flowing towards the point. This means that the point is a "sink".

2. The divergence of electric field $\vec{E}$ over the surface of a unit volume element surrounding a point in an electric field is given by

$$\text{div. } \vec{E} = \lim_{V \to 0} \frac{1}{V} \iint_S \vec{E} \cdot d\vec{S}$$

where S is the closed surface surrounding the point and V is the volume of the element surrounding it.

The divergence of electric field at any point gives us the charge density (scalar function) at that point.

$$\vec{\nabla} \cdot \vec{E} = \frac{\rho}{\varepsilon_0}$$

6.13 CURL OF A VECTOR FIELD

Consider a differentiable vector field $\vec{A}$. Let A_x, A_y and A_z be the rectangular components of $\vec{A}$.

The curl of $\vec{A}$ is defined as :

$$\text{Curl } \vec{A} = \begin{vmatrix} \hat{i} & \hat{j} & \hat{k} \\ \frac{\partial}{\partial x} & \frac{\partial}{\partial y} & \frac{\partial}{\partial z} \\ A_x & A_y & A_z \end{vmatrix}$$

or

$$\text{Curl } \vec{A} = \vec{\nabla} \times \vec{A}$$

Clearly, curl $\vec{A}$ is a vector quantity.

$$\vec{\nabla} \times \vec{A} = \hat{i}\left(\frac{\partial A_z}{\partial y} - \frac{\partial A_y}{\partial z}\right) + \hat{j}\left(\frac{\partial A_x}{\partial z} - \frac{\partial A_z}{\partial x}\right) + \hat{k}\left(\frac{\partial A_y}{\partial x} - \frac{\partial A_x}{\partial y}\right)$$

Curl of a vector field is related to the rotation of a vector field in hydrodynamics. As an illustration, if the velocity $\vec{v}$ of the fluid is taken as vector field $\vec{A}$, then $\vec{\nabla} \times \vec{v}$ gives the rotation of the fluid at a point where curl $\vec{v}$ is evaluated.

If the curl of a vector field vanishes, then the field is called irrotational.

For irrotational field, $\vec{\nabla} \times \vec{A} = \vec{0}$

The curl of a rotational or non-conservative vector at any point in a vector field is a vector whose magnitude is equal to the line integral per unit area along the boundary of an infinitesimal area drawn around that point. Its direction is along the normal to the area.

Physical Significance

(*i*) Curl measures the twist of a vector.

(*ii*) A region of high curl is a whirlpool.

(*iii*) Curl is an important property of those vector fields for which closed path-line integral is non-zero.

(*iv*) The curl of magnetic field at a point near a current-carrying conductor represents the current flowing per unit area to that point. Clearly, curl of magnetic field $\vec{B}$ represents magnetomotive force.

> *The closed path line integral is called the circulation of the vector field around the closed path.* It is generally denoted by C.
>
> $\therefore \qquad C = \oint \vec{A} \cdot \vec{dl}.$

(*v*) Consider a rigid body rotating about a fixed axis. The curl of the linear velocity of a point on the body represents twice its angular velocity.

6.14 SOME USEFUL RELATIONS INVOLVING GRADIENT, DIVERGENCE AND CURL

(*i*) $\vec{\nabla}(\phi + \psi) = \vec{\nabla}\phi + \vec{\nabla}\psi$

Here, ϕ and ψ are differentiable scalar functions of (x, y, z).

(*ii*) $\vec{\nabla}(\phi\psi) = \phi\vec{\nabla}\psi + \psi\vec{\nabla}\phi$

(*iii*) $\vec{\nabla} \cdot (\vec{A} + \vec{B}) = \vec{\nabla} \cdot \vec{A} + \vec{\nabla} \cdot \vec{B}$

Here $\vec{A}$ and $\vec{B}$ are differentiable vector functions.

(*iv*) $\vec{\nabla} \cdot (\phi\vec{A}) = \phi\vec{\nabla} \cdot \vec{A} + \vec{\nabla}\phi \cdot \vec{A}$

(*v*) $\vec{\nabla} \times (\vec{A} + \vec{B}) = \vec{\nabla} \times \vec{A} + \vec{\nabla} \times \vec{B}$

(*vi*) $\vec{\nabla} \times (\phi\vec{A}) = \phi\vec{\nabla} \times \vec{A} + \vec{\nabla}\phi \times \vec{A}$

(*vii*) $\vec{\nabla} \cdot (\vec{\nabla} \times \vec{A}) = 0$

The divergence of a curl of a vector is always zero.

(*viii*) $\vec{\nabla} \times \vec{\nabla} V = \vec{0}$

The curl of grad. V is always zero.

6.15 INTEGRATION OF VECTORS

The integration of vector field functions may be carried out in one dimension, in two dimensions or in three dimensions. The corresponding integrals are respectively named as line integral, surface integral and volume integral.

The symbolic representations of (*i*) open line integral (*ii*) closed line integral (*iii*) open surface integral (*iv*) closed surface integral and (*v*) volume integral are as under.

(*i*) $\int$ (*ii*) $\oint$ (*iii*) $\iint$ (*iv*) $\oiint$ (*v*) $\iiint$

The **line integral** of a vector $\vec{A}$ along the curve PQ is given by :

$$\int_P^Q \vec{A} \cdot \vec{dl} = \underset{dl_i \to 0}{\text{Lt.}} \quad \vec{A}_i \cdot \vec{dl}_i$$

If the line integral of a vector field around a closed path is zero, then the field is conservative. For a conservative field, $\oint \vec{A} \cdot \vec{dl} = 0$. The circle on the integral signifies that the path is closed.

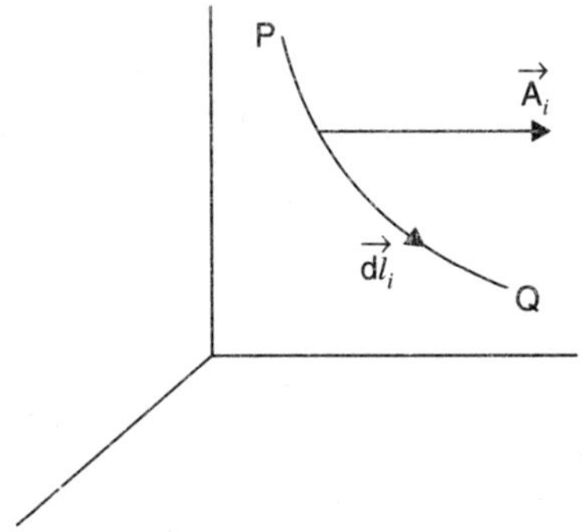

Fig. 6.3

If $\vec{A}$ represents force, then the line integral represents the work done.

If $\vec{A}$ represents electric field, then the line integral represents the work done.

If $\vec{A}$ represents gravitational field, then the line integral represents gravitational potential energy difference.

Let us now discuss **surface integral.**

Consider a surface S in a vector field $\vec{A} = \vec{A}(x, y, z)$. Consider an area element dS upon this surface. Let $\hat{n}$ be a unit vector along the outward drawn normal to the surface element dS.

Component of $\vec{A}$ normal to dS is $\vec{A} \cdot \hat{n} = A \cos \theta$, where θ is the angle between the directions of $\hat{n}$ and $\vec{A}$. The scalar product $\vec{A} \cdot \hat{n}\ dS$ is called outward flux of $\vec{A}$ through the area element dS.

Integration of $\vec{A} \cdot \hat{n}\ dS$ taken over the entire surface S is defined as the total outward flux or surface integral of $\vec{A}$ through whole surface S.

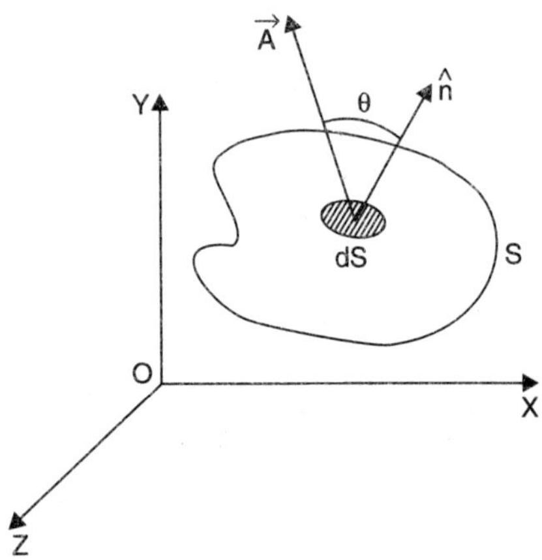

Fig. 6.4

$$\iint_S \vec{A} \cdot \hat{n}\ dS = \iint_S A \cos \theta\ dS = \iint_S A_x\ dS_x + A_y\ dS_y + A_z\ dS_z$$

If the integration is over a closed surface S, then it is denoted by $\oiint_S \vec{A} \cdot \hat{n}\ dS$.

or $$\oiint_S \vec{A} \cdot d\vec{S}.$$

Let us now consider **volume integral.**

Consider a closed surface enclosing volume V in a vector field $\vec{A}$. Let $dV = dx\,dy\,dz$ represent a small volume element inside the closed surface.

Then, $\iiint_V \vec{A}\,dV$ represents the volume integral or space integral of the vector field $\vec{A}$ for entire volume V.

In Cartesian components,

$$\iiint_V \vec{A}\,dV = \int_x \int_y \int_z (A_x\hat{i} + A_y\hat{j} + A_z\hat{k})\,dx\,dy\,dz$$

6.16 GAUSS DIVERGENCE THEOREM

This theorem is used to transform a volume integral into surface integral.

Statement. The surface integral of a vector field $\vec{A}$ over the closed surface S is equal to the volume integral of the divergence of the vector field $\vec{A}$ over the volume enclosed by the surface. Mathematically, it can be expressed as

$$\oiint_S \vec{A} \cdot d\vec{S} = \iiint_V \vec{\nabla} \cdot \vec{A}\,dV.$$

Proof. Consider a closed surface S of arbitrary shape in a vector field $\vec{A}$. Suppose the surface S encloses a volume V.

Consider a surface element dS on surface S. Let $\hat{n}$ be a unit vector along the outward drawn normal upon dS. Flux of $\vec{A}$ through the surface element $dS = (A\cos\theta)\,dS = (\vec{A}\cdot\hat{n})\,dS = \vec{A}\cdot d\vec{S}$ where $d\vec{S}$ is the area vector along $\hat{n}$.

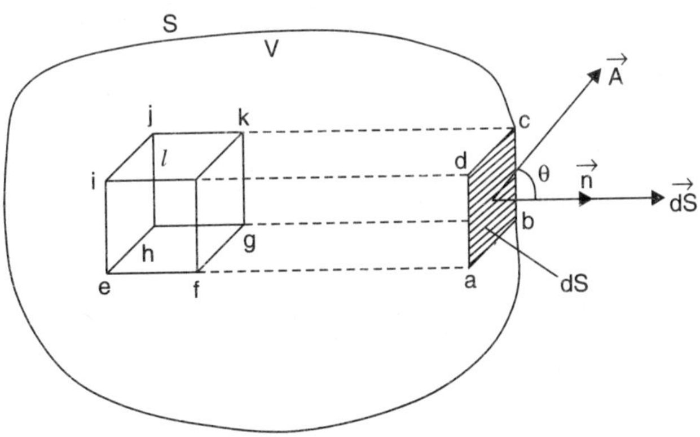

Fig. 6.5

Total flux through the closed surface $S = \oiint_S \vec{A} \cdot d\vec{S}$ ...(i)

Let *efghijkl* be a small cubical volume element of volume dV within the surface S.

The amount of flux diverging per unit volume is represented by div. $\vec{A}$. The amount of flux diverging from the volume element dV is div. $\vec{A}\, dV$. The total flux diverging through the entire volume V enclosed by the entire surface S = $\iiint_V \text{div.} \vec{A}\, dV$. ...(*ii*)

Since the total flux through the entire surface S must be equal to the total flux diverging from volume V enclosed by the surface S.

Equating (*i*) and (*ii*), we get

$$\oiint_S \vec{A} \cdot d\vec{S} = \iiint_V \vec{\nabla} \cdot \vec{A}\, dV.$$

6.17 STOKES' THEOREM

It is an important theorem which relates a line integral to a surface integral over any surface of which the line integral path is a boundary.

Statement. The line integral of a vector field $\vec{A}$ around any closed curve C is equal to the surface integral of the curl of $\vec{A}$ over an open surface S bounded by curve C.

Mathematically, it is expressed as :

$$\oint_C \vec{A} \cdot d\vec{l} = \iint_S (\vec{\nabla} \times \vec{A}) \cdot d\vec{S}$$

Proof. Let us consider an open surface S with boundary C. This surface is in a differentiable vector field $\vec{A}$.

The line integral of $\vec{A}$ around a closed curve C forming the periphery of the surface is

$$\oint_C \vec{A} \cdot d\vec{l} \qquad \text{...(1)}$$

Figure shows the surface area divided into a large number of infinitesimally small rectangular elements. Let us consider one such element of area $d\vec{S}$. If $\hat{n}$ is a unit vector along the outward drawn normal upon dS, then

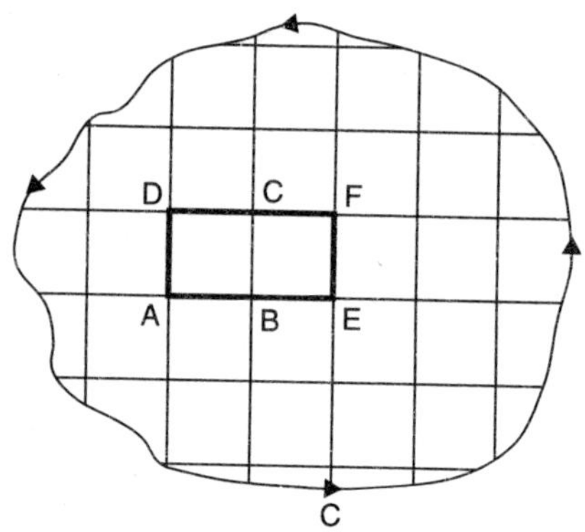

Fig. 6.6

$$d\vec{S} = dS\, \hat{n}$$

The curl of a vector field $\vec{A}$ at any point in the field is the maximum line integral of the vector per unit area along the boundary of an infinitesimally small area around that point. It is in a direction normal to the surface.

The line integral of $\vec{A}$ along the boundary of the infinitesimal rectangular element of area dS is :

$$(\text{curl}\ \vec{A}) \cdot d\vec{S}$$

The total line integral of $\vec{A}$ around the closed curve C forming the periphery of the surface S is :

$$\iint_S (\text{curl}\ \vec{A}) \cdot d\vec{S}$$

The sides of every rectangular element inside the curve are common for two continuous elements. The line integrals along the common sides of the continuous elements mutually cancel each other. This is because they are traversed in opposite directions. Clearly, only the sides of the elements which lie on the periphery of the surface contribute to the line integral. Thus, the surface integral $\iint_S (\text{curl}\ \vec{A}) \cdot d\vec{S}$ represents the line integral of $\vec{A}$ along the boundary C of the surface. It is equal to $\oint_C \vec{A} \cdot \vec{dl}$.

$$\iint_S (\text{curl}\ \vec{A}) \cdot d\vec{S} = \oint_C \vec{A} \cdot \vec{dl}$$

$$\iint_S (\vec{\nabla} \times \vec{A}) \cdot d\vec{S} = \oint_C \vec{A} \cdot \vec{dl}.$$

6.18 GAUSS'S LAW IN ELECTROSTATICS

$$\oint_S \vec{E} \cdot d\vec{S} = \frac{1}{\varepsilon_0} q$$

But $q = \int_V \rho\, dV$, where ρ is the volume charge density.

$$\therefore \qquad \oint_S \vec{E} \cdot d\vec{S} = \frac{1}{\varepsilon_0} \int_V \rho\, dV$$

or

$$\oint_S \varepsilon_0\, \vec{E} \cdot d\vec{S} = \int_V \rho\, dV$$

But Displacement current, $\vec{D} = \varepsilon_0 \vec{E}$

$$\therefore \qquad \oint_S \vec{D} \cdot d\vec{S} = \int_V \rho\, dV$$

Using Gauss's divergence theorem,

$$\int_V (\text{div}\,.\,\vec{D}) dV = \int_V \rho\, dV$$

or

$$\boxed{\text{div}.\ \vec{D} = \rho}$$

6.19 GAUSS'S LAW IN MAGNETOSTATICS

We know that the magnetic monopoles do not exist.

So, the magnetic flux density $\vec{B}$ is given by

$$\oint_S \vec{B}.d\vec{S} = 0$$

Using Gauss's divergence theorem,

$$\int_V (\text{div}.\vec{B})dV = 0 \qquad \text{or} \qquad \boxed{\text{div}.\ \vec{B} = 0}$$

6.20 FARADAY'S LAW OF ELECTROMAGNETIC INDUCTION

$$\vec{\nabla} \times \vec{E} = -\frac{\partial \vec{B}}{\partial t}$$

or

$$\text{curl}\ \vec{E} = -\frac{\partial \vec{B}}{\partial t}$$

6.21 AMPERE'S LAW

If $\vec{H}$ is the magnetic field intensity and $\vec{J}$ is the current density, then

$$\oint_C \vec{H}.d\vec{l} = \int_S \vec{J}.\ d\vec{S}$$

Using Stoke's theorem,

$$\int_S \text{curl}\ \vec{H}.d\vec{S} = \int_S \vec{J}.d\vec{S} \qquad \text{or} \qquad \text{curl}\ \vec{H} = \vec{J}.$$

6.22 THE CONTINUITY EQUATION

It is well known that total charge is always constant. It neither created nor lost, it is always conserved. If the total charge in some volume changes then some charge must be flowing out or in through the surface S of volume V. If charge flows out of a volume V then rate of flow of charge is the current which is equal to the rate at which charge decreases inside the volume. If $\vec{J}$ is the surface current density, then current flows through the surface S is

$$I = \int_S \vec{J}.d\vec{S} = -\frac{dQ_{in}}{dt}$$

Where Q_{in} is the charge inside the volume, but

$$Q_{in} = \int_V \rho\, dV$$

ρ is volume charge density, therefore

$$I = -\frac{d}{dt}\int_V \rho\, dV$$

$$= -\int_V \frac{d\rho}{dt} dV$$

using divergence theorem $\int \vec{J}.d\vec{S} = \int (\vec{\nabla}.\vec{J})dV$

$$I = \int_V \left(\vec{\nabla}.\vec{J}\right) dV = -\int_V \frac{d\rho}{dt} dV$$

Removing volume integral on both sides, we get

$$\vec{\nabla}.\vec{J} = -\frac{d\rho}{dt}$$

$$\boxed{\vec{\nabla}.\vec{J} + \frac{d\rho}{dt} = 0}$$

This is an important equation regarding charge flow called continuity equation and describe the fact that charge is conserved.

In steady state case when time derivative term is zero $\frac{d\rho}{dt} = 0$ then

$$\vec{\nabla}.\vec{J} = 0$$

or

$$\oint \vec{J}.\vec{ds} = 0$$

i.e., the net flux of current through any close surface is zero.

6.23 ELECTROMAGNETIC BEFORE MAXWELL

There are four basic laws of electromagnetic theory which relates electric and magnetic field with charge motion. These are

(*i*) Gauss's law for electrostatics

$$\oint \vec{E}.\vec{ds} = \frac{q}{\epsilon_0} \text{ or div } \vec{E} = \frac{\rho}{\epsilon_0}$$

(*ii*) Gauss's law for magnetostatics

$$\oint \vec{B}.\vec{ds} = 0 \text{ or div } \vec{B} = 0$$

(*iii*) Faraday's law of induction

$$\oint \vec{E}.\vec{dl} = -\frac{d\phi_B}{dt} \text{ or curl } \vec{E} = -\frac{d\vec{B}}{dt}$$

(*iv*) Ampere's law

$$\oint \vec{B}.\vec{dl} = \mu_0 I \text{ or curl } \vec{B} = \mu_0 \vec{J}$$

These equations represent the state of electromagnetic theory long ago, when Maxwell started his work. The in-consistency in the (*iv*) equation of Ampere's law was observed which will be explained in following section.

6.24 NEED FOR MODIFICATION OF AMPERE'S LAW FOR VARYING CURRENTS

The differential form of Ampere's law is

$$\text{curl } \vec{B} = \vec{\nabla} \times \vec{B} = \mu_0 \vec{J}$$

Taking divergence of both the sides, we get

$$\vec{\nabla} . (\vec{\nabla} \times \vec{B}) = \vec{\nabla} . \mu_0 \vec{J}$$

or

$$\vec{\nabla} . (\vec{\nabla} \times \vec{B}) = \mu_0 \vec{\nabla} . \vec{J}$$

But the divergence of curl of vector is zero.

$$\therefore \qquad \mu_0 \vec{\nabla} . \vec{J} = 0 \quad \text{or} \quad \vec{\nabla} . \vec{J} = 0$$

This is the equation of continuity for steady currents. So, we conclude that the Ampere's law given by equation $\vec{\nabla} \times \vec{B} = \mu_0 \vec{J}$ is valid only for steady currents.

Figure shows a charged parallel plate capacitor connected to a resistance R. When the capacitor discharges, a varying current flows through the circuit.

Let us assume that Ampere's law holds good for varying currents.

$$\vec{\nabla} \times \vec{B} = \mu_0 \vec{J} \qquad ...(1)$$

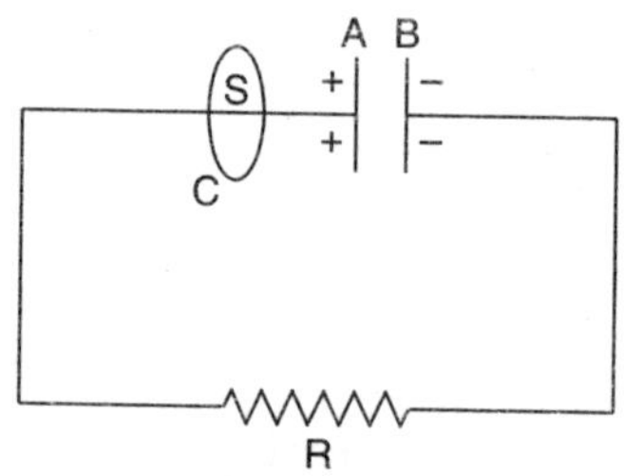

Fig. 6.7

Now two cases arise :

Case (*i*) Consider a plane surface S enclosing the wire. Let C represent its boundary. Taking the dot product of both sides of equation (1) with $d\vec{S}$, we get

$$\vec{\nabla} \times \vec{B} . d\vec{S} = \mu_0 \vec{J} . d\vec{S}$$

Integrating over the surface S, we get

$$\iint_S \vec{\nabla} \times \vec{B} . d\vec{S} = \mu_0 \iint_S \vec{J} . d\vec{S}$$

Since the current I in the circuit passes through surface S, $\therefore \iint_S \vec{J} . d\vec{S} = I$

$$\therefore \qquad \iint_S \vec{\nabla} \times \vec{B} . d\vec{S} = \mu_0 I$$

Using stokes' theorem, we get

$$\oint_C \vec{B} . d\vec{r} = \mu_0 I \qquad ...(2)$$

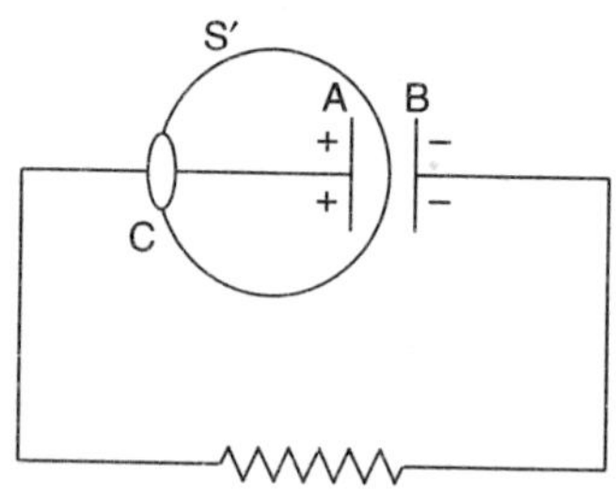

Fig. 6.8

Case (*ii*) Let us now consider another hemispherical surface S′ such that its boundary C is identical with that of the surface S. A part of this hemispherical surface passes through the space between the plates of the capacitor.

$$\text{Now,} \qquad \iint_{S'} \vec{\nabla} \times \vec{B} . d\vec{S} = \mu_0 \iint_{S'} \vec{J} . d\vec{S}$$

No current passes through S′. This is because the surface S′ does not enclose the wire.

$$\therefore \qquad \iint_{S'} \vec{J} . d\vec{S} = 0$$

$$\therefore \qquad \iint_{S'} \vec{\nabla} \times \vec{B} . d\vec{S} = 0$$

Applying Stokes' theorem, we get

$$\oint_C \vec{B} . \vec{dr} = 0 \qquad ...(3)$$

A comparison of equations (2) and (3) shows that the line integral of magnetic field around the same path C give different values. So, we conclude that the equation $\vec{\nabla} \times \vec{B} = \mu_0 \vec{J}$ is not valid for varying currents. Thus, Ampere's law needs modification for varying currents.

6.25 MODIFICATION OF AMPERE'S LAW

Maxwell modified Ampere's law by introducing the cencept of displacement current as explained below.

The differential form of Gauss's law is :

$$\vec{\nabla} . \vec{E} = \frac{\rho}{\varepsilon_0} \quad \text{or} \quad \rho = \varepsilon_0 \vec{\nabla} . \vec{E}$$

Differentiating both sides with respect to t partially, we get

$$\frac{\partial \rho}{\partial t} = \frac{\partial}{\partial t}(\varepsilon_0 \vec{\nabla} . \vec{E}) \quad \text{or} \quad \frac{\partial \rho}{\partial t} = \varepsilon_0 \frac{\partial}{\partial t}(\vec{\nabla} . \vec{E})$$

or

$$\frac{\partial \rho}{\partial t} = \varepsilon_0 \vec{\nabla} . \frac{\partial \vec{E}}{\partial t}$$

Adding $\vec{\nabla} . \vec{J}$ to both sides, we get

$$\vec{\nabla} . \vec{J} + \frac{\partial \rho}{\partial t} = \vec{\nabla} . \vec{J} + \varepsilon_0 \vec{\nabla} . \frac{\partial \vec{E}}{\partial t}$$

The equation of continuity for varying currents is :

$$\vec{\nabla} . \vec{J} + \frac{\partial \rho}{\partial t} = 0$$

$$\therefore \qquad \vec{\nabla} . \vec{J} + \varepsilon_0 \vec{\nabla} . \frac{\partial \vec{E}}{\partial t} = 0 \quad \text{or} \quad \vec{\nabla} . \left(\vec{J} + \varepsilon_0 \frac{\partial \vec{E}}{\partial t} \right) = 0$$

This is another form of the equation of continuity for varying currents.

Maxwell, therefore, proposed that for varying currents, $\vec{J}$ in Ampere law be replaced by

$$\vec{J} + \varepsilon_0 \frac{\partial \vec{E}}{\partial t}$$

So, Ampere's law becomes

$$\text{curl } \vec{B} = \mu_0 \left(\vec{J} + \varepsilon_0 \frac{\partial \vec{E}}{\partial t} \right) \quad \text{or} \quad \vec{\nabla} \times \vec{B} = \mu_0 \left(\vec{J} + \varepsilon_0 \frac{\partial \vec{E}}{\partial t} \right)$$

This is the modified form of Ampere's law. This is applicable to both the varying as well as steady currents.

The term $\varepsilon_0 \frac{\partial \vec{E}}{\partial t}$ has the dimensions of current density. It is vacuum displacement current density. It is denoted by $\vec{J}_D$.

For steady currents, the term $\varepsilon_0 \frac{\partial \vec{E}}{\partial t}$ becomes zero.

In empty space where there are no currents,

$$\vec{J} = 0$$

$$\therefore \quad \vec{\nabla} \times \vec{B} = \mu_0 \varepsilon_0 \frac{\partial \vec{E}}{\partial t}$$

The term $\frac{\partial \vec{E}}{\partial t}$ is the rate at which electric field is changing between the plates of the capacitor. The equation $\vec{\nabla} \times \vec{B} = \mu_0 \varepsilon_0 \frac{\partial \vec{E}}{\partial t}$ that a varying electric field between the plates of capacitor gives rise to a magnetic field. So, we conclude that the varying electric field is another source of magnetic field.

6.26 DISPLACEMENT CURRENT

Term $\varepsilon_0 \frac{\partial \vec{E}}{\partial t}$ is called displacement current density in vacuum. This name was given by Maxwell as the concept of displacement current was first introduced by him. Maxwell postulated that not only the current in conductor produces magnetic field, but a change in electric field in vacuum or in dielectric also produces magnetic field. So changing electric field is equivalent to current and exist as long as the field is changing. It produces same effect of magnetic field as due to conduction current and also equal in magnitude to conduction current. This can be shown with following example.

Consider a parallel plate capacitor of plate area A and separation d. Due to applied potential a charge q is lying on each plate which produces an electric field E between the plate as given (fig. 6.9)

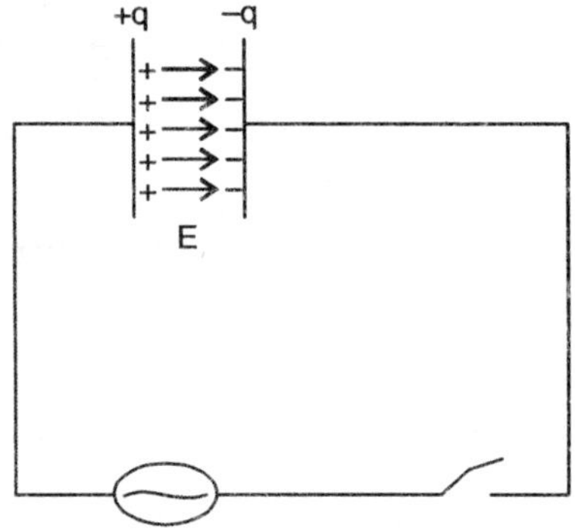

Fig. 6.9

$$E = \frac{q}{\varepsilon_0 A}$$

differentiating the above equation with time

$$\frac{dE}{dt} = \frac{1}{\varepsilon_0 A}\frac{dq}{dt} = \frac{1}{\varepsilon_0 A} I$$

or

$$I = \varepsilon_0 A \frac{dE}{dt}$$

but the displacement current $I_D = \varepsilon_0 \frac{d\phi_E}{dt}$ where $\phi_E = EA$

$$I_D = \varepsilon_0 A \frac{dE}{dt} = I$$

so

$$\boxed{I_D = I}$$

> $\varepsilon_0 \frac{\partial \vec{E}}{\partial t}$ *acts like a current density in the sense that it is a source of magnetic field.*

The displacement current in between the plates of a capacitor is equal to the conduction current flow through the wire due to actual flow of charge. This displacement current make the current continuous across the discontinuity in a conduction current. It is negligible in comparison to conduction current in good conductors at frequencies less than 10^{15}Hz.

6.27 MAXWELL'S EQUATIONS IN INTEGRAL FORM

1. Maxwell's first equation is $\vec{\nabla} . \vec{D} = \rho$

Integrating over an arbitrary volume V, we get

$$\int_V \vec{\nabla} . \vec{D}\, dV = \int_V \rho\, dV$$

Using Gauss' divergence theorem, we get

$$\int_S \vec{D} . d\vec{S} = \int_V \rho\, dV \qquad \text{...}(i)$$

Where S is the surface which bounds volume V

Physical significance

$$\int_V \rho\, dV = q$$

Here, q is the net charge contained in volume V. The following statement gives the physical significance of Maxwell's first equation.

"The net outward flux of electric displacement vector through the surface enclosing a volume is equal to the net charge contained within that volume".

2. Maxwell's second equation is :

$$\vec{\nabla} . \vec{B} = 0$$

Integrating over an arbitrary volume V, we get

$$\int_V \vec{\nabla} . \vec{B} = 0$$

Using Gauss's divergence theorem,

$$\int_S \vec{B} . d\vec{S} = 0$$

Here S is the surface which bounds volume V.

Physical significance. The net outward flux of magnetic induction B through any closed surface is equal to zero.

3. Maxwell's third equation is $\vec{\nabla} \times \vec{E} = -\dfrac{\partial \vec{B}}{\partial t}$

Integrating over surface S bounded by curve C, we get

$$\int_S (\vec{\nabla} \times \vec{E}) . d\vec{S} = -\int_S \frac{\partial \vec{B}}{\partial t}\, d\vec{S}$$

Using Stokes theorem,

$$\int_C \vec{E} . d\vec{l} = -\frac{\partial}{\partial t}\int_S \vec{B} . d\vec{S}$$

Physical significance

$\int_C \vec{E} \cdot \vec{dl}$ represents emf. $\int_S \vec{B} \cdot \vec{dS}$ represents magnetic flux.

The electromotive force around a closed path is equal to negative rate of change of magnetic flux linked with the path.

4. Maxwell's fourth equation is $\vec{\nabla} \times \vec{H} = \vec{J} + \frac{\partial \vec{D}}{\partial t}$

Integrating over surface S bounded by curve C, we get

$$\int_S \vec{\nabla} \times \vec{H} \cdot \vec{dS} = \int_S \left(\vec{J} + \frac{\partial \vec{D}}{\partial t} \right) \cdot \vec{dS}$$

Using Stoke's theorem, we get

$$\oint_C \vec{H} \cdot \vec{dl} = \int_S \left(\vec{J} + \frac{\partial \vec{D}}{\partial t} \right) \cdot \vec{dS}$$

Physical significance

$\oint_C \vec{H} \cdot \vec{dl}$ represents magnetomotive force around a closed path.

So, the physical significance of Maxwell's fourth equation is as follows.

The magnetomotive force around a closed path is equal to the conduction current plus displacement current through any surface bounded by the path.

6.28 MAXWELL'S DIFFERENTIAL EQUATIONS AND THEIR DERIVATION

Maxwell's first equation

$$\vec{\nabla} \cdot \vec{D} = \rho \quad \text{or} \quad \text{div. } \vec{D} = \rho$$

Since $\vec{D} = \varepsilon \vec{E}$ for a medium, therefore, Maxwell's first equation may be written as follows.

$$\vec{\nabla} \cdot \vec{E} = \frac{\rho}{\varepsilon}$$

Derivation. In electrostatics, Gauss's law is given as

$$\oint_S \vec{E} \cdot \vec{dS} = \frac{1}{\varepsilon_0} q$$

If ρ be the volume charge density, then

$$q = \int_V \rho \, dV$$

$$\therefore \quad \oint_S \vec{E} \cdot \vec{dS} = \frac{1}{\varepsilon_0} \int_V \rho \, dV$$

or

$$\oint_S \varepsilon \vec{E} \cdot \vec{dS} = \int_V \rho \, dV \quad \text{or} \quad \oint_S \vec{D} \cdot \vec{dS} = \int_V \rho \, dV$$

The surface integral can be changed to volume integral by the use of Gauss divergence theorem.

$$\oint_S \vec{D} \cdot d\vec{S} = \int_V (\text{div.} \vec{D})\, dV$$

$$\therefore \quad \int_V (\text{div.} \vec{D})\, dV = \int_V \rho\, dV$$

or $$\int_V (\text{div.} \vec{D} - \rho)\, dV = 0 \quad \text{or} \quad \text{div.} \vec{D} - \rho = 0$$

or $$\text{div.} \vec{D} = \rho \quad \text{or} \quad \vec{\nabla} \cdot \vec{D} = \rho$$

This is Maxwell's first equation.

In free space, the volume charge density ρ is zero. So, Maxwell's first equation in free space is :

$$\vec{\nabla} \cdot \vec{D} = 0$$

Maxwell's second equation

$$\vec{\nabla} \cdot \vec{B} = 0 \quad \text{or} \quad \text{div.} \vec{B} = 0$$

Derivation. Using Gauss's law in magnetism,

$$\oint_S \vec{B} \cdot d\vec{S} = 0$$

It is based on the fact that a megnetic monopole does not exist. So, magnetic induction across any closed surface is always zero.

Using Gauss divergence theorem,

$$\int_V (\text{div.} \vec{B})\, dV = 0$$

or $$\text{div.} \vec{B} = 0 \quad \text{or} \quad \vec{\nabla} \cdot \vec{B} = 0$$

This is Maxwell's second equation.

Maxwell's third equation

$$\text{Curl } \vec{E} = \vec{\nabla} \times \vec{E} = -\frac{\partial \vec{B}}{\partial t}$$

Derivation. The magnetic flux through any closed surface in terms of magnetic flux density $\vec{B}$ is given by

$$\phi_B = \oint_S \vec{B} \cdot d\vec{S} \qquad \ldots(i)$$

Using Faraday's law of electromagnetic induction, the induced emf is given by

$$e = -\frac{\partial \phi_B}{\partial t}$$

Using equation (*i*), $$e = -\frac{\partial}{\partial t}\oint_S \vec{B}\,.\,d\vec{S}$$

or $$e = -\oint_S \frac{\partial \vec{B}}{\partial t}\,.\,d\vec{S}$$

The emf is defined as the work done in carrying a unit charge over a closed cycle.

$$\therefore \quad e = \oint_C \vec{E}\,.\,d\vec{l} \quad \therefore \quad \oint_C \vec{E}\,.\,d\vec{l} = -\oint_S \frac{\partial \vec{B}}{\partial t}\,.\,d\vec{S}$$

Using Stokes' theorem,

$$\oint_S (\text{curl}\,\vec{E})\,.\,d\vec{S} = -\oint_S \frac{\partial \vec{B}}{\partial t}\,.\,d\vec{S}$$

or $$\text{curl}\,\vec{E} + \frac{\partial \vec{B}}{\partial t} = 0 \quad \text{or} \quad \text{curl}\,\vec{E} = -\frac{\partial \vec{B}}{\partial t} \quad \text{or} \quad \vec{\nabla}\times\vec{E} = -\frac{\partial \vec{B}}{\partial t}$$

This is maxwell's third equation.

Maxwell's fourth equation

$$\text{curl}\,\vec{H} = \vec{J} + \frac{\partial \vec{D}}{\partial t} \quad \text{or} \quad \text{curl}\,\vec{B} = \mu_0\left(\vec{J} + \varepsilon_0 \frac{\partial \vec{E}}{\partial t}\right)$$

Derivation. Consider a region in which there is a steady flow of current. Consider a closed path in this region bounded by surface S.

The total current I enclosed by this path is given by

$$I = \oint_S \vec{J}\,.\,d\vec{S}$$

According to Ampere's circuital law,

$$\oint \vec{B}\,.\,d\vec{l} = \mu_0 I \quad \text{or} \quad \oint \vec{B}\,.\,d\vec{l} = \mu_0 \int_S \vec{J}\,.\,d\vec{S}$$

Using Stoke's theorem,

$$\oint_S \text{curl}\,\vec{B}\,.\,d\vec{S} = \mu_0 \oint_S \vec{J}\,.\,d\vec{S}$$

or $$\oint_S (\text{curl}\,\vec{B} - \mu_0\,\vec{J})\,.\,d\vec{S} = 0 \quad \text{or} \quad \text{curl}\,\vec{B} - \mu_0\,\vec{J} = 0$$

or $$\text{curl}\,\vec{B} = \mu_0\,\vec{J} \qquad \ldots(1)$$

For time-dependent charge distribution,

$$\text{div.}\,\vec{J} + \frac{\partial \rho}{\partial t} = 0$$

$$\because \quad \frac{\partial \rho}{\partial t} \neq 0, \quad \therefore \quad \text{div.}\,\vec{J} \neq 0 \qquad \ldots(2)$$

From equation (1), $\vec{J} = \frac{1}{\mu_0}$ curl $\vec{B}$

Taking divergence, div. $\vec{J}$ = div. $\left(\frac{1}{\mu_0} \text{curl } \vec{B}\right)$

or $$\text{div. } \vec{J} = \frac{1}{\mu_0} \text{div.} \left(\text{curl } \vec{B}\right) = 0 \quad ...(3)$$

The contradiction between equations (2) and (3) shows that the equation curl $\vec{B} = \mu_0 \vec{J}$ representing Ampere's law is valid only for steady currents. It is not valid for time-varying fields. So, for time-varying fields, Ampere's law must be modified to include time-dependent term. Maxwell modified Ampere's law by stating that changing electric field is equivalent to a current called displacement current I_d. This current flows only so long as the electric field is changing. On the analogy of* Faraday's law, Maxwell included the term $\mu_0 \varepsilon_0 \frac{\partial \vec{E}}{\partial t}$ in the expression of curl $\vec{B}$.

$$\therefore \quad \text{curl } \vec{B} = \mu_0 \vec{J} + \mu_0 \varepsilon_0 \frac{\partial \vec{E}}{\partial t}$$

or $$\text{curl } \vec{B} = \mu_0 \left(\vec{J} + \varepsilon_0 \frac{\partial \vec{E}}{\partial t}\right) \quad \text{or} \quad \text{curl } \vec{H} = \left(\vec{J} + \frac{\partial \vec{D}}{\partial t}\right)$$

This is Maxwell's fourth equation.

6.29 POYNTING VECTOR

The rate at which energy is transmitted through unit area perpendicular to the direction of propagation of energy is called Poynting vector.

Poynting vector is also called the flux vector. It is represented by $\vec{P}$ or $\vec{S}$.

The cross product of the electric field vector $\vec{E}$ and the magnetic field vector $\vec{H}$ is known as a Poynting vector.

Poynting vector $\quad \vec{S} = \vec{E} \times \vec{H}$

The direction of propagation of electromagnetic wave is in the direction of the Poynting vector *i.e.*, in the direction of $\vec{E} \times \vec{H}$.

The poynting vector measures the **electrical power per unit area.

The SI unit of Poynting vector is W m^{-2}.

---

*According to this law, changing magnetic field produces electric field.

$$^{**}\vec{E} \times \vec{H} = \frac{\text{volt}}{\text{Length}} \times \frac{\text{current}}{\text{Length}} = \frac{\text{Power}}{\text{Area}}$$

6.30 POYNTING THEOREM OR CONSERVATION OF ENERGY IN ELECTROMAGNETIC WAVE

As the electromagnetic wave propagates from one point to another, there is a transfer of electromagnetic energy. As an example, the electromagnetic waves from the sun transport energy. With the help of Maxwell's equations, we can obtain an equation showing the energy propagation in an electromagnetic wave.

In differential form, Maxwell's third and fourth equations are given by

$$\vec{\nabla} \times \vec{E} = -\frac{\partial \vec{B}}{\partial t} \quad \ldots(i)$$

and

$$\vec{\nabla} \times \vec{H} = \vec{J} + \frac{\partial \vec{D}}{\partial t} \quad \ldots(ii)$$

Taking the dot product of $\vec{H}$ with equation (*i*) and that of $\vec{E}$ with equation (*ii*), we have

$$\vec{H} \cdot (\vec{\nabla} \times \vec{E}) = -\vec{H} \cdot \frac{\partial \vec{B}}{\partial t} \quad \ldots(iii)$$

and

$$\vec{E} \cdot (\vec{\nabla} \times \vec{H}) = \vec{E} \cdot \vec{J} + \vec{E} \cdot \frac{\partial \vec{D}}{\partial t} \quad \ldots(iv)$$

Subtracting equation (*iv*) from equation (*iii*), we get

$$\vec{H} \cdot (\vec{\nabla} \times \vec{E}) - \vec{E} \cdot (\vec{\nabla} \times \vec{H}) = -\vec{H} \cdot \frac{\partial \vec{B}}{\partial t} - \vec{E} \cdot \vec{J} - \vec{E} \cdot \frac{\partial \vec{D}}{\partial t}$$

$$= -\left[\vec{H} \cdot \frac{\partial \vec{B}}{\partial t} + \vec{E} \cdot \frac{\partial \vec{D}}{\partial t}\right] - \vec{E} \cdot \vec{J} \quad \ldots(v)$$

But

$$\vec{\nabla} \cdot (\vec{E} \times \vec{H}) = \vec{H} \cdot (\vec{\nabla} \times \vec{E}) - \vec{E} \cdot (\vec{\nabla} \times \vec{H}) \quad \ldots(vi)$$

From equation (*v*) and (*vi*), we get

$$\vec{\nabla} \cdot (\vec{E} \times \vec{H}) = -\left[\vec{H} \cdot \frac{\partial \vec{B}}{\partial t} + \vec{E} \cdot \frac{\partial \vec{D}}{\partial t}\right] - \vec{E} \cdot \vec{J}$$

But

$$\vec{B} = \mu \vec{H} \quad \text{and} \quad \vec{D} = \varepsilon \vec{E}$$

$\therefore$

$$\vec{\nabla} \cdot (\vec{E} \times \vec{H}) = -\left[\vec{H} \cdot \frac{\partial}{\partial t}(\mu \vec{H}) + \vec{E} \cdot \frac{\partial}{\partial t}(\varepsilon \vec{E})\right] - \vec{E} \cdot \vec{J}$$

$$= -\frac{1}{2}\left[\frac{\partial}{\partial t}(\vec{H} \cdot \mu \vec{H}) + \frac{\partial}{\partial t}(\vec{E} \cdot \varepsilon \vec{E})\right] - \vec{E} \cdot \vec{J}$$

$$= -\frac{\partial}{\partial t}\left[\frac{1}{2}(\vec{H} \cdot \vec{B} + \vec{E} \cdot \vec{D})\right] - \vec{E} \cdot \vec{J}$$

Taking the volume integral over a volume V enclosed by surface S, we get

$$\int_V [\vec{\nabla} \cdot (\vec{E} \times \vec{H})]\, dV = -\int_V \frac{\partial}{\partial t}\left\{\frac{1}{2}(\vec{H} \cdot \vec{B} + \vec{E} \cdot \vec{D})\right\} dV - \int_V (\vec{E} \cdot \vec{J})\, dV$$

Using Gauss's divergence theorem,

$$\oint_S (\vec{E} \times \vec{H}) \cdot \vec{dS} = -\frac{\partial}{\partial t} \int_V \frac{1}{2}(\vec{H} \cdot \vec{B} + \vec{E} \cdot \vec{D})\, dV - \int_V (\vec{E} \times \vec{J}) \cdot \vec{dV}$$

or
$$\int_V (\vec{E} \cdot \vec{J})\, dV = -\frac{\partial}{\partial t} \int_V \frac{1}{2}(\vec{H} \cdot \vec{B} + \vec{E} \cdot \vec{D})\, dV - \oint_S (\vec{E} \times \vec{H}) \cdot \vec{dS}$$

or
$$-\int_V (\vec{E} \cdot \vec{J})\, dV = \frac{\partial}{\partial t} \int_V \frac{1}{2}(\vec{H} \cdot \vec{B} + \vec{E} \cdot \vec{D})\, dV + \oint_S (\vec{E} \times \vec{H}) \cdot \vec{dS}$$

Rearranging terms, we get

$$-\int_V (\vec{J} \cdot \vec{E})\, dV = \frac{\partial}{\partial t} \int_V \frac{1}{2}(\vec{E} \cdot \vec{D} + \vec{H} \cdot \vec{B})\, dV + \oint_S (\vec{E} \times \vec{H}) \cdot \vec{dS}$$

or
$$-\int_V (\vec{J} \cdot \vec{E})\, dV - \frac{\partial}{\partial t} \int_V \frac{1}{2}(\vec{E} \cdot \vec{D} + \vec{H} \cdot \vec{B})\, dV = \oint_S (\vec{E} \times \vec{H}) \cdot \vec{dS}$$

The term $-\int_V (\vec{J} \cdot \vec{E})\, dV$ represents the rate of transfer of electromagnetic energy due to the motion of the charge. It is the rate at which electromagnetic energy is lost through Joule heating.

The term $-\frac{\partial}{\partial t} \int_V \frac{1}{2}(\vec{E} \cdot \vec{D} + \vec{H} \cdot \vec{B})\, dV$ represents the rate of decrease of energy stored in volume V due to electric and magnetic fields.

Clearly, $\oint_S (\vec{E} \times \vec{H}) \cdot \vec{dS}$ represents the rate of flow of energy over the surface S enclosing the volume V.

Clearly, $\vec{E} \times \vec{H}$ represents the rate of flow of energy through unit area enclosing the volume V.

$$\vec{S} = \vec{E} \times \vec{H}$$

The direction of $\vec{S}$ is perpendicular to $\vec{E}$ and $\vec{H}$.

6.31 ELECTROMAGNETIC WAVE PROPAGATION IN FREE SPACE

One of the important applications of Maxwell's equations is to derive electromagnetic wave equations in a homogeneous linear medium.

In regions of space where there is no charge or current, Maxwell's equations read

(*i*) $\vec{\nabla} \cdot \vec{E} = 0$ (*ii*) $\vec{\nabla} \cdot \vec{B} = 0$

(*iii*) $\vec{\nabla} \times \vec{E} = -\dfrac{\partial \vec{B}}{\partial t}$ (*iv*) $\vec{\nabla} \times \vec{B} = \mu_0 \varepsilon_0 \dfrac{\partial \vec{E}}{\partial t}$

These equations constitute a set of coupled, first order partial differential equations for $\vec{E}$ and $\vec{B}$. They can be decoupled by applying the curl to equations (*iii*) and (*iv*).

Applying curl to equation (*iii*), we get

$$\vec{\nabla} \times (\vec{\nabla} \times \vec{E}) = \vec{\nabla} \times \left(-\frac{\partial \vec{B}}{\partial t} \right)$$

Using vector product rule, we get

or $$\vec{\nabla}(\vec{\nabla} \cdot \vec{E}) - \nabla^2 \vec{E} = -\frac{\partial}{\partial t}(\vec{\nabla} \times \vec{B})$$

Using equation (*i*), we get

$$-\nabla^2 \vec{E} = -\frac{\partial}{\partial t}(\vec{\nabla} \times \vec{B})$$

or $$\nabla^2 \vec{E} = \frac{\partial}{\partial t}(\vec{\nabla} \times \vec{B})$$

Using equation (*iv*), we get

$$\boxed{\nabla^2 \vec{E} = \mu_0 \varepsilon_0 \frac{\partial^2 \vec{E}}{\partial t^2}} \quad \ldots(v)$$

Now, applying curl to equation (*iv*), we get

$$\vec{\nabla} \times (\vec{\nabla} \times \vec{B}) = \vec{\nabla} \times \left(\mu_0 \varepsilon_0 \frac{\partial \vec{E}}{\partial t} \right)$$

Using vector product rule, we get

$$\vec{\nabla}(\vec{\nabla} \cdot \vec{B}) - \nabla^2 \vec{B} = \mu_0 \varepsilon_0 \frac{\partial}{\partial t}(\vec{\nabla} \times \vec{E})$$

Using equation (*ii*), we get

$$-\nabla^2 \vec{B} = \mu_0 \varepsilon_0 \frac{\partial}{\partial t}(\vec{\nabla} \times \vec{E})$$

Using equation (*iii*), we get

$$-\nabla^2 \vec{B} = -\mu_0 \varepsilon_0 \frac{\partial^2 \vec{B}}{\partial t^2} \quad \text{or} \quad \boxed{\nabla^2 \vec{B} = \mu_0 \varepsilon_0 \frac{\partial^2 \vec{B}}{\partial t^2}} \quad \ldots(vi)$$

We now have separate equations (*v*) and (*vi*) for $\vec{E}$ and $\vec{B}$. These equations are of second order. Clearly, equations (*v*) and (*vi*) represent the relation between the space and time variation of electric field $\vec{E}$ and magnetic field $\vec{B}$ respectively in three dimensional case. These equations are called wave equations for $\vec{E}$ and $\vec{B}$ respectively.

> **An interesting fact**
> *If you decouple coupled equations, you get second order equations.*

In vacuum, each Cartesian component of $\vec{E}$ and $\vec{B}$ satisfies the three-dimensional wave equation,

$$\nabla^2 f = \frac{1}{v^2} \frac{\partial^2 f}{\partial t^2}$$

So, Maxwell's equations imply that empty space supports the propagation of electromagnetic waves travelling at a speed

$$v = \frac{1}{\sqrt{\mu_0 \varepsilon_0}} = 3.00 \times 10^8 \text{ m s}^{-1}$$

This happens to be precisely the velocity c of light.

Discussion

1. Let us dimensionally analyse $\nabla^2 \vec{B} = \mu_0\varepsilon_0 \frac{\partial^2 \vec{E}}{\partial t^2}$

$$\frac{1}{\text{metre}^2} \cdot \frac{\text{weber}}{\text{metre}^2} = \mu_0\varepsilon_0 \frac{1}{\text{second}^2} \cdot \frac{\text{weber}}{\text{metre}^2}$$

or
$$\mu_0\varepsilon_0 = \frac{\text{second}^2}{\text{metre}^2} \quad \text{or} \quad \frac{1}{\sqrt{\mu_0 \varepsilon_0}} = \frac{\text{metre}}{\text{second}}.$$

2. In Maxwell's time, it was astounding to conclude that light is an electromagnetic wave.

3. ε_0 and μ_0 are constants in Coulomb's law and Biot-Savart law respectively. They are measured in experiments involving charged pith balls, batteries and wires—experiments having nothing whatever to do with light. And yet, according to Maxwell's theory, you can calculate *c* from μ_0 and ε_0.

4. Notice the crucial role played by Maxwell's contribution to Ampere's law $\left(\mu_0\varepsilon_0 \frac{\partial \vec{E}}{\partial t}\right)$. Without it, the wave equation would not emerge and there would be no electromagnetic theory of light.

6.32 TRANSVERSE NATURE OF ELECTROMAGNETIC WAVES

The wave equations for $\vec{E}$ and $\vec{H}$ are given by

$$\nabla^2\vec{E} - \frac{1}{c^2}\frac{\partial^2 \vec{E}}{\partial t^2} = 0 \qquad \ldots(i)$$

$$\nabla^2\vec{H} - \frac{1}{c^2}\frac{\partial^2 \vec{H}}{\partial t^2} = 0 \qquad \ldots(ii)$$

The general form of the plane wave solutions of these equations are :

$$\vec{E}(\vec{r},t) = \vec{E}_0\, e^{i\vec{k}\cdot\vec{r} - i\omega t} \qquad \ldots(iii)$$

and
$$\vec{H}(\vec{r},t) = \vec{H}_0\, e^{i\vec{k}\cdot\vec{r} - i\omega t} \qquad \ldots(iv)$$

Here $\vec{E}_0$ and $\vec{H}_0$ represent the amplitudes of electric and magnetic fields respectively. These are constant in space and time. The wave propagation vector is given by

$$\vec{k} = \frac{2\pi}{\lambda}\hat{n}$$

Here $\hat{n}$ is a unit vector in the direction of propagation of electromagnetic wave.

Now,
$$\vec{\nabla}\cdot\vec{E} = \left(\frac{\partial}{\partial x}\hat{i} + \frac{\partial}{\partial y}\hat{j} + \frac{\partial}{\partial z}\hat{k}\right)\cdot\left\{E_0 e^{i(\vec{k}\cdot\vec{r} - \omega t)}\right\}$$

Now, $$\vec{k}\cdot\vec{r} = (k_x\hat{i} + k_y\hat{j} + k_z\hat{k})\cdot(x\hat{i} + y\hat{j} + z\hat{k})$$
$$= k_x x + k_y y + k_z z.$$

$$\therefore \quad \vec{\nabla}\cdot\vec{E} = \left(\frac{\partial}{\partial x}\hat{i} + \frac{\partial}{\partial y}\hat{j} + \frac{\partial}{\partial z}\hat{k}\right).[(E_{0x}\hat{i} + E_{0y}\hat{j} + E_{0z}\hat{k})\,e^{i(k_x x + k_y y + k_z z - \omega t)}]$$
$$= E_0 x\, e^{i[(k_x x + k_y y + k_z z) - \omega t]}(ikx) + E_0 y e^{i[(k_x x + k_y y + k_z z) - \omega t]}(ik_y) + E_0 z e^{i[(k_x x + k_y y + k_z z) - \omega t]}(ik_z)$$
$$= i[k_x E_{0x} + k_y E_{0y} + k_z E_{0z}]\, e^{i[(k_n \bar{x} + k_y y + k_z z) - \omega t]}$$
$$= i(\vec{k}\cdot\vec{E}_0)\, e^{i(kr - \omega t)} \quad \text{or} \quad \vec{\nabla}\cdot\vec{E} = i(\vec{k}\cdot\vec{E})$$

For free space, Maxwell's first equation is

$$\vec{\nabla}\cdot\vec{E} = 0$$
$$\therefore \quad \vec{k}\cdot\vec{E} = 0 \qquad \ldots(v)$$

Proceeding in the same manner, we shall get

$$\vec{k}\cdot\vec{H} = 0 \qquad \ldots(vi)$$

We conclude from equations (*v*) and (*vi*) that both the electric vector $\vec{E}$ and magnetic vector $\vec{H}$ in an electromagnetic wave are perpendicular to the direction of the propagation vector. In other words, $\vec{E}$ and $\vec{H}$ are perpendicular to the direction of the propagation of electromagnetic wave.

Conclusion. Electromagnetic waves are transverse in nature.

Now, according to Maxwell's third and fourh equation,

$$\vec{\nabla}\times\vec{E} = -\mu_0\frac{\partial\vec{H}}{\partial t} \qquad \ldots(vii)$$
$$\vec{\nabla}\times\vec{H} = \varepsilon_0\frac{\partial\vec{E}}{\partial t} \qquad \ldots(viii)$$

We know that for any vector $\vec{V} = V_x\hat{i} + V_y\hat{j} + V_z\hat{k}$

$$\text{curl}\ \vec{V} = \hat{i}\left(\frac{\partial}{\partial y}V_x - \frac{\partial}{\partial z}V_z\right) + \hat{j}\left(\frac{\partial}{\partial z}V_x - \frac{\partial}{\partial x}V_z\right) + \hat{k}\left(\frac{\partial}{\partial x}V_y - \frac{\partial}{\partial y}V_x\right)$$

So $$\text{curl}\ \vec{E} = \vec{\Delta}\times\vec{E} = \vec{\Delta}\times\vec{E}_0\, e^{i(\vec{k}\cdot\vec{r} - wt)}$$
$$= \hat{i}\left[\frac{\partial}{\partial y}\left\{E_{0z}e^{i(k_x x + k_y y + k_z z) - iwt}\right\} - \frac{\partial}{\partial z}\left\{E_{0y}\,e^{i(k_x x + k_y y + k_z z) - iwt}\right\}\right]$$
$$+\hat{k}\left[\frac{\partial}{\partial x}\left\{E_{0y}e^{i(k_x x + k_y y + k_z z) - iwt}\right\} - \frac{\partial}{\partial y}\left\{E_{0x}\,e^{i(k_x x + k_y y + k_z z) - iwt}\right\}\right]$$

$$+\hat{j}\left[\frac{\partial}{\partial z}\left\{E_{0x}e^{i(k_x x+k_y y+k_z z)-iwt}\right\}-\frac{\partial}{\partial z}\left\{E_{0z}\,e^{i(k_x x+k_y y+k_z z)-iwt}\right\}\right]$$

$$=\hat{i}\left[i\mathrm{E}_{0z}\,k_y e^{i(\vec{k}.\vec{r}-wt)}-i\mathrm{E}_{0y}\,k_z e^{i(\vec{k}.\vec{r}-wt)}\right]$$

$$+j\left[i\mathrm{E}_{0x}\,k_z e^{i(\vec{k}.\vec{r}-wt)}-i\mathrm{E}_{0z}\,k_x e^{i(\vec{k}.\vec{r}-wt)}\right]$$

$$+\hat{k}\left[i\mathrm{E}_{0y}\,k_x e^{i(\vec{k}.\vec{r}-wt)}-i\mathrm{E}_{0x}\,k_y e^{i(\vec{k}.\vec{r}-wt)}\right]$$

$$=i\left[i\{\mathrm{E}_{0z}\,k_y-\mathrm{E}_{oy}\,k_z\}+j\{\mathrm{E}_{0x}\,k_z-\mathrm{E}_{oz}\,k_x\}\right.$$

$$\left.+\hat{k}\left\{\mathrm{E}_{0y}\,k_x-\mathrm{E}_{ox}\,k_y\right\}\right]e^{i(\vec{k}.\vec{r}-wt)}$$

$$=i\left[\hat{k}\times\vec{\mathrm{E}}_0\right]e^{i(\vec{k}.\vec{r}-wt)}$$

$$=i\left[\hat{k}\times\vec{\mathrm{E}}\right].$$

Using equation (*vii*), we can arrive at the following.

$$\vec{\nabla}\times\vec{\mathrm{E}}=i(\vec{k}\times\vec{\mathrm{E}})=-\mu_0(-i\omega\vec{\mathrm{H}})$$

or
$$\vec{k}\times\vec{\mathrm{E}}=-\mu_0\omega\vec{\mathrm{H}} \qquad \ldots(ix)$$

Similarly using equation (*viii*) we can have

$$\vec{\nabla}\times\vec{\mathrm{H}}=i(\vec{k}\times\vec{\mathrm{H}})=\varepsilon_0(-i\omega\vec{\mathrm{E}})$$

or
$$\vec{k}\times\vec{\mathrm{H}}=-\varepsilon_0\omega\vec{\mathrm{E}} \qquad \ldots(x)$$

It follows from equation (*ix*) that $\vec{\mathrm{H}}$ is perpendicular to both $\vec{k}$ and $\vec{\mathrm{E}}$. Similarly, it follows from equation (*x*) that $\vec{\mathrm{E}}$ is perpendicular to both $\vec{k}$ and $\vec{\mathrm{H}}$.

Conclusion. $\vec{\mathrm{E}}, \vec{\mathrm{H}}$ and $\vec{k}$ are mutually perpendicular to each other. They can be taken along the three rectangular axes as shown in Fig. 6.10.

Now, from equation (*ix*),

$$\vec{\mathrm{H}}=\frac{1}{\mu_0\,\omega}(\vec{k}\times\vec{\mathrm{E}})$$

or
$$\vec{\mathrm{H}}=\frac{1}{\mu_0\,\omega}(k\hat{n}\times\vec{\mathrm{E}})$$

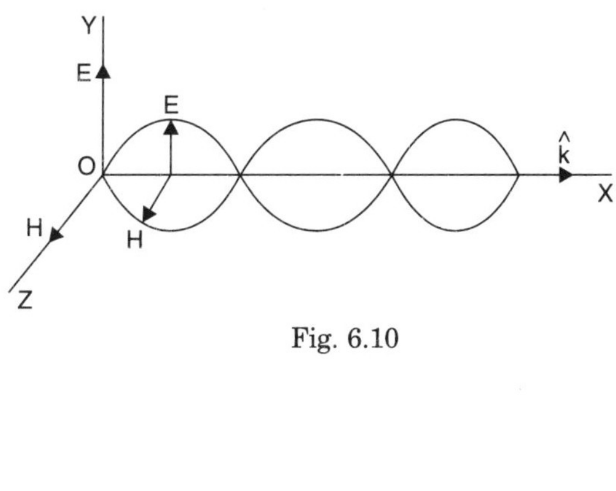

Fig. 6.10

or $$\vec{H} = \frac{k}{\mu_0 \omega}(\hat{n} \times \vec{E}) \quad \text{or} \quad |\vec{H}| = \frac{1}{\mu_0 c}|\vec{E}|$$

Here $$c = \frac{\omega}{k}.$$

If $$Z_0 = \frac{|\vec{E}|}{|\vec{H}|}$$

then $$Z_0 = \frac{E_0}{H_0} = \mu_0 c$$

But speed of electromagnetic waves, c

$$= \frac{1}{\sqrt{\mu_0 \varepsilon_0}}$$

$$\therefore \quad Z_0 = \frac{E_0}{H_0} = \frac{\mu_0}{\sqrt{\mu_0 \varepsilon_0}} \quad \text{or} \quad \frac{E_0}{H_0} = \sqrt{\frac{\mu_0}{\varepsilon_0}}$$

Also, $$\frac{E}{H} = \sqrt{\frac{\mu_0}{\varepsilon_0}}.$$

$$\frac{E}{H} = \sqrt{\frac{\mu_0}{\varepsilon_0}} = \sqrt{\frac{4\pi \times 10^{-7}\ \text{web/amp-mt}}{8.85 \times 10^{-12}\ \text{coulomb}^2/\text{N} - \text{mt}^2}}$$

$$= \mathbf{376.7\ ohms.}$$

This shows that ratio of magnitude of $\vec{E}$ to the magnitude of $\vec{H}$ is a constant and has a dimension of electrical resistance. This constant is known as characteristic impedance or intrinsic impedance of free space. Since ratio of E and H is constant hence in free space both fields are in same phase.

6.33 ELECTROMAGNETIC WAVES IN CONDUCTING MEDIUM

Consider a conducting medium having finite permeability μ, permittivity ε and conductivity σ. If the medium is free from a charge source, then $\rho = 0$. In such a case, the Maxwell's equations for $\vec{E}$ and $\vec{H}$ are :

$$\varepsilon_0\, \vec{\nabla} . \vec{E} = 0 \quad \text{or} \quad \vec{\nabla} . \vec{E} = 0 \qquad ...(i)$$

$$\mu \vec{\nabla} . \vec{H} = 0 \quad \text{or} \quad \vec{\nabla} . \vec{H} = 0 \qquad ...(ii)$$

$$\vec{\nabla} \times \vec{E} = -\mu \frac{\partial \vec{H}}{\partial t} \qquad ...(iii)$$

$$\vec{\nabla} \times \vec{H} = \vec{J} + \varepsilon \frac{\partial \vec{E}}{\partial t}$$

or $$\vec{\nabla} \times \vec{H} = \sigma \vec{E} + \varepsilon \frac{\partial \vec{E}}{\partial t} \qquad \vec{J} = \sigma \vec{E} \qquad ...(iv)$$

Taking curl of equation (*iii*),

$$\vec{\nabla} \times (\vec{\nabla} \times \vec{E}) = \vec{\nabla} \times \left(-\mu \frac{\partial \vec{H}}{\partial t} \right) = -\mu \left(\vec{\nabla} \times \frac{\partial \vec{H}}{\partial t} \right) = -\mu \frac{\partial}{\partial t} (\vec{\nabla} \times \vec{H})$$

Using equation (*iv*), we get

$$\vec{\nabla} \times (\vec{\nabla} \times \vec{E}) = -\mu \frac{\partial}{\partial t} \left(\sigma \vec{E} \times \varepsilon \frac{\partial \vec{E}}{\partial t} \right) = -\mu\sigma \frac{\partial \vec{E}}{\partial t} - \mu\varepsilon \frac{\partial^2 \vec{E}}{\partial t^2}$$

Applying the triple vector identity

$$\vec{A} \times (\vec{B} \times \vec{C}) = \vec{B}(\vec{A} \cdot \vec{C}) - \vec{C}(\vec{A} \cdot \vec{B}),$$

$$\vec{\nabla}(\vec{\nabla} \cdot \vec{E}) - \vec{E}(\vec{\nabla} \cdot \vec{\nabla}) = -\mu\sigma \frac{\partial \vec{E}}{\partial t} - \mu\varepsilon \frac{\partial^2 \vec{E}}{\partial t^2}$$

But $\qquad \vec{\nabla} \cdot \vec{E} = 0$

$$\therefore \qquad -\vec{E}(\vec{\nabla} \cdot \vec{\nabla}) = -\mu\sigma \frac{\partial \vec{E}}{\partial t} - \mu\varepsilon \frac{\partial^2 \vec{E}}{\partial t^2}$$

or

$$\vec{E}(\vec{\nabla} \cdot \vec{\nabla}) = \mu\sigma \frac{\partial \vec{E}}{\partial t} + \mu\varepsilon \frac{\partial^2 \vec{E}}{\partial t^2}$$

or

$$\boxed{\nabla^2 \vec{E} = \mu\sigma \frac{\partial \vec{E}}{\partial t} + \mu\varepsilon \frac{\partial^2 \vec{E}}{\partial t^2}} \qquad \ldots(v)$$

For a non-conducting medium, $\sigma = 0$

$$\therefore \qquad \nabla^2 \vec{E} = \mu\varepsilon \frac{\partial^2 \vec{E}}{\partial t^2} \qquad \ldots(vi)$$

Comparing equations (*v*) and (*vi*), we find that $\mu\sigma \frac{\partial \vec{E}}{\partial t}$ is the dissipative term which allows the current to flow.

Now, taking the curl of equation (*iv*),

$$\vec{\nabla} \times (\vec{\nabla} \times \vec{H}) = \vec{\nabla} \times \left(\sigma \vec{E} + \varepsilon \frac{\partial \vec{E}}{\partial t} \right) = \sigma(\vec{\nabla} \times \vec{E}) + \varepsilon \frac{\partial}{\partial t} (\vec{\nabla} \times \vec{E})$$

Using equation (*iii*), we get

$$\vec{\nabla} \times (\vec{\nabla} \times \vec{H}) = -\mu\sigma \frac{\partial \vec{H}}{\partial t} - \varepsilon\,\mu \frac{\partial^2 \vec{H}}{\partial t^2}$$

Using the triple vector identity

$$\vec{A} \times (\vec{B} \times \vec{C}) = \vec{B}(\vec{A} \cdot \vec{C}) - \vec{C}(\vec{A} \cdot \vec{B}), \text{ we get}$$

$$\vec{\nabla}(\vec{\nabla}\cdot\vec{H}) - \vec{H}(\vec{\nabla}\cdot\vec{\nabla}) = -\mu\sigma\frac{\partial \vec{H}}{\partial t} - \varepsilon\mu\frac{\partial^2 \vec{H}}{\partial t^2}$$

Using equation (*ii*), we get

$$-\vec{H}(\vec{\nabla}\cdot\vec{\nabla}) = -\mu\sigma\frac{\partial \vec{H}}{\partial t} - \varepsilon\mu\frac{\partial^2 \vec{H}}{\partial t^2}$$

or

$$\vec{H}(\vec{\nabla}\cdot\vec{\nabla}) = \mu\sigma\frac{\partial \vec{H}}{\partial t} + \varepsilon\mu\frac{\partial^2 \vec{H}}{\partial t^2}$$

or

$$\boxed{\nabla^2\vec{H} = \mu\sigma\frac{\partial \vec{H}}{\partial t} + \varepsilon\mu\frac{\partial^2 \vec{H}}{\partial t^2}} \quad \ldots(vii)$$

For non-conducting medium, $\sigma = 0$

$$\nabla^2\vec{H} = \varepsilon\mu\frac{\partial^2 \vec{H}}{\partial t^2} \quad \ldots(viii)$$

Comparing equations (*vii*) and (*viii*), we find that $\mu\sigma\frac{\partial \vec{H}}{\partial t}$ is the dissipative term which allows the current to flow.

Electromagnetic wave equations for a charge-free conducting medium having finite values of permeability and permittivity.

For a charge-free ($\rho = 0$) conducting medium ($\sigma \neq 0$), three of Maxwell's equations in component form remain the same but the fourth relation shall include an additional term due to $\vec{J} = \sigma\vec{E}$ where σ represents the conductivity of the medium.

Let us now express $\vec{\nabla}\times\vec{H} = \sigma\vec{E} + \varepsilon\frac{\partial \vec{E}}{\partial t}$ in component-form.

$$\sigma E_x + \varepsilon\frac{\partial E_x}{\partial t} = \frac{\partial H_z}{\partial y} - \frac{\partial H_y}{\partial z} \quad \ldots(ix)$$

$$\sigma E_y + \varepsilon\frac{\partial E_y}{\partial t} = \frac{\partial H_x}{\partial z} - \frac{\partial H_z}{\partial x} \quad \ldots(x)$$

$$\sigma E_z + \varepsilon\frac{\partial E_z}{\partial t} = \frac{\partial H_y}{\partial x} - \frac{\partial H_x}{\partial y} \quad \ldots(xi)$$

Since we are considering only the oscillatory nature of either $\vec{E}$ or $\vec{H}$ therefore a constant E_z or H_z will have no effect on the wave propagation.

Putting $E_z = H_z = 0$ in equations (*ix*) and (*x*), we get

$$\sigma E_x + \varepsilon\frac{\partial E_x}{\partial t} = -\frac{\partial H_y}{\partial z} \quad \ldots(xii)$$

$$\sigma E_y + \varepsilon\frac{\partial E_y}{\partial t} = \frac{\partial H_x}{\partial z} \quad \ldots(xiii)$$

If we consider a plane polarised wave travelling along the $+z$ direction, we can choose the electric field vector $\vec{E}$ to be zero along X-axis or Y-axis. If we take $\vec{E}$ along X-axis, then $E_y = 0$. In this case, the magnetic field vector will be zero along X-axis *i.e.*, $H_x = 0$.

Putting $E_y = H_x = 0$, we find that equation (*xiii*) vanishes.

Taking $\frac{\partial}{\partial t}$ of equation (*xii*), we get

$$\sigma \frac{\partial E_x}{\partial t} + \varepsilon \frac{\partial^2 E_x}{\partial t^2} = -\frac{\partial^2 H_y}{\partial z\, \partial t} \qquad \ldots(xiv)$$

Now,
$$-\mu \frac{\partial \vec{H}}{\partial t} = \vec{\nabla} \times \vec{E}$$

In component form,

$$-\mu \frac{\partial H_x}{\partial t} = \frac{\partial E_z}{\partial y} - \frac{\partial E_y}{\partial z}$$

$$-\mu \frac{\partial H_y}{\partial t} = \frac{\partial E_x}{\partial z} - \frac{\partial E_z}{\partial x}$$

$$-\mu \frac{\partial H_z}{\partial t} = \frac{\partial E_y}{\partial x} - \frac{\partial E_x}{\partial y}$$

If we consider a system of plane waves and assume that the waves propagate in positive z-direction, then the wave properties are constant in the xy plane. Clearly, $\vec{E}$ and $\vec{H}$ are independent of x and y. Their partial derivatives with respect to x and y are zero. Moreover, the z-components of either $\vec{E}$ or $\vec{H}$ do not vary with space or time.

$\therefore$
$$-\mu \frac{\partial H_y}{\partial t} = \frac{\partial E_x}{\partial z} \qquad \ldots(xv)$$

Taking $\frac{\partial}{\partial z}$ of the above equation, we get

$$-\mu \frac{\partial^2 H_y}{\partial z \partial t} = \frac{\partial^2 E_x}{\partial z^2}$$

or
$$-\frac{\partial^2 H_y}{\partial z \partial t} = \frac{1}{\mu} \frac{\partial^2 E_x}{\partial z^2} \qquad \ldots(xvi)$$

Combining equations (*xiv*) and (*xvi*), we get

$$\frac{1}{\mu} \frac{\partial^2 E_x}{\partial z^2} = \sigma \frac{\partial E_x}{\partial t} + \varepsilon \frac{\partial^2 E_x}{\partial t^2}$$

or
$$\boxed{\frac{\partial^2 E_x}{\partial z^2} = \mu\sigma \frac{\partial E_x}{\partial t} + \mu\varepsilon \frac{\partial^2 E_x}{\partial t^2}}$$

This is the wave equation for E_x for a charge-free conducting medium.

Now, taking $\frac{\partial}{\partial t}$ of equation (*xv*), we get

$$-\mu \frac{\partial^2 H_y}{\partial t^2} = \frac{\partial^2 E_x}{\partial z \partial t} \qquad \ldots(xvii)$$

Again, taking $\frac{\partial}{\partial z}$ of equation (*xii*), we get

$$\sigma\frac{\partial E_x}{\partial z}+\varepsilon\frac{\partial^2 E_x}{\partial t\partial z}=-\frac{\partial^2 Hy}{\partial z^2} \quad ...(xviii)$$

combining equations (*xvii*) and (*xviii*), we get

$$\sigma\frac{\partial E_x}{\partial z}-\mu\varepsilon\frac{\partial^2 H_y}{\partial t^2}=-\frac{\partial^2 H_y}{\partial z^2} \quad \text{or} \quad -\mu\varepsilon\frac{\partial^2 H_y}{\partial t^2}=-\frac{\partial^2 H_y}{\partial z^2}-\sigma\frac{\partial E_x}{\partial z}$$

Using equation (*xv*), we get

$$-\mu\varepsilon\frac{\partial^2 H_y}{\partial t^2}=-\frac{\partial^2 H_y}{\partial z^2}+\sigma\mu\frac{\partial H_y}{\partial t}$$

or

$$\boxed{\frac{\partial^2 H_y}{\partial z^2}=\mu\varepsilon\frac{\partial^2 H_y}{\partial t^2}+\mu\sigma\frac{\partial H_y}{\partial t}}$$

This is the wave equation for H_y for a charge-free conducting medium.

6.34 ATTENUATION OF WAVE IN CONDUCTING MEDIUM

The solutions for wave equation for E_x and H_y are

$$E_x(z,t) = E_{0x}\,e^{j(kz-wt)} \text{ and} \quad ...(i)$$

$$H_y(z,t) = H_{0y}\,e^{j(kz-wt)} \quad ...(ii)$$

Substituting these values of E_x as

$$\frac{\partial^2 E_x}{\partial z^2} = -k^2E_x;\ \frac{\partial E_x}{\partial t}=-jwE_x;\ \frac{\partial^2 E_x}{\partial t^2}=-w^2E_x$$

in wave equation for E, then

$$-k^2 = -jw\mu\sigma - w^2\mu\varepsilon$$

$$k^2 = w^2\mu\varepsilon + jw\mu\sigma \quad ...(iii)$$

it is clear that k is a complex quantity, so let it be as

$$k = \alpha + j\beta \quad ...(iv)$$

then equation (*iii*) becomes

$$\alpha^2-\beta^2+2j\alpha\beta = w^2\mu\varepsilon + jw\mu\sigma$$

comparing the real and imaginary components

$$\alpha^2-\beta^2 = w^2\mu\varepsilon \text{ and}$$

$$2\alpha\beta = w\mu\sigma$$

from the above two equations α and β can be determined. Which will be

$$\alpha = w\sqrt{\frac{\varepsilon\mu}{2}}\left[\left(1+\frac{\sigma^2}{\varepsilon^2w^2}\right)^{1/2}+1\right]^{1/2} \quad ...(v)$$

and

$$\beta = w\sqrt{\frac{\varepsilon\mu}{2}}\left[\left(1+\frac{\sigma^2}{\varepsilon^2w^2}\right)^{1/2}-1\right]^{1/2} \quad ...(vi)$$

α is the real part of k which determine wavelength, the propagation speed (v) and refractive index (n) in usual way are

$$\lambda = \frac{2\pi}{\alpha}; \; v = \frac{w}{\alpha}; \; n = \frac{c\alpha}{w}$$

β, the imaginary part of k is an attenuation of wave (decreasing amplitude with increasing z) and can be shown as

$$E_x(z, t) = E_{0x} \, e^{j[z(\alpha + j\beta) - wt]}$$
$$= E_{0x} \, e^{-\beta z} \, e^{j[\alpha z - wt]} \quad \text{...(vii)}$$

$e^{-\beta z}$ is responsible for attenuation of wave since β and z are both positive.

6.35 SKIN DEPTH

Skin depth is an important parameter when electromagnetic waves penetrating in conducting medium. It is the depth in conducting medium in which the amplitude of electromagnetic wave is reduced $\frac{1}{e}$ times its original value.

In equation, $E_x = E_{ox} \, e^{-\beta z} \, e^{j[\alpha z - wt]}$, the attenuation factor is $e^{-\beta z}$, therefore E_x should be $\frac{1}{e}$ times its original value if $\beta z = 1$ or $z = \frac{1}{\beta}$, in this case z becomes skin depth and denoted by d as shown in figure 6.11 $d = \frac{1}{\beta}$.

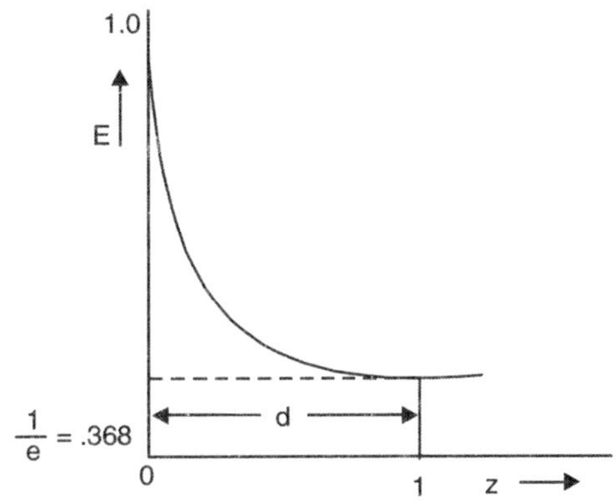

Fig. 6.11

Since $\beta = w\sqrt{\frac{\mu\varepsilon}{2}\left[\left(1+\frac{\sigma^2}{w^2\varepsilon^2}\right)^{1/2} - 1\right]}$

For good conductor ($\sigma >> \varepsilon w$)

$$\beta = w\sqrt{\frac{\mu\varepsilon}{2}\left[\frac{\sigma}{w\varepsilon}\right]}$$
$$= \sqrt{\frac{w\mu\sigma}{2}}$$

or skin depth $d = \frac{1}{\beta} = \sqrt{\frac{2}{w\mu\sigma}} = \sqrt{\frac{2}{2\pi f\mu\sigma}}$

where f is frequency.

Hence skin depth or depth of penetrations is inversely proportional to frequency of wave. For very high frequency wave the depth of penetration is very small in conducting medium.

For poor conductor ($\sigma << \varepsilon w$)

$$\beta = w\sqrt{\frac{\mu\varepsilon}{2}\left[\left(1+\frac{1}{2}\frac{\sigma^2}{w^2\varepsilon^2}\right) - 1\right]}$$
$$= w\sqrt{\frac{\mu\varepsilon}{2} \times \frac{\sigma^2}{2w^2\varepsilon^2}}$$
$$= \frac{\sigma}{2}\sqrt{\frac{\mu}{\varepsilon}}$$

Skin depth for poor conductor is independent of frequency of wave.

SOLVED EXAMPLES BASED ON VECTOR CALCULUS, MAXWELL'S EQUATIONS AND ELECTROMAGNETIC WAVES

■ **Example 6.01.** *Prove that* $div.(\vec{A}+\vec{B}) = (div.\vec{A} + div.\vec{B})$.

Solution. $\text{div.}(\vec{A}+\vec{B}) = \vec{\nabla}.(\vec{A}+\vec{B})$

$$= \frac{\partial}{\partial x}(A_x + B_x) + \frac{\partial}{\partial y}(A_y + B_y) + \frac{\partial}{\partial z}(A_z + B_z)$$

$$= \frac{\partial A_x}{\partial x} + \frac{\partial B_x}{\partial x} + \frac{\partial A_y}{\partial y} + \frac{\partial B_y}{\partial y} + \frac{\partial A_z}{\partial z} + \frac{\partial B_z}{\partial z}$$

$$= \left(\frac{\partial A_x}{\partial x} + \frac{\partial A_y}{\partial y} + \frac{\partial A_z}{\partial z}\right) + \left(\frac{\partial B_x}{\partial x} + \frac{\partial B_y}{\partial y} + \frac{\partial B_z}{\partial z}\right)$$

$$= \text{div.}\vec{A} + \text{div.}\vec{B}.$$

■ **Example 6.02.** *Prove that*

$$div.(\vec{A}\times\vec{B}) = \vec{B}.\ curl\ \vec{A} - \vec{A}\ curl\ \vec{B}.$$

[UPTU, B. Tech. II Sem. 2001]

Solution. $\text{div.}\vec{A} = \vec{\nabla}.\vec{A}$

$$= \frac{\partial A_x}{\partial x} + \frac{\partial A_y}{\partial y} + \frac{\partial A_z}{\partial z}$$

$$\vec{A}\times\vec{B} = \begin{vmatrix} \hat{i} & \hat{j} & \hat{k} \\ A_x & A_y & A_z \\ B_x & B_y & B_z \end{vmatrix}$$

$$= \hat{i}(A_yB_z - A_zB_y) + \hat{j}(A_zB_x - A_xB_z) + \hat{k}(A_xB_y - A_yB_x)\ \vec{\nabla}.(\vec{A}\times\vec{B})$$

$$= \left(\hat{i}\frac{\partial}{\partial x} + \hat{j}\frac{\partial}{\partial y} + \hat{k}\frac{\partial}{\partial z}\right).[\hat{i}(A_yB_z - A_zB_y) + \hat{j}(A_zB_x - A_xB_z) + \hat{k}(A_xB_y - A_yB_x)]$$

$$= \frac{\partial}{\partial x}(A_yB_z - A_zB_y) + \frac{\partial}{\partial y}(A_zB_x - A_xB_z) + \frac{\partial}{\partial z}(A_xB_y - A_yB_x)$$

$$= A_y\frac{\partial B_z}{\partial x} + B_z\frac{\partial A_y}{\partial x} - A_z\frac{\partial B_y}{\partial x} - B_y\frac{\partial A_z}{\partial x} + A_z\frac{\partial B_x}{\partial y} + B_x\frac{\partial A_z}{\partial y} - A_x\frac{\partial B_z}{\partial y} - B_z\frac{\partial A_x}{\partial y} + A_x\frac{\partial B_y}{\partial z} + B_y\frac{\partial A_x}{\partial z} - A_y\frac{\partial B_x}{\partial z} - B_x\frac{\partial A_y}{\partial z}$$

$$= B_x\left(\frac{\partial A_z}{\partial y} - \frac{\partial A_y}{\partial z}\right) + B_y\left(\frac{\partial A_x}{\partial z} - \frac{\partial A_z}{\partial x}\right) + B_z\left(\frac{\partial A_y}{\partial x} - \frac{\partial A_x}{\partial y}\right) - A_x\left(\frac{\partial B_z}{\partial y} - \frac{\partial B_y}{\partial z}\right) - A_y\left(\frac{\partial B_x}{\partial z} - \frac{\partial B_z}{\partial x}\right) - A_z\left(\frac{\partial B_y}{\partial x} - \frac{\partial B_x}{\partial y}\right)$$

$$= (\hat{i}B_x + \hat{j}B_y + \hat{k}B_z).\left[\hat{i}\left(\frac{\partial A_z}{\partial y} - \frac{\partial A_y}{\partial z}\right) + \hat{j}\left(\frac{\partial A_x}{\partial z} - \frac{\partial A_z}{\partial x}\right) + \hat{k}\left(\frac{\partial A_y}{\partial x} - \frac{\partial A_x}{\partial y}\right)\right] - (\hat{i}A_x + \hat{j}A_y + \hat{k}A_z).\left[\hat{i}\left(\frac{\partial B_z}{\partial y} - \frac{\partial B_y}{\partial z}\right) + \hat{j}\left(\frac{\partial B_x}{\partial z} - \frac{\partial B_z}{\partial x}\right) + \hat{k}\left(\frac{\partial B_y}{\partial x} - \frac{\partial B_x}{\partial y}\right)\right]$$

$$= \vec{B}.\begin{vmatrix} \hat{i} & \hat{j} & \hat{k} \\ \frac{\partial}{\partial x} & \frac{\partial}{\partial y} & \frac{\partial}{\partial z} \\ A_x & A_y & A_z \end{vmatrix} - \vec{A}.\begin{vmatrix} \hat{i} & \hat{j} & \hat{k} \\ \frac{\partial}{\partial x} & \frac{\partial}{\partial y} & \frac{\partial}{\partial z} \\ B_x & B_y & B_z \end{vmatrix}$$

$$= \vec{B}.\text{curl}\ \vec{A} - \vec{A}.\text{curl}\ \vec{B}.$$

■ **Example 6.03.** *If* $\vec{r}$ *is the position vector of a point, then show that* $div.\left(\frac{\vec{r}}{r^3}\right) = 0$.

[UPTU, B.Tech. I Sem. 2001]

Solution. $\dfrac{\vec{r}}{r^3} = \dfrac{x\hat{i} + y\hat{j} + z\hat{k}}{(x^2+y^2+z^2)^{3/2}} = \vec{A}$ (say)

$$A_x = \frac{x}{(x^2+y^2+z^2)^{3/2}},$$

$$A_y = \frac{y}{(x^2+y^2+z^2)^{3/2}}$$

and $$A_z = \frac{z}{(x^2+y^2+z^2)^{3/2}}$$

$$\text{div. } \vec{A} = \text{div.}\left(\frac{\vec{r}}{r^3}\right)$$

$$= \frac{\partial}{\partial x}\left[\frac{x}{(x^2+y^2+z^2)^{3/2}}\right] + \frac{\partial}{\partial y}\left[\frac{y}{(x^2+y^2+z^2)^{3/2}}\right] + \frac{\partial}{\partial z}\left[\frac{z}{(x^2+y^2+z^2)^{3/2}}\right]$$

$$= x\frac{\partial}{\partial x}(x^2+y^2+z^2)^{-3/2} + (x^2+y^2+z^2)^{-3/2} + y\frac{\partial}{\partial y}(x^2+y^2+z^2)^{-3/2} + (x^2+y^2+z^2)^{-3/2} + z\frac{\partial}{\partial z}(x^2+y^2+z^2)^{-3/2} + (x^2+y^2+z^2)^{-3/2}$$

$$= x\left(-\frac{3}{2}\right)(x^2+y^2+z^2)^{-5/2}(2x) + (x^2+y^2+z^2)^{-3/2} + y\left(-\frac{3}{2}\right)(x^2+y^2+z^2)^{-5/2}(2y) + (x^2+y^2+z^2)^{-3/2} + z\left(-\frac{3}{2}\right)(x^2+y^2+z^2)^{-5/2}(2z) + (x^2+y^2+z^2)^{-3/2}$$

$$= -3(x^2+y^2+z^2)(x^2+y^2+z^2)^{-5/2} + 3(x^2+y^2+z^2)^{-3/2}$$

$$= -3(x^2+y^2+z^2)^{-3/2} + 3(x^2+y^2+z^2)^{-3/2}$$

$$= 0.$$

Example 6.04. *Prove that grad* $\left(\frac{1}{r}\right) = -\frac{\vec{r}}{r^3}$, *where* $\vec{r}$ *is a position vector.*

[UPTU, B. Tech. I Sem. 2001]

Solution. $\text{grad}\left(\frac{1}{r}\right) = \vec{\nabla}\left[\frac{1}{r}\right]$

$$= \left(\hat{i}\frac{\partial}{\partial x} + \hat{j}\frac{\partial}{\partial y} + \hat{k}\frac{\partial}{\partial z}\right)\left[\frac{1}{(x^2+y^2+z^2)^{1/2}}\right]$$

$$= \hat{i}\frac{\partial}{\partial x}(x^2+y^2+z^2)^{-1/2} + \hat{j}\frac{\partial}{\partial y}(x^2+y^2+z^2)^{-1/2} + \hat{k}\frac{\partial}{\partial z}(x^2+y^2+z^2)^{-1/2} \quad ...(i)$$

Now, $\frac{\partial}{\partial x}(x^2+y^2+z^2)^{-1/2}$

$$= -\frac{1}{2}(x^2+y^2+z^2)^{-1/2-1}(2x)$$

$$= -\frac{x}{(x^2+y^2+z^2)^{3/2}}, \quad ...(ii)$$

$$\frac{\partial}{\partial y}(x^2+y^2+z^2)^{-1/2} = -\frac{y}{(x^2+y^2+z^2)^{3/2}} \quad ...(iii)$$

and $$\frac{\partial}{\partial z}(x^2+y^2+z^2)^{-1/2} = -\frac{z}{(x^2+y^2+z^2)^{3/2}} \quad ...(iv)$$

From equations (*i*), (*ii*), (*iii*) and (*iv*)

$$\text{grad}\left(\frac{1}{r}\right) = -\frac{x\hat{i}}{(x^2+y^2+z^2)^{3/2}} - \frac{y\hat{j}}{(x^2+y^2+z^2)^{3/2}} - \frac{z\hat{k}}{(x^2+y^2+z^2)^{3/2}}$$

$$= -\left[\frac{x\hat{i}+y\hat{j}+z\hat{k}}{(x^2+y^2+z^2)^{3/2}}\right] = -\frac{\vec{r}}{(x^2)^{3/2}} = -\frac{\vec{\mathbf{r}}}{\mathbf{r}^3}.$$

Example 6.05. *Find grad* r^n *where r is the distance of any point (x, y, z) from the origin (0, 0, 0).* **[UPTU, B.Tech. I, Q. Bank, 2001]**

Solution. $\text{grad}\,\phi = \vec{\nabla}\phi$

$$= \left(\hat{i}\frac{\partial}{\partial x} + \hat{j}\frac{\partial}{\partial y} + \hat{k}\frac{\partial}{\partial z}\right)\phi$$

$$\text{grad}\, r^n = \left(\hat{i}\frac{\partial}{\partial x} + \hat{j}\frac{\partial}{\partial y} + \hat{k}\frac{\partial}{\partial z}\right)r^n$$

Now, $\vec{r} = x\hat{i} + y\hat{j} + z\hat{k}$

or $r = (x^2+y^2+z^2)^{1/2}$

or $r^n = (x^2+y^2+z^2)^{n/2}$

$$\therefore \quad \text{grad}\, r^n = \left(\hat{i}\frac{\partial}{\partial x} + \hat{j}\frac{\partial}{\partial y} + \hat{k}\frac{\partial}{\partial z}\right)(x^2+y^2+z^2)^{n/2} \quad ...(i)$$

Now, $\frac{\partial}{\partial x}(x^2+y^2+z^2)^{n/2}$

$$= \frac{n}{2}(x^2+y^2+z^2)^{\frac{n}{2}-1}(2x)$$

$$= nx(x^2+y^2+z^2)^{\frac{n-2}{2}}$$

Similarly, $\frac{\partial}{\partial y}(x^2+y^2+z^2)^{n/2}$

$$= ny(x^2+y^2+z^2)^{\frac{n-2}{2}}$$

and $\frac{\partial}{\partial z}(x^2+y^2+z^2)^{n/2}$

$$= nz(x^2+y^2+z^2)^{\frac{n-2}{2}}$$

$\therefore$ grad r^n

$$= n(x^2+y^2+z^2)^{\frac{n-2}{2}}(x\hat{i}+y\hat{j}+z\hat{k})$$

$$= n(r^2)^{\frac{n-2}{2}}\vec{r} = \mathbf{nr^{n-2}\vec{r}}.$$

Example 6.06. *If $\phi(x, y, z) = 3(x^2y - y^2x)$, calculate grad at the point $(1, -2, 1)$.*

[UPTU, B. Tech. I, Q. Bank 2001]

Solution. grad $\phi = \left(\hat{i}\frac{\partial}{\partial x}+\hat{j}\frac{\partial}{\partial y}+\hat{k}\frac{\partial}{\partial z}\right)\phi$

Here $\phi(x,y,z) = 3x^2y - 3y^2x$

$$\therefore \text{grad } \phi = \hat{i}\frac{\partial}{\partial x}(3x^2y-3y^2x) + \hat{j}\frac{\partial}{\partial y}(3x^2y-3y^2x) + \hat{k}\frac{\partial}{\partial z}(3x^2y-3y^2x)$$

$$= (6xy-3y^2)\hat{i} + (3x^2-6yx)\hat{j} + 0\hat{k}$$

$$= (6xy-3y^2)\hat{i} + (3x^2-6yx)\hat{j}$$

At the point $(1, -2, -1)$, we have

$$\text{grad } \phi = [6(1)(-2)-3(-2)^2]\hat{i} + [3(1)^2-6(-2)(1)]\hat{j}$$

$$= \mathbf{-24\,\hat{i} + 15\hat{j}}.$$

Example 6.07. *Evaluate curl grad ϕ.*

[UPTU, B. Tech. II Sem. 2001]

Solution. Curl grad $\phi = \vec{\nabla}\times\vec{\nabla}\phi$

$$= \left(\hat{i}\frac{\partial}{\partial x}+\hat{j}\frac{\partial}{\partial y}+\hat{k}\frac{\partial}{\partial z}\right) \times \left(\hat{i}\frac{\partial\phi}{\partial x}+\hat{j}\frac{\partial\phi}{\partial y}+\hat{k}\frac{\partial\phi}{\partial z}\right)$$

$$= \begin{vmatrix} \hat{i} & \hat{j} & \hat{k} \\ \frac{\partial}{\partial x} & \frac{\partial}{\partial y} & \frac{\partial}{\partial z} \\ \frac{\partial\phi}{\partial x} & \frac{\partial\phi}{\partial y} & \frac{\partial\phi}{\partial z} \end{vmatrix}$$

$$= \hat{i}\left[\frac{\partial}{\partial y}\left(\frac{\partial\phi}{\partial z}\right)-\frac{\partial}{\partial z}\left(\frac{\partial\phi}{\partial y}\right)\right] + \hat{j}\left[\frac{\partial}{\partial z}\left(\frac{\partial\phi}{\partial x}\right)-\frac{\partial}{\partial x}\left(\frac{\partial\phi}{\partial z}\right)\right] + \hat{k}\left[\frac{\partial}{\partial x}\left(\frac{\partial\phi}{\partial y}\right)-\frac{\partial}{\partial y}\left(\frac{\partial\phi}{\partial x}\right)\right]$$

$$= \hat{i}[0-0] + \hat{j}[0-0] + \hat{k}[0-0]$$

$$= \mathbf{0}.$$

Example 6.08. *Electric potential at any point in the xy plane is given by*

$$\phi = 5x(x^2+y^2)^{1/2} + y(x^2+y^2)^{-1/2}$$

Find the cartesian components of the electric field intensity at that point.

[UPTU, B.Tech. I, Q. Bank 2001]

Solution. $\vec{E} = -\text{grad } \phi$

$$E_x\hat{i} + E_y\hat{j} + E_z\hat{k} = -\hat{i}\frac{\partial\phi}{\partial x} - \hat{j}\frac{\partial\phi}{\partial y} - \hat{k}\frac{\partial\phi}{\partial z}$$

$$E_x = -\frac{\partial\phi}{\partial x},\; E_y = -\frac{\partial\phi}{\partial y},\; E_z = -\frac{\partial\phi}{\partial z}$$

$$E_x = -\frac{\partial}{\partial x}[5x(x^2+y^2)^{1/2} + y(x^2+y^2)^{-1/2}]$$

$$E_x = -\left[(5x)\frac{\partial}{\partial x}(x^2+y^2)^{1/2} + (x^2+y^2)^{1/2}\frac{\partial}{\partial x}(5x) + y\frac{\partial}{\partial x}(x^2+y^2)^{-1/2} + (x^2+y^2)^{-1/2}\frac{\partial y}{\partial x}\right]$$

$$= -\left[(5x)\left(\frac{1}{2}\right)(x^2+y^2)^{\frac{1}{2}-1}\frac{\partial}{\partial x}(x^2+y^2) + 5(x^2+y^2)^{1/2} + y\left(-\frac{1}{2}\right)(x^2+y^2)^{-\frac{1}{2}-1}\frac{\partial}{\partial x}(x^2+y^2) + (x^2+y^2)^{-\frac{1}{2}}\times 0\right]$$

$$= -[5x^2(x^2+y^2)^{-1/2} + 5(x^2+y^2)^{1/2} - xy(x^2+y^2)^{-3/2}]$$

Again, $E_y = -\frac{\partial}{\partial y}[5x(x^2+y^2)^{1/2} + y(x^2+y^2)^{-1/2}]$

$$= -\left[(5x)\left(\frac{1}{2}\right)(x^2+y^2)^{\frac{1}{2}-1}(2y) + y\left(-\frac{1}{2}\right)(x^2+y^2)^{-3/2}(2y) + (x^2+y^2)^{-\frac{1}{2}}\right]$$

$$= -[5xy(x^2+y^2)^{-1/2} - y^2(x^2+y^2)^{3/2} + (x^2+y^2)^{-1/2}].$$

■ Example 6.09. *A region is specified by the potential function given by*

$$\phi = 4x^2 + 3y^2 - 9z^2.$$

Calculate electric field strength at any point (3, 4, 5) in this region.

[UPTU, B. Tech. I, Q. Bank, 2001]

Solution. $\vec{E} = -\text{grad}\,\phi$

$$= -\left(\hat{i}\frac{\partial\phi}{\partial x} + \hat{j}\frac{\partial\phi}{\partial y} + \hat{k}\frac{\partial\phi}{\partial z}\right)$$

Now, $\frac{\partial\phi}{\partial y} = \frac{\partial}{\partial x}[4x^2 + 3y^2 - 9z^2] = 8x$

Similarly, $\frac{\partial\phi}{\partial y} = 6y$ and $\frac{\partial\phi}{\partial z} = -18z$

$$\vec{E} = -(8x\hat{i} + 6y\hat{j} - 18z\hat{k})$$

At point (3, 4, 5),

$$\vec{E} = -(8\times 3\hat{i} + 6\times 4\hat{j} - 18\times 5\hat{k})$$

$$\vec{\mathbf{E}} = -\mathbf{24\hat{i} - 24\,\hat{j} + 90\hat{k}}.$$

■ Example 6.10. *If the earth receives 2 cal min^{-1} cm^{-2} solar energy, what are the amplitudes of electric and magnetic field of radiation ?*

[II Sem 2002, UPTU, B. Tech. I, Q. Bank, 2001]

Solution. Energy flux per unit area per second,

$$|\vec{S}| = |\vec{E}\times\vec{H}|$$

or $S = EH\sin 90°$

or $S = EH$

Energy flux per unit area per second at earth

$$= 2 \text{ cal min}^{-1}\text{ cm}^{-2}$$

$$= \frac{2\times 4.2\times 10^4}{60}\text{ J m}^{-2}\text{ s}^{-1}$$

Clearly, $EH = \frac{2\times 4.2\times 10^4}{60}$

or $EH = 1400$ ...(i)

Again, $\frac{E}{H} = \sqrt{\frac{\mu_0}{\varepsilon_0}}$

$$\frac{E}{H} = \sqrt{\frac{4\pi\times 10^{-7}}{8.85\times 10^{-12}}}$$

$$= 376.8\ \Omega \quad \text{...(ii)}$$

Multiplying (i) and (ii), we get

$$EH\times\frac{E}{H} = 1400\times 376.8$$

or $E^2 = 5.2752\times 10^5$

or $E = 726.3\text{ Vm}^{-1}$

From equation (i), $H = \frac{1400}{E}$

or $H = \frac{1400}{726.3}$

or **H = 1.93 ampere-turn/metre.**

The amplitudes of electric and magnetic fields are $E_0 = \sqrt{2}\,E = 1.414\times 726.3\text{ Vm}^{-1} =$ **1026.988 Vm^{-1}** and $H_0 = \sqrt{2}\,H = 1.414\times 1.93$ ampere-turn/metre = **2.729 ampere-turn/metre.**

■ Example 6.11. *If the magnitude of $\vec{H}$ in a plane wave is 1 ampere / metre, find the magnitude of $\vec{E}$ for plane wave in free space.*

[UPTU, B. Tech. I, Q. Bank, 2001]

Solution. $\frac{H_0}{E_0} = \sqrt{\frac{\varepsilon_0}{\mu_0}}$ or $\frac{E_0}{H_0} = \sqrt{\frac{\mu_0}{\varepsilon_0}}$

or $E_0 = \sqrt{\frac{\mu_0}{\varepsilon_0}}\,H_0$

or $E_0 = \sqrt{\frac{4\pi\times 10^{-7}}{8.85\times 10^{-12}}}\text{ Vm}^{-1}$

$$= \mathbf{376.84\text{ Vm}^{-1}}.$$

Example 6.12. *The relative permittivity of distilled water is 81. Calculate refractive index and velocity of light in it.*

[UPTU, B. Tech. I, Q Bank, 2001]

Solution. $\mu = \sqrt{K} = \sqrt{81} = 9$

$$v = \frac{c}{\mu} = \frac{3\times 10^8}{9}\text{ m s}^{-1} = \mathbf{3.33\times 10^7\text{ m s}^{-1}}.$$

Example 6.13. *An alternate potential is applied to a parallel plate capacitor. Show that conduction current is the same as displacement current.*

Solution. Let an alternate potential across a capacitor C is

$$V = V_0 \sin \omega t$$

V_0 is amplitude and ω is frequency then conduction current I is

$$\frac{dq}{dt} = I = \frac{C\,dV}{dt} = CV_0\,\omega \cos \omega t \quad ...(i)$$

The capacitance of capacitor $C = \frac{\varepsilon_0 A}{d}$ where A is the area of plates and d is separation. The electric field between the plates will be

$$E = \frac{V}{d} = \frac{V_0 \sin \omega t}{d}$$

The displacement current

$$I_d = A\varepsilon_0 \frac{dE}{dt} = \frac{\varepsilon_0}{d} A V_0 \omega \cos \omega t$$

$$I_d = C V_0\, \omega \cos \omega t \quad ...(ii)$$

which is equal to eq. (*i*) hence

$$I = I_d$$

Example 6.14. *If earth received a solar energy of 1400 watt/m², then calculate the amplitudes of electric and magnetic field vectors.*

Solution. The solar energy is the energy transported by electromagnetic wave which is defined by poynting vector.

$$\vec{S} = \vec{E} \times \vec{H} \text{ its magnitude is}$$

$$|\vec{S}| = |\vec{E} \times \vec{H}| = EH \sin 90° = EH$$

Give that $|\vec{S}| = 1400$ watt/m²

or EH = 1400

$$\frac{E}{H} = \sqrt{\frac{\mu_0}{\varepsilon_0}} = \sqrt{\frac{4\pi\times10^{-7}}{8.86\times10^{-12}}} = 376.72 \text{ ohm}$$

then $\frac{E}{H} \times EH = 376.72 \times 1400$

or $E^2 = 5.274 \times 10^5$, E = 726.2 volt/m.

and $H = \frac{1400}{726.2} = 1.928$ amp-turn/m

The amplitudes of electric and magnetic fields of radiated *e.m.* energy are

$$E_0 = E\sqrt{2} = 726.2\sqrt{2} = 1026.8 \text{ volt/m}$$

and $H_0 = H\sqrt{2} = 1.928\sqrt{2}$

$$= 2.726 \text{ amp-turn/m}$$

Example 6.15. *If a bulb radiate energy 1000 watt uniformally in all directions, then calculate the average value of electric and magnetic fields at distance of 2 meter from the bulb.*

Solution. The magnitude of poynting vector $\vec{S}$ is

$$S = \frac{1000}{4\pi r^2} = \frac{1000}{4\pi(2)^2} \text{ watt/m}^2$$

also $S = EH$ and $\frac{E}{H} = 376.72$

then $EH \times \frac{E}{H} = E^2 = 376.72 \times \frac{1000}{4\pi(2)^2}$

$$E = \frac{\sqrt{376.72}}{16\pi} = 86.59 \text{ volt/m}$$

and $H = \frac{1000}{16\pi\times86.59} = 0.23$ amp-turn/m

Example 6.16. *Calculate the skin depth for the wave of frequency 71.6 MHz in aluminium. The conductivity for aluminium σ is 3.54×10^7 mho/m and permeability $\mu = 4\pi \times 10^{-7}$ N/amp².*

Solution. Skin depth d is defined as

$$d = \sqrt{\frac{2}{w\sigma\mu}}$$

putting the volume

$$= \sqrt{\frac{2}{2\pi\times71.6\times10^6\times3.54\times10^7\times4\pi\times10^{-7}}}$$

$$= 10.0 \times 10^{-6} \text{ m} = 10\ \mu\text{m}.$$

Example 6.17. *Find the attenuation constant for sea water for its following data.*

$\mu = \mu_0 = 4\pi \times 10^{-7}$ N/Amp²

$\varepsilon = 70\,\varepsilon_0$

$\sigma = 5$ mho/m.

Solution. Sea water is a poor conductor for which skin depth is independent of frequency and given as

$$d = \frac{2}{\sigma}\sqrt{\frac{\varepsilon}{\mu}}$$

$$= \frac{2}{5}\sqrt{\frac{70\times8.86\times10^{-12}}{4\pi\times10^{-7}}} = 0.0089 \text{ m}$$

The attenuation constant $\alpha = \frac{1}{d}$

so $\alpha = \frac{1}{0.0089} = 112.36$ N/m.

EXERCISES

1. Prove that grad $r = \hat{n}$ where $\hat{n}$ is a unit vector.
2. A plane electromagnetic wave propagating in the x-direction has a wavelength 7.0 mm. The electric field is in the y-direction and its maximum magnitude is 42 Vm^{-1}. Write suitable equations for the electric and magnetic fields as a function of x and t.

$$\left[\textbf{Ans. } 42 \sin\left[\frac{2\pi}{7}(ct - x)\right] \text{ and } 1.4 \times 10^{-7} \sin\left[\frac{2\pi}{7}(ct - x)\right]\right]$$

3. Assuming that all the energy from a 1000 W lamp is rediated uniformly, calculate the average values of the intensities of electric and magnetic fields of radiation at a distance of 2m from the lamp.

[**Ans.** 86.59 Vm^{-1}, 0.23 ampere-turn/m]

SHORT QUESTIONS

1. Compare Gauss's law and Ampere's law.
2. Which equation shows that isolated magnetic poles do not exist ?
3. Name the work-energy theorem of electrodynamics.
4. What is represented by Poynting vector ?

ANSWERS

1.

	Gauss's Law	*Ampere's Law*
1.	It is used to determine electric field due to stationary symmetric charge distributions.	It is used to evaluate magnetic field due to symmetric steady current distributions.
2.	It involves a surface integral.	It involves a line integral.
3.	For applying this law, we need to choose a Gauss's an surface.	For applying this law, we need to choose a path.

2. The equation $\vec{\nabla} . \vec{B} = 0$ shows that isolated magnetic poles do not exist.
3. Poynting's theorem.
4. It is the energy per unit time per unit area transported by the fields.

MULTIPLE CHOICE QUESTIONS

1. Which of the following represents Gauss's theorem in differential form ?

   (*a*) curl $\vec{E} = 0$ (*b*) div. curl $\vec{E} = 0$ (*c*) div. $\vec{E} = 0$ (*d*) div. $\vec{E} = \frac{\rho}{\varepsilon_0}$.

2. The ratio of the electric field E and the magnetic field H has the dimensions of

   (*a*) inductance (*b*) capacitance (*c*) resistance (*d*) velocity.

3. The direction of propagation of the electromagnetic wave is given by

   (*a*) $\vec{E} \times \vec{B}$ (*b*) $\vec{E} . \vec{B}$ (*c*) $\vec{B}$ (*d*) $\vec{E}$.

4. Displacement current is represented by

(*a*) $\varepsilon_0 \dfrac{d\phi_E}{dt}$ (*b*) $\varepsilon_0 \dfrac{d\phi_B}{dt}$ (*c*) $\mu_0 \dfrac{d\phi_B}{dt}$ (*d*) $\mu_0 \dfrac{d\phi_E}{dt}$.

5. Electromagnetic waves are produced by

(*a*) static charge (*b*) moving charge (*c*) Accelerated charge

(*d*) Oscilating megnetic field.

6. Which relation is true for an electromagnetic wave?

(*a*) $B_0 = \mu_0 \varepsilon_0 c E_0$ (*b*) $B_0 = cE_0$ (*c*) $E_0 = \mu_0 \varepsilon_0 B_0$ (*d*) $c = \sqrt{\dfrac{\mu_0}{\varepsilon_0}}$

7. Poynting vector is defined as

(*a*) $\vec{S} = \dfrac{1}{\mu_0}(\vec{E} \times \vec{B})$ (*b*) $\vec{S} = (\vec{E} \times \vec{B})$ (*c*) $\vec{S} = \mu_0 (\vec{E} \times \vec{B})$ (*d*) $\vec{S} = \dfrac{1}{\varepsilon_0}(\vec{E} \times \vec{B})$

8. The plane electromagnetic wave in free space is represented by the equation

(*a*) $\nabla^2 E - \mu_0 \varepsilon_0 \dfrac{\delta^2 B}{\delta t^2} = 0$ (*b*) $\nabla^2 B - \dfrac{1}{\mu_0 \varepsilon_0} \dfrac{\delta^2 B}{\delta t^2} = 0$

(*c*) $\nabla^2 E - \mu_0 \varepsilon_0 \dfrac{\delta^2 B}{\delta t^2} = 0$ (*d*) $\nabla^2 E - \mu\sigma \dfrac{\delta^2 E}{\delta t} + \mu\varepsilon \dfrac{\delta^2 E}{\delta t^2} = 0$

9. For the electromagnetic wave in conducting medium

(*a*) E and B fields are in phase. (*b*) E and B fields attenuated

(*c*) E field attenuated but not B (*d*) B field attenuated but not E.

10. Skin depth or depth of penetration for good conductor is given as

(*a*) $\sqrt{\dfrac{2}{w\varepsilon\sigma}}$ (*b*) $\sqrt{\dfrac{2}{w\mu\sigma}}$ (*c*) $\sqrt{\dfrac{w\mu\sigma}{2}}$ (*d*) $\dfrac{\sigma}{2}\sqrt{\dfrac{\varepsilon}{\mu}}$.

ANSWERS

1. (*d*)	**2.** (*c*)	**3.** (*a*)	**4.** (*a*)	**5.** (c)
6. (*a*)	**7.** (*a*)	**8.** (*c*)	**9.** (*b*)	**10.** (b).

LONG ANSWER QUESTIONS

1. Show that the Faraday's law of electromagnetic induction can be expressed as, $\vec{\nabla} \times \vec{E} = -\dfrac{\partial \vec{B}}{\partial t}$. Write down its integral form. *(UPTU, B.Tech. II Sem. 2001)*
2. Show that if $\vec{B}$ is a magnetic field and its divergence $\vec{B} = 0$ everywhere, it means that an isolated magnetic pole is not possible. *(UPTU, B.Tech. I, Q. Bank, 2001)*
3. Write the Maxwell's equations. Show that the velocity of plane electromagnetic wave in the free space is given by $c = 1/\sqrt{(\mu_0 \varepsilon_0)}$ *(UPTU, B.Tech. I Sem.* 2001)
4. Deduce four Maxwell's equations is free space. *(UPTU, B.Tech. I, Q. Bank, 2001)*
5. Deduce the equation for the propagation of plane electromagnetic wave in the free space. Show that the electric, megnetic vectors are normal to each other and to the direction of the propagation of the wave. *(UPTU, B.Tech.I, Q. Bank, 2001)*

6. Show that the wave equation for the electric field in free space is given by $\nabla^2 \vec{E} = \mu_0\varepsilon_0 \frac{\partial^2 \vec{E}}{\partial t^2}$.
(UPTU, B.Tech. II Sem 2001)

7. Prove that the velocity of plane electromagnetic wave in the vacuum is given by $c = 1/\sqrt{(\mu_0\varepsilon_0)}$.
(UPTU, B. Tech. I, Q Bank, 2001)

8. Show that the energy stored in a magnetic field per unit volume in free space is $B^2/2\mu_0$.
(UPTU, B. Tech. I, Q. Bank, 2001)

9. Show that the equation for electric field E is given by

$$\nabla^2\vec{E} = \mu_0\varepsilon_0\left(\frac{\partial^2 \vec{E}}{\partial^2 t}\right)$$
(UPTU, B. Tech. I, Q. Bank, 2001)

10. Deduce the Maxwell's equations for free space and prove that the electromagnetic waves and transverse. *(UPTU, B.Tech. I, Q Bank, 2001)*
11. Discuss the propagation of electromagnetic waves in free space.*(UPTU, B. Tech. I, Q. Bank, 2001)*
12. Deduce poynting theorem for the flow of energy in an electromagnetic field.
(UPTU, B.Tech. I, Q. Bank, 2001)
13. Show that the plane electromagnetic waves follow the laws of reflection just like light radiations.
(UPTU, B.Tech. I, Q. Bank, 2001)
14. Write the Maxwell equation in conducting medium and deduce wave equation from it.
15. Explain the attenuation of wave in conducting medium.
16. Define skin depth and show that it is independent for poor conductor ($\sigma << \varepsilon\,\omega$).
17. Write the wave equation in conducting medium and discuss its propagation.
18. Define poynting vector. How it represent the flow of electromagnetic energy density?
19. What do you understand by displacement current? How it resolve the paradox in Ampere's law?
20. Describe continuity equation. How it is used to modify the Ampere's law?
21. Define displacement current and show by an example of a capacitor that this is same as conduction current.

UNIT IV

SUPERCONDUCTIVITY AND SCIENCE AND TECHNOLOGY OF NANOMATERIALS

This unit contains the following chapters:

7

Superconductivity

7.01 TEMPERATURE DEPENDENCE OF ELECTRICAL RESISTIVITY

The electrical resistivity of many metals and alloys suddenly drops to zero when they are cooled to a sufficient low temperature, often a temperature in liquid helium range. It was discovered by a Dutch physicist Heik Kamarlingh Onnes in 1911 first time that electrical resistance of mercury vanishes suddenly at 4.2 K temperature. As this transitions occurs over a very small temperature range of about 0.05 K as shown in Fig. 7.1, the phenomenon given the name superconductivity and the material which exhibit this phenomenon are called superconductors. Thus superconductors are the material which possess infinitive conductivity *i.e.*, perfect electrical conductors under special condition. The conduction electron require no energy to move in conductor and no energy is lost in friction to sustain the current in conductor under this special conditions.

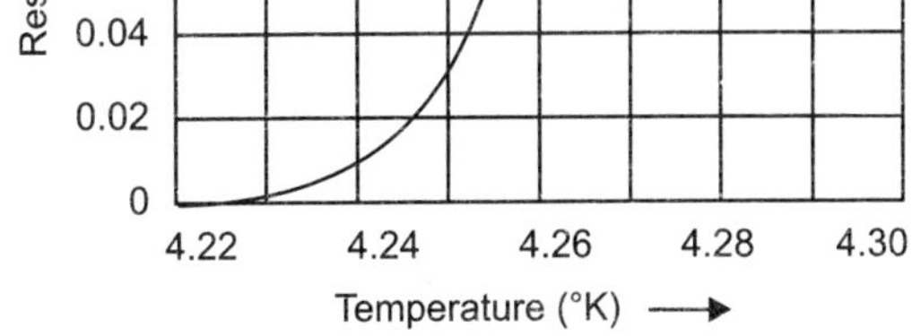

Fig. 7.1. Resistance of mercury as a function of temperature: critical temperature T_C = 4.2°K.

The temperature below which the electrical resistivity of the material become zero in absence of any magnetic field is called the critical (transition) temperature T_C. The critical temperature T_C separates the superconducting state from the normal state as shown in Fig. 7.2. Above T_C the substance is in normal state with finite resistivity. However, below T_C, it is in superconducting state with infinite conductivity. The superconducting state is a distinct phase of matter having characteristic electrical, magnetic, thermodynamic and other physical properties. The most easily observed characteristic of bulk superconductors are the zero electrical resistance and the perfect diamagnetism. Not only pure metals, there are large number of alloys exhibiting this phenomenon of superconductivity. It is possible for a alloy to be a super conductor even if its components are not superconductors. The conductors which are not metals such as semiconducting mixed oxide of barium, lead and bismuth is a superconductor. The conducting polymer, polysulphur nitride (SN) are behave as superconductor at 3K temperature.

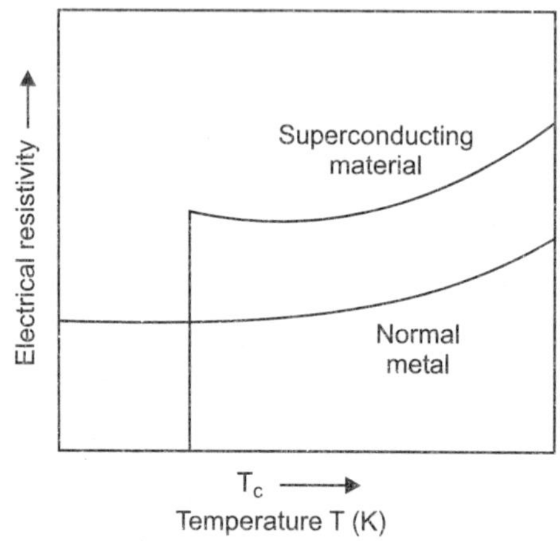

Fig. 7.2. Temperature dependence of the electrical resistivity.

7.02 TEMPERATURE DEPENDENCE OF MAGNETIC FIELD

On application of a sufficient strong magnetic field, the superconducting behaviour of a conductor disappear and it restore its normal conducting state. The critical value of the applied magnetic field necessary to restore the normal conductivity is denoted by H_C (T) is a function of temperature and defined by the following relation,

$$H_C = H_O \left(1 - \frac{T^2}{T_C^2}\right)$$

where H_O is the critical field at 0° K and has a specific value for each material. Higher is the temperature lower is the value of critical field. At critical temperature ($T = T_C$) the critical field become zero.

The phase diagram for normal and superconducting state is shown in Fig. 7.3. Inside the curve, material is in superconducting phase (zero resistivity) and outside curve, the material is in its normal state of conductivity (normal resistivity is restored).

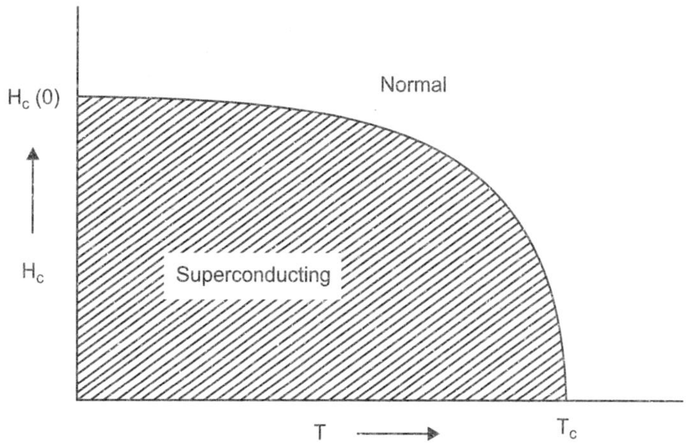

Fig. 7.3. Variation of critical field H_C with temperature.

7.03 EFFECT OF MAGNETIC FIELD : MEISSNER EFFECT

The magnetic properties exhibits by superconductors are as dramatic as their electrical properties. The magnetic properties cannot be accounted for by the assumption that a superconductor is a normal conductor with zero electrical resistivity.

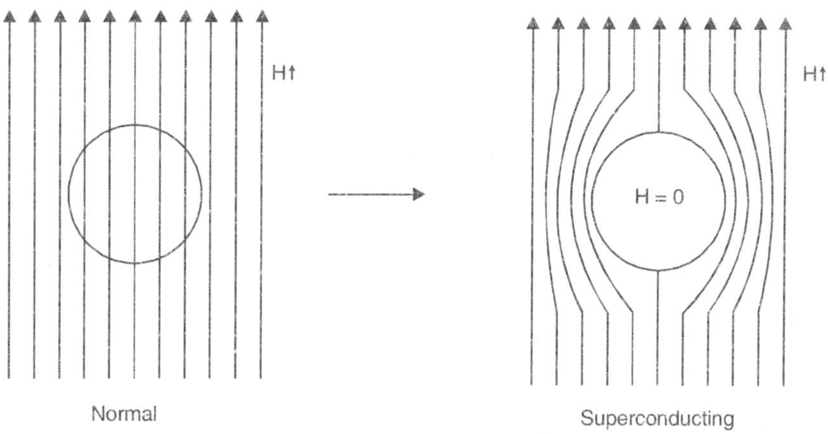

Fig. 7.4. Meissner effect in a sphere cooled in a magnetic field: note the ejection of lines of magnetic induction at the transition of normal state into superconducting state.

It is an experimental fact that a bulk superconductor in a weak magnetic field will act as a perfect diamagnet, with zero magnetic induction in the interior part. When a sample is placed in a magnetic field and is then cooled beyond the transition (critical) temperature for superconductivity, the magnetic field which was present in the sample is pushed out from it. This is called Meissner effect. This effect remain valid in solid sample *i.e.*, solid sphere, solid cylinder or rod (having no hole inside the sample) irrespective of the order of cooling below T_C and the presence of magnetic field. The sequence of the events for the experiment on a sherical conductor is shown in Fig. 7.4.

When field become zero; B = H + M = 0, $\chi = -1$ that is sample behave as perfect diamagnetic. Further from ohm's law $\vec{E} = \rho\vec{J}$, E will be zero as $\rho = 0$ (resistivity zero) then from Maxwell's equation $\nabla \times E = -\frac{\partial \vec{B}}{\partial t}$, $\frac{dB}{dt} = 0$ or B = constant.

This results that magnetic flux remain the same inside the specimen and cannot change on cooling through the transition, which contradict the Meissner effect. Thus, perfect diamagnetism and zero resistivity are two independent essential properties of superconducting state.

7.04 TYPE I AND TYPE II SUPERCONDUCTORS

Superconducting materials are classified as soft (Type-I) and hard (Type-II) superconductors. Type-I superconductors are pure specimen of some element with a low value of critical field H_C. These type of superconductors exhibit a complete Meissner effect *i.e.*, a complete diamagnetic and hence the flux is completely excluded. The magnetization curve for Type-I superconductors is shown in Fig. 7.5(*a*).

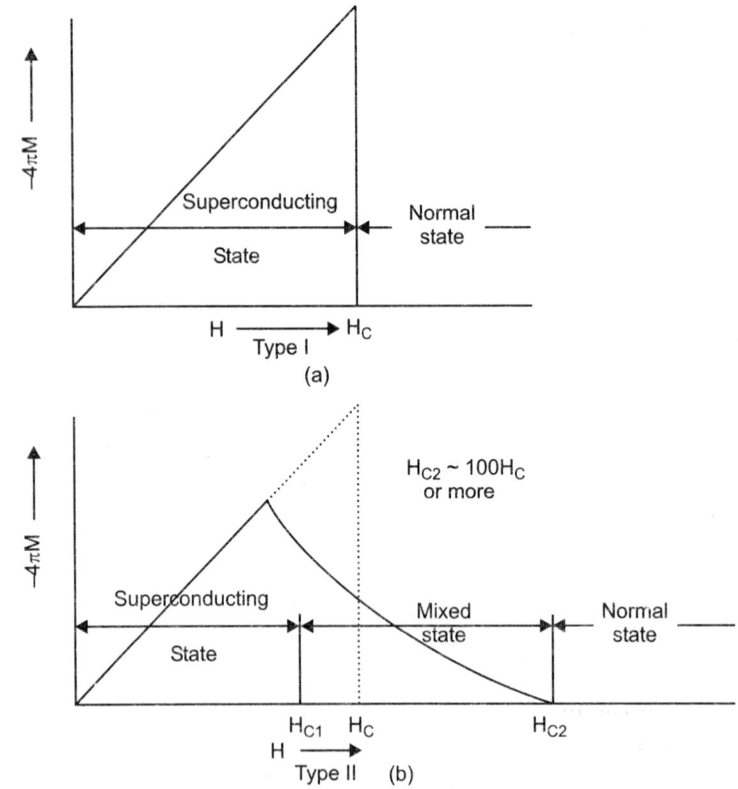

Fig. 7.5. Magnetization curve for Type-I and Type-II material.

The magnetization curve of Type-II superconductor is shown in Fig. 7.5(*b*) in which material behave in different way. In this Type-II, the material is diamagnetic and flux is excluded below the lower applied critical field H_{C1}. At H_{C1} magnetic flux begins to enter the superconductor and an upper critical magnetic field H_{C2} at which superconductivity disappear and magnetization vanishes. There is gradual fall of magnetization in Type-II superconductors where as abrupt fall in case of Type-I. The value of H_C for Type-II may be 100 times more or even higher than that for Type-I material.

A material can change from Type-I to Type-II on the substitution of some impurities. For example, lead is a Type-I superconductor with $H_C = 4.8 \times 10^4$ A/*m* at 4° K. When 2 % indium is added to it, it becomes a Type-II superconductor with $H_C = 3.18 \times 10^4$ A/m and $H_{C2} = 7.96 \times 10^4$ A/m. On adding 20 % of indium in lead, $H_{C1} = 5.6 \times 10^3$ A/m and $H_{C2} = 2.9 \times 10^5$ A/m. Magnetization curves for both Type-I and Type-II materials are reversible in case of the ideal superconductors.

Table 7.1 Critical temperature T_C and critical magnetic field H_O at 0° K of some superconducting material.

Element	T_C *(K)*	H_O *(A/m)*	*Type*
Aluminium	1.196	0.79×10^4	I
Cadmium	0.52	0.22×10^4	I
Gallium	1.09	0.4×10^4	I
Indium	3.4	2.2×10^4	I
Tin	3.72	2.4×10^4	I
Mercury	4.12	3.3×10^4	I
Zinc	0.9	0.42×10^4	I
Tantalum	4.5	6.6×10^4	II
Thalium	2.4	1.4×10^4	II
Niobium	9.3	15.6×10^4	II
Zirconium	0.8	0.37×10^4	II
Nb – 44% Ti	10.5	9.5×10^6	II
NbN	16	8×10^6	II
Nb_3 Sn	18.5	1.6×10^7	II
Nb_3 Al	18	3.2×10^7	II

7.05 BCS THEORY OF SUPERCONDUCTIVITY

John Bardean, Leon Cooper and John Schreiffer (BCS) put forward a microscopic theory in 1957, which provides the better quantum explanation of superconductivity and accounts very well for all the properties exhibited by the superconductors. The other fundamentally different theories of superconductors works within the framework of this general theory and thus are of little importance. BCS theory basically involves the electron intraction through phonons as mediator. Infact, the idea underlying this theory was originally conceived by Frohlich and Bardean in 1950, who pointed out that an electron moving through a crystal lattice has a self energy accompanied by a virtual phonon, although could not explain the superconductivity.

BCS showed that basis of intraction responsible for superconductivity to be a pair of electron by means of interchange of virtual phonons. Usually the electrons repel each other by the coulombic force but in special case when temperature is lower than critical value, the attraction force between the two electron is stronger than electron coulombic force. The attraction force can be explained in following way. Suppose an electron approaches a positive core. It suffers attractive coulumb interaction. Due to this attraction, ion core is set in motion and consequently distorts the lattice. Smaller the mass of the positive ion core, the greater will be the distortion. Let another electron comes towards that and sees this distorted lattice. Then the interactions between distorted lattice and electron taken place which effects the lowering energy of second electron. Thus, the two electron interact via lattice distortion or the phonon field resulting in the lowering of energy of the electrons. This lowering energy of electron implies that the force between two electron is attractive. This type of interaction is called electron-lattice-electron interaction. This interaction is strongest when the two electrons have equal and opposit spin and momenta.

The superconductivity occurs when attractive interaction between two electrons by means of a phonon exchange dominate the usual repulsive interaction. The two attracted electron make a pair and called cooper pair. The energy of cooper pair is less than the energy of electron in free state by an amount of binding energy. That is the binding energy is needed to break the pair at critical temperature. Below critical temperature the motion of several cooper pair with same velocity in same direction causes an electric current in the material. A single wave therefore be regarded as collectively representing these cooper pairs. Since the momentum of cooper pair is small corresponding velocity of wave is small consequently wavelength of wave is very large. Due to this large wavelength, scattering from imperfections rarely occurs, and the conductivity tends to infinite to be a superconductor.

7.06 HIGH-TEMPERATURE SUPERCONDUCTORS

The discovery of high temperature superconductors has been hailed as one of the most important scientific events of the 20th century. High temperature superconductor is a fiekle term. It used to mean any superconductor with a transition temperature above 10 K. Now, it probably means anything with a transition temperature above 40 K or may be even 90 K.

The high temperature superconductors are not metals or intermetallic compounds at all; they are oxides. According to crystal chemists, the compounds are classified as ceramics. Aluminium oxide, quartz, and even ordinary table salt are examples of ceramic materials, so defined by virtue of the kind of bonding that holds their constituent atoms together. For most of us, the term ceramic makes us think of property that is fired in a Kiln, and in fact, the new high temperature superconductors were first made in just this way.

The new superconductors have a far more complicated unit cell. The cell contains one atom of a rare earth metal, two barium atoms, three copper atoms and seven oxygen atoms. The number of atoms of each metal element in the superconductor gives these compounds the name '1-2-3'.

Several new oxide superconductors with above 90 K transition temperatures were discovered in the year 1988. These materials do not even have the same crystal structure as 1-2-3. The newer compounds are still layered copper oxide materials, but they do not contain rare earth elements. In place of yttrium or other rare earths, the compounds contain metals like bismuth or thallium. Thus superconductors contain four metallic elements instead of three. Transition temperatures at 120 K and above mean that these superconductors can actually be used at liquid nitrogen temperature.

7.07 CHARACTERISTIC PROPERTIES OF SUPERCONDUCTING STATE

(*i*) Perfect Diamagnetism

Meissner showed that magnetic flux is ejected out (Fig. 7.4) as soon as the material passes over to the superconducting phase. Meissner effect contradicts Maxwell's equation and suggests that perfect diamagnetism is an essential property of the superconducting state.

The Meissner effect shows that a superconductor behaves as if inside the specimen we have B = 0. Consequently from the relation.

$$B = \mu_0 (H + M) = 0$$

where M is magnetization and $\mu_0 = 4\pi \times 10^{-7}$

$$\text{Thus, the susceptibility } \chi = \frac{M}{H} = -1$$

Thus, the superconductor acts as perfect diamagnet that excludes the flux from its interior by means of surface current.

(*ii*) Persistent Current Setup

Since in superconducting state resistivity is zero. The current due to induction in the specimen persists as along as the field is on. Due to this persistent current, specimen acquires a negative moment, producing a field in a direction opposite to the direction of applied field. Due to the negative magnetic moment, the specimen behaves as a diamagnetic one and when the external field is removed persistent current goes off.

It is observed that persistent current can be setup only in the presence of applied field—a concept that does not support fully the occurrence of superconducting state for we must get persistent current irrespective of the presence or absence of applied magnetic field. The latter is really true, was predicted by Meissner who showed that persistent current can be setup in presence or absence of applied field. Persistent currents are maintained in the superconductor—a fact that goes in favour of occurrence of superconductivity.

(*iii*) Penetration Depth of the Magnetic Field

In superconducting state the supercurrents flowing over the surface of the conductor with their associated magnetic fields shielding the interior of the specimen. Applying the conditions.

$$E = 0 \text{ and } B = 0$$

Characterising the superconducting state, with Maxwell's electromagnetic equation in one dimension, *i.e.*,

$$\frac{d^2 J_s}{dx^2} = \frac{J_s}{\lambda^2}$$

where J_s is the supercurrent density and x is the distance from the surface into the superconducting material. The parameter λ is called the penetration depth and is given by

$$\lambda^2 = \frac{m}{\mu_o n_g e^2}$$

where m the mass, e the charge of electron, and n_g is the number of superconducting electrons per unit volume. λ varies with temperature and quite large at critical temperature. It also changes with applied magnetic field.

(*iv*) Isotope Effect

The transition temperature T_C is found to be proportional to the reciprocal of the square root of their respective isotopic masses *i.e.*,

$$T_C \propto M^{-1/2}$$

$$M^{1/2}\ T_C = \text{Const.}$$

A Heavier Isotopic Mass Lowers the Lattice Vibrations. Lattice vibrations or electron phonon interactions may be important for the occurrence of superconductivity.

7.08 APPLICATIONS OF SUPERCONDUCTORS

1. Superconductors are used for producing very strong magnetic field of about 20–50 Tesla.
2. Magnetic energy can be stored in large superconductors and drawn as required to counter the voltage fluctuations during peak loading.
3. Superconductors are also being used in medical industry as superconducting quantum interferometer (SQUID).
4. Superconductors are also used in computers and information processing *i.e.*, semiconductor-superconductor hybrids (A-D converters). Active superconducting elements (FETs), voltage standard, opto electronics.

SOLVED EXAMPLE

1. *A superconducting tin has a critical temperature of 3.7 K in zero magnetic field and a critical field of 0.0306 Tesla at 0 K. Find the critical field at 2 K.*

Sol. We know that

$$H_C = H_O \ \ 1 - (T/T_C)^2$$

$$= 0.0306 \ \ 1 - \frac{2}{3.7} \ ^2$$

$$= \mathbf{0.0217\ T.} \qquad \textbf{Ans.}$$

2. *The critical field for niobium is 1 × 10⁵ Amp/m at temperature 6 K and 2 × 10⁵ Amp/m at 0 K. Calculate the transition temperature of niobium.*

Sol. We know that

$$H_C = H_O\left[1-\left(\frac{T}{T_C}\right)^2\right]$$

here

$$H_C = 1 \times 10^5 \text{ Amp/m}$$

$$H_O = 2 \times 10^5 \text{ Amp/m}$$

$$T = 6 \text{ K}$$

putting the value

$$1 \times 10^5 = 2 \times 10^5\left[1-\left(\frac{6}{T_C}\right)^2\right]$$

$$\frac{1}{2} = 1-\left(\frac{6}{T_C}\right)^2$$

$$\frac{6}{T_C} = \sqrt{1-\frac{1}{2}} = \frac{1}{\sqrt{2}}$$

$$T_C = 6\sqrt{2} = \mathbf{8.46\ K.\ Ans.}$$

MULTIPLE CHOICE QUESTIONS

1. Superconductors are generally.
   (*a*) Ferromagnetic and antiferomagnetic metals.
   (*b*) Monovalent metals.
   (*c*) Amorphous thin films of Be and Bi.
   (*d*) Thin films of barium titnate.
2. Hard Superconductors observe
   (*a*) Break down of silsbee's rule
   (*b*) Incomplete Meissner effect
   (*c*) High critical field and transition temperature
   (*d*) All of these
3. The temperature at which a conductor becomes a superconductor is called
   (*a*) Superconducting temperature (*b*) Curie temperature
   (*c*) Onne's temperature (*d*) Transition temperature
4. Soft superconductors observe
   (*a*) Silsbee's rule (*b*) Meissner effect
   (*c*) Both (*a*) and (*b*) (*d*) None of these

5. The transition temperature of most superconducting elements lie in the range
   (a) Zero to 10K (b) 10K to 20K
   (c) 20K to 50K (d) Above 50K
6. A superconducting material when placed in a magnetic field will
   (a) Attract the magnetic field towards its centre.
   (b) Repell all the magnetic lines of force passing through it
   (c) Attract the magnetic field but transfer it into a concentrated zone
   (d) Not influence the magnetic field
7. A superconducting material on being subjected to the critical field changes to
   (a) Critical conductivity
   (b) Normal state
   (c) Superconducting which is independent of temperature
   (d) Remains uninfluenced
8. The width of the energy gap of a superconductor is maximum at
   (a) 0 K (b) Transition temperature
   (c) Room temperature (d) None of these.

ANSWERS

1. (c)	2. (d)	3. (d)	4. (c)	5. (a)
6. (b)	7. (c)	8. (a)		

LONG QUESTIONS AND NUMERICAL PROBLEMS

1. Lead in superconducting state has critical temperature of 6.2 K. at zero magnetic field and a critical field of 8.5 T at 0°K. Determine the critical field at 4.2 °K.
2. The critical temperature, T_C for Hg with isotopic mass 199.5 is 4.185 K. Calculate its T_C when its isotopic mass changes to 203.4.
3. A long thin superconducting wire of a metal produces a magnetic field 105×10^3 amp/m on its surface due to current through it at a certain temperature T. The critical field of the metal is 150×10^3 amp/m at T = 0 K. The critical temperature T_C of the metal is 9.2 K. What is the value of T?
4. What is superconductivity? Describe the effect of magnetic field on a superconductor.
5. Explain Meissner effect.
6. How does critical field vary with temperature in Type-II superconductors?
7. Define critical magnetic field. Explain the effect of isotopes on superconductor.
8. Give the BCS theory of superconductivity.

8

Nanotechnology

8.01 INTRODUCTION

Nanotechnology is a field whose theme is to control of matter on an atomic and molecular scale. It deals generally with structures of 100 nanometers or smaller, and involve developing materials or devices within that size. A number of physical phenomena become pronounced as the size of the system devices. These includes statistical mechanical effects, as well as quantum mechanical effects. For example, the quantum size effect, where the electronic properties of solids are altered with great reduction in particle size. This effect does not come into play by going from macro to micro dimensions. However, it becomes dominant when the nanometer size range is reached. Additionally, a number of physical, like mechanical, electrical, optical etc., properties changes on moving from macroscopic systems to nanoscale.

The increase in surface area to volume ratio altering mechanical, thermal and catalytic properties of the materials. It has the potential to create many new materials and devices with wide ranging applications such as in medicine, electronics and energy production. This technology also concern about the toxicity and invironmental impact of nanomaterials and their potential effect on global economics, as well as speculation about various doomsday scenarios.

8.02 NANOSCIENCE AND NANOTECHNOLOGY

Nanotechnology is an advance field of technology which deals with small structure or nanomaterials. Nanotechnology is extremely diverse, ranging from novel extension of conventional device physics, to completely new approach based upon molecular self-assembly to develop new materials with dimensions on the nanoscale. The size is ranging from 1 nanometer to 100 nanometers. One nanometer is one thousandth of a micrometer which is approximately the length of equivalent to 10 hydrogen or 5 silicon atoms aligned in a line. The nano size can be understood by Fig. 8.1 which compare the nano size with dimension of others. The comparative size of a nanometer to a meter is the same as that of a marble to the size of the earth.

Nanostructure science and technology is a interdisciplinary field of research and development activity that has been growing explosively worldwide in past few years. It can be thought of as an extension of traditional disciplines towards the explicit consideration of these properties. The traditional disciplines can be reinterpreted as specific application of nanotechnology. Broadly speaking, nanotechnology is the synthesis and application of ideas from science and engineering towards the understanding and production of novel materials and devices. These products generally make copious use of physical properties associated with small scales.

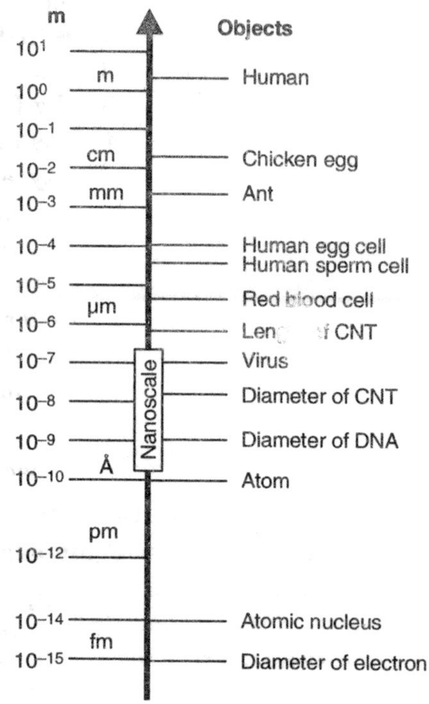

Fig. 8.1. Nanoscale comparison with other scales.

Materials in microscale generally exhibit similar physical properties that of bulk form, but exhibits remarkable distinct physical properties when reduces to nano-scale. For example copper material which opaque at macroscale become transparent at nanoscale, the platinum which is inert at macro attain catalytic properties at nanoscale, the silicon insulator at bulk form become conductor at nanoscale, gold is solid in bulk form become liquid at room temperature when reduce to nanoscale. Material such as gold, which is chemically inert at normal scales, can serve as a chemical catalyst at nanoscale. An aspect of the nanotechnology is the vastly increased ratio of surface area to volume which present in many nanomaterials and makes possible new quantum mechanical effects. Because of this effect the electronic properties of solid are changed with reduction in particle size. Further this effect does not come into picture from macro to micro dimensions but becomes more pronounced when reduced to nonometer scale. The catalytic activity of nanomaterials also opens potential risks in their interaction with biomaterials.

Nanotechnology can be understood as a technology to design, fabricate and application of nanostructure and nanomaterials. It includes fundamental understanding of physical properties and phenomena of nanomaterials. The study of nanomaterials is the study of fundamental relationship between physical properties and phenomena of material dimension in the nanoscale.

8.03 NANOMATERIALS

Nanomaterial is the structure of material with improved properties through controlled synthesis and assembly of the material at nanoscale level. There are unique defining feature of nanomaterials having higher surface area than conventional materials. The high surface area can be attained either by fabricating small particles or clusters where the surface to volume ratio of each particle is high. This unique property of surface to volume ratio leads to greater chemical reactivity and effect their strength. These materials may have a significantly lower melting point due to a huge fraction of surface atoms in the total amount of atoms. The electrical conductivity of nanomaterials also enhanced appreciably due to better ordering in microstructure. Many such properties of size dependent. In other words, properties of nanostructure material can be tuned considerably simply by adjusting the size, shape or extent of agglomeration. Interface and colloid science has given many materials which may be useful in nanotechnology, such as carbon nanotubes and other fullerenes, and various nanoparticles and nanorods.

There are two main approaches are used to get the material at nanoscale.

(*i*) **Bottomup approach.** In this approach atoms and molecules are assembled together chemically and can stop the growth process at desired size of material.

(*ii*) **Top to bottom approach.** In this approach materials are cut using high energy ballomill, to a designed scale. The process of cutting material is called the milling of material.

8.04 CARBON BUCKYBALLS AND NANOTUBES

Carbon is a unique material and can be a good metallic conductor in the form of graphite. A wide band gap semiconductor in the form of diamond or a polymer when reacted with hydrogen. Carbon provide material at nanoscale like buckyballs and nanotubes which are constructed by strong bonding carbon atoms.

Buckyballs

A buckyball or buck minister fullerence is the spherical fullerence commonly refer to a molecule with 60 carbon atoms C_{60} and with an icosahedral symmetry. It also includes larger molecular weight C_n ($n > 60$). The 60 carbon atoms in C_{60} are located at the vertices of a regular trancated icosahedron and every carbon site on C_{60} is equivalent to every other site* (Fig. 8.2). The average nearest neighbor C–C distance in C_{60} is 1.44 Å is almost equal to that in graphite. Each carbon atom in C_{60} is trigonally bonded to other carbon atom and most of the faces are hexagonal. There are 20 hexagonal faces and 12 pentagonal faces in C_{60} molecule. The pentagonal faces consists of alternating single and double bonds, where the pentagonal faces are defined by single bond. The length of single bond is 1.46 Å which is longer than the average bond length 1.44 Å, while the double bond are shorter 1.40 Å. The structure of buckyballs like C_{70}, C_{80} can be considered as modification of C_{60} by varying the number of hexagonal faces.

---

* Which resembles a socer ball.

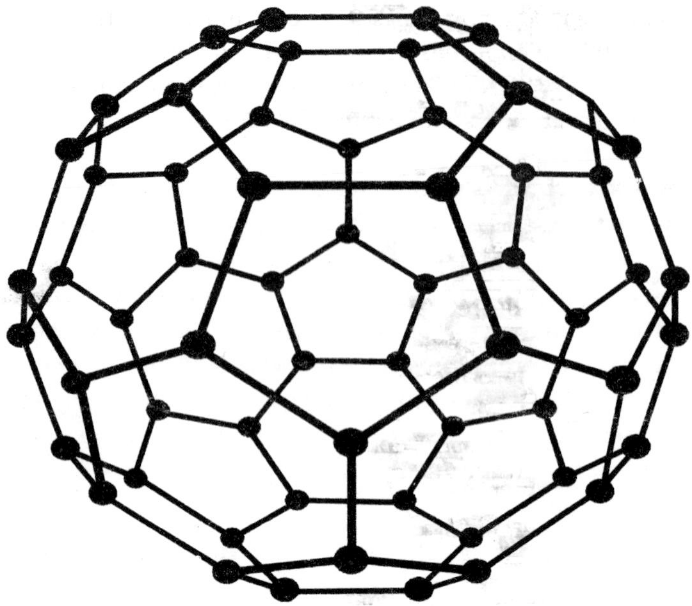

Fig. 8.2. Buckminster fullerence (Buckyballs) C_{60}.

Creation of Buckyballs

The buckyballs usually created by using an arc discharge between graphite electrodes in approximately 200 Ton of He gas. The heat generated at constant point between the electrodes in the reaction chamber filled with low pressure helium. The carbon is evaporated to form soot and buckyballs which condensed on the water cooled walls of chamber. Because of uniqueness of C_{60} buckyball molecule and its ability to have a variety of chemical reactions, its chemistry become a very active research field.

Use of Buckyballs

Because of unique nature of buckyballs there have variety of uses. One of the important use of buckyballs is in medical field. It can be used as antioxidant to counteract free redical in human body. A free redical is a molecule has unpaired electron and become very active. When another electron provided by antioxidant to free redical, it become inactive. Buckyballs unique cage like structure might allow it to take place of other molecules shuttling toxic metal substances through human body during MRI scan.

The superconducting properties are also observed by fitting a potassium ion in buckyball and therefore used for superconducting need.

Buckyballs can also used as catalyst by attaching metals on to the surface of buckyballs.

Carbon Nanotubes

Carbon Nanotubes (CNT) are the cylindrical fullerence, the cylindrical structure of carbon atoms, in which one carbon atom is covalently bonded to three other carbon atoms. These carbon cylindrical tubes are usually a few nanometer wide but can range from less than a micrometer to several millimetres in length. Generally these CNT are closed ends, but can found open end also. Their name is derived from their size, the length to diameter ratio greater than 1,000,000. Nanotubes are classified as single-walled nanotubes (SWCNT) and multi walled nanotubes (MWCNT). The fundamental carbon nanotube is SWCNT. However MWCNT are consists of several nested coaxial SWCNT.

Structure of Carbon Nanotube

The structure of nanotube can be conceptualised by wrapping a one atom thick layer of graphite sheets called graphene into seamless cylinder in which each carbon atom covalently bonded with three neighbours similar atoms. The structure can be specified by a chiral vector (n, m) where n and m are the integers of the vector equation.

$$C_h = na_1 + ma_2$$

a_1 and a_2 are the unit vectors for the honey comb lattice. In Fig. 8.3, we see that point O and A are crystallographically equivalent. The point O and A can be connected by a vector C_h. We can draw normal to C_h at point O and A to obtain lines OT and AT′. If we now superimpose OT and AT′, we obtain a cylinder of carbon atoms that constitute carbon nanotube when properly capped at both end with half of buckyball. Depending on the value of n and m, the carbon nanotube are devided into three category. If $m = 0$, the nanotube are called "zigzag" [Fig. 8.4(*a*)]. If $n = m$, the nanotube are called "armchair" [Fig. 8.4(*b*)]. Otherwise they are called "chiral" [Fig. 8.4(*c*)].

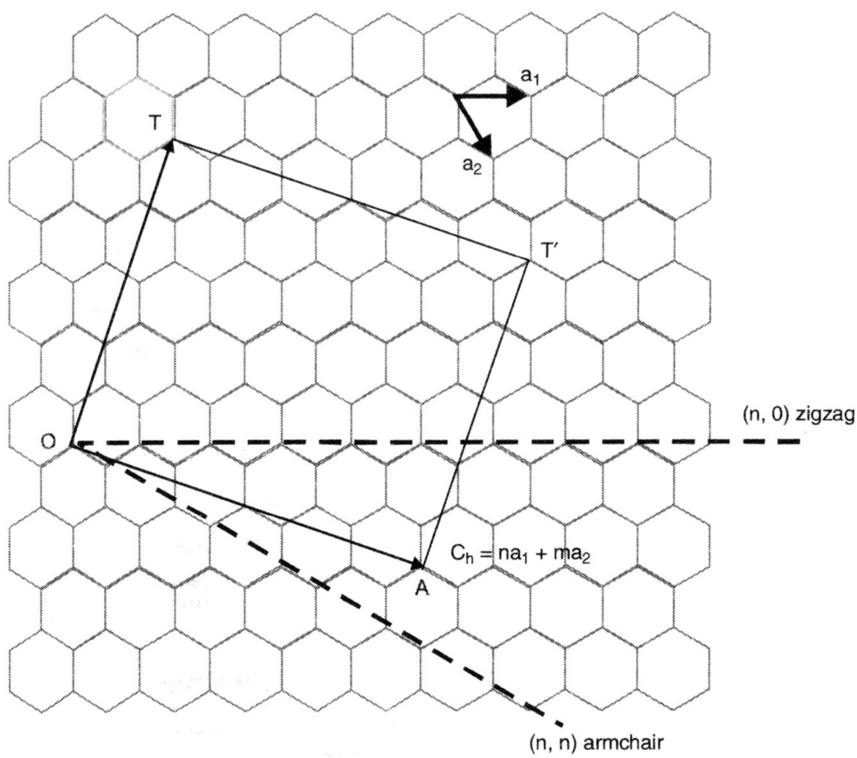

Fig. 8.3. Structure of graphite sheet which turn in nanotube by chiral vector $(n, m)^3$.

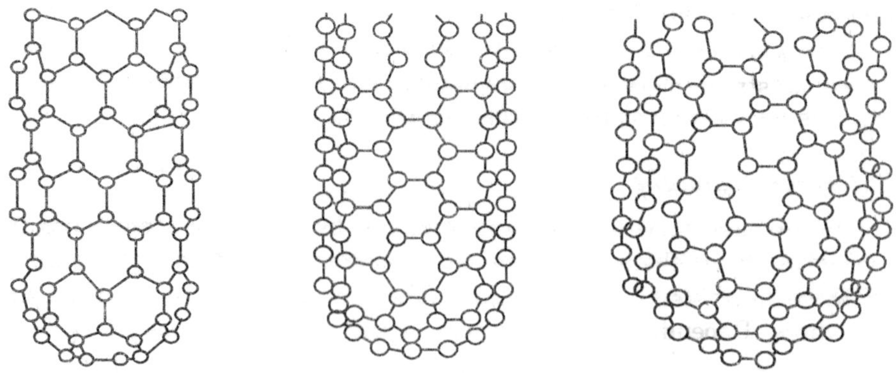

Fig. 8.4. Nanotubes described through chiral vector $C_h = na_1 + ma_2$.

Synthesis of Carbon Nanotubes

There have been different techniques developed to produce nanotubes in sizeable quantities, including arc discharge, laser ablation, high pressure carbon monoxide (HiPCO), and chemical vapour deposition (CVD).

Arc Discharge

From this method nanotubes were observed in 1991 in the carbon soot of graphite electrodes during an arc discharge. During this process the carbon contained in the negative electrode evaporate because of the high temperature caused by the discharge. Since nanotubes were initially discovered using this technique, therefore this has been the widely used method of nanotube synthesis. This method produces both single and multi walled nanotubes with lengths upto 50 micrometers.

Laser Ablation

In this process of laser ablation, a pulsed laser vaporizes a graphite target in a high temperature reactor while an inert gas is bled into chamber. The nanotubes develop on the cooler surfaces of the reactor, as the vaporized carbon condenses. A water cooled surface may be included in the system to collect the nanotubes. This method can produce about 70% primarily SWCNT with controlled diameter determined by reaction temperature. However this process is more expensive than arc discharge method.

Chemical Vapour Deposition (CVD)

CVD is a common method for commercial production of carbon nanotubes. In this method two gases, a process gas such as ammonia, nitrogen, hydrogen and a carbon-containing gas such as methane are bled into chamber. The carbon containing gas is broken into carbon atoms at surface of catalyst particle such as iron, and carbon is deposited to the edges of the iron particle forming nanotubes. The catalyst particles can stay at the tips of the growing nanotubes during the growth process or remain at the nanotube base depending on the adhesion between the catalyst particle and the substrate.

Properties of Carbon Nanotubes

1. Carbon nanotubes are the strongest and stiffert materials on the earth. The tensile strength of carbon nanotube is about 100 times greater than steel. These are much softer in the radial direction than along the tube axis.
2. MWCNT are multiple concentric nanotubes nested within one another, exhibit a striking telescoping property where the inner nanotube may slide almost without friction within its outer nanotube shell. This property has already been utilized to create the world's smallest rotational moter.
3. The unique electronic structure of graphena, affects its electrical properties strongly. Depending value of n and m its electrical properties changing. For $n = m$ the metallic nanotube have electrical current density more than 1000 times greater than bulk metal such as silver and copper.
4. All nanotube are good thermal conductor along the axis but good insulator along radial.
5. If there is defect in the CNT due to atomic vacancies, its stencil strength and electrical properties are very much affected.
6. It is commonly referred as one dimensional as electron transport in CNT takes place along the axis of the tube.
7. Under certain conditions, CNT can cross the membrane barriers and reach the organs to make harm.
8. Wide use of CNT may lead to lung cancer.

Uses of Carbon Nanotubes

High strength, flexibility and electrical conductance of CNT makes them of potential use in various fields.

1. Because of the great mechanical properties of the carbon nanotube, a variety of structures have been proposed ranging from everyday items like cloths and sports gear to combat jacket and space elevators.
2. Nanotube based transistors have been made which operates at room temperature and are capable of digital switching using single electron.
3. Depending upon the structure of surface it can be used as plain conductor or semi conductor.
4. In cancer related incidents, the CNT are used as a vessel for transporting drugs into the body. The drug either attached to the side or trail behind or can actually be placed inside the nanotube.
5. Bulk carbon nanotubes have been used as composite fibers in polymers to improve the mechanical, thermal and electrical properties of the bulk.
6. Carbon nanotube are used in construction of handle bars for mountain bikes.
7. Carbon nanotubes are used in solar cells as copper wires to make flow electron or current.
8. It can be used as radio receiver.
9. A sheet of nanotubes can operate as a loudspeaker if an alternating current

is applied. The sound is produced through thermoacoustically and not by vibrations.

10. Carbon nanotube have the strength of diamond, therefore research is going on to wearing CNT into cloths to create stab-proof and bullet proof clothing.

8.05 APPLICATIONS OF NANOTECHNOLOGY

Nanotechnology has potential to develop various types of material and devices whose physical properties changes as sizes are shrunk. Nanoparticles take advantage of their dramatically increased surface area to volume ratio and influence the mechanical properties of material strongly like strength and elasticity. Traditional polymers can be reinforced by nanoparticles to obtain novel materials which are light in weight and can replace the metals. Nanotechnology reduces weight and increase stability and reactivity of materials. In the light of above, there are many applications of nanotechnology, some of them are described here:

1. Medicine

The size of nanomaterials is similar to that of most biological molecules and structure, therefore nanomaterials can be useful for both in vivo and in vitro biomedical research and applications. The integration of nanomaterials with biology has led to the development of diagnostic devices, contrast agents, analytical tools, physical therapy applications and drug delivery vehicles.

Biological test, measuring the presence or activity of selected substances, become quicker, more sensitive and more flexible when nanoscale particles are put to work as tags or labels. For example gold nanoparticles tagged with short segment of DNA, can be used for detection of genetic sequence in a sample. The overall drug consumption and side effects can be lowered significantly by depositing the active agent in the morbid region only and no higher dose then needed. As drug delivery vehicle, a nanotube can transport a small drug molecule to the desired location which reduces the drug consumption and treatment expenses, such as in cancer treatment. This results in an overall societal benefit by reducing cost to the public health system.

Nanotechnology can help to reproduce or to repair damaged tissues and bones which might replace todays conventional treatment like organ transplants or artificial implants.

2. Chemistry

Nanotechnology plays an important role in the techniques of chemical catalysis and filtration. All chemical synthesis can be understood in terms of nanotechnology, because of its ability to manufacture certain molecules, polymers as well as clusters and nanoparticles.

Chemical catalysis from nanoparticles is very important because of extremely large surface to volume ratio. There is potential application of nanoparticles in catalysis ranging from fuel cell to catalytic converters and photo catalytic devices.

Nanochemistry is applicable on waste water treatment, air purification and energy storage devices as filtration. The filtration techniques are based on the use of membranes with suitable hole sizes, where liquid is pressed through the membranes. Such nanoporous membranes are suitable for a mechanical filtration with extremely small pores, smaller than 10 nm and may be composed of nanotubes. Nanofiltration is mainly used for removal of ions or the separation of different fluids. On a large scale, the membrane filtration technique is named as ultrafiltration, which works between 10 and 100 nm and used for medical purpose in renal dialysis. Some water treatment devices incorporating nanotechnology are already available in the market.

3. Energy

Nanotechnology can be applied to improve the systems for energy storage, energy conversion, energy saving, and renewable energy sources. Energy consumption can be reduced by making better insulation systems, more efficient lighting or combustion systems and use of lighter and stronger materials in the transportation sector. Nanotechnological approach, like light emitting diodes or quantum caged atoms, could lead to a strong reduction of energy consumption for illumination.

From energy conversion systems, presently the best solar cells absorb light at different energies and manage to use only 40 percent of the sun's energy and works at very lower efficiencies (15–20%). The use of nanostructures with a continuum bandgaps could help in increasing efficiency of nanotech solar cells and other light conversion devices. The nanotech solar cell can be manufacture at significantly lower cost than conventional solar cells.

4. Fuel cells

Nanotechnology is being used to reduce the cost of catalyst used in fuel cells to produce hydrogen ions from fuel such as methanol. This also improve the efficiency of membranes used in fuel cells to separate hydrogen ions from gases.

5. Aerospace

The manufacturing of aircraft and spacecraft with lighter and stronger material will lead to increase the performance as weight is a major factor. Nanotechnology also help to reduce the size of equipments used in the aircraft and spacecraft.

6. Laser

The nanoscaled quantum dots can be used for the construction of lasers. The advantage of quantum dot laser over the traditional semiconductor laser is that their emitted wavelength depends on the diameter of the dot. Quantum dot laser are cheaper and offer a higher beam quality than conventional laser diodes.

7. Food

Nanotechnology can be applied in the production, processing, safety and packing of food. A nanocomposit coating process could improve food packing by placing antimicrobial agents directly on the surface of the coated film. Research is in progress to apply technology for the detection of chemical and biological substances for biochemical changes in foods.

8. Textile

The use of engineered nanofibers makes clothes water and stain repellent or wrinkle free. Textile with a nanotechnological finish can be washed less frequently and at lower temperatures. Nanotechnology has been used to integrate tiny carbon particles membrane which gives full surface protection to the wearer from electrostatic charges.

9. Chemical Sensors

Nanotechnology can enable sensors to detect very small amounts of chemical vapours. Various types of detecting elements, such as carbon nanotubes, zinc oxide, nanowires can be used for sensors. Due to their small size, a few gas molecules are sufficient to charge the electrical properties of the sensing elements.

10. Consumer Goods

Nanotechnology is already impacting the field of consumer goods such as cosmetics, sunscreens, water filters, glare filters, ink, stain resistance clothing, more durable tennis balls, more light weight tennis rackets, dressing for burn or injuries etc. However, we have only begun to see what can be achieved through nanotechnology and its continuous development will lead to changes in the world that will redefine life.

MULTIPLE CHOICE QUESTIONS

1. Nanotechnology deals with particles size of
   (*a*) 1 to 100 mm (*b*) 1 to 100 nm
   (*c*) 1 to 100 μm (*d*) 1 to 100 cm.
2. Properties of materials changes remarkably when their size approaches to
   (*a*) microscale (*b*) macroscale
   (*c*) nanoscale (*d*) vernier scale.
3. A buckyball commonly refer to a molecule with 60 carbon atoms have
   (*a*) 20 pentagonal faces and 12 hexagonal faces
   (*b*) 12 pentagonal faces and 20 hexagonal faces
   (*c*) 22 pentagonal faces and 10 hexagonal faces
   (*d*) 26 pentagonal faces and 12 hexagonal faces.
4. The average nearest neighbour C-C distance in C_{60} buckyball is
   (*a*) 1.0 Å (*b*) 2.04 Å
   (*c*) 1.44 Å (*d*) 1.04 Å
5. Carbon nanotubes are about 100 times stronger than steel but its weight is
   (*a*) One tenth of steel (*b*) One sixth of steel
   (*c*) equal to steel (*d*) five times of steel
6. In nanotechnology, Chemical Vapour Deposition (CVD) is a technique used for
   (*a*) identification of nanoparticle
   (*b*) obtaining the size of nanoparticles

(*c*) synthesis of carbon nanatubes

(*d*) Characterization of nanoparticles

7. Zigzag, Armchair and Chiral are the classification of

(*a*) Single walled nanotubes

(*b*) Double walled nanotubes

(*c*) Multi walled nanotubes

(*d*) None of the above

8. In Chiral vector $C_n = na_1 + ma_2$, the relation between n and m for Armchair nanotube is

(*a*) $n > m$ (*b*) $n < m$

(*c*) $n = m$ (*d*) $m = 0, n \neq 0$.

ANSWERS

1. (*b*)	**2.** (*c*)	**3.** (*b*)	**4.** (*c*)	**5.** (*b*)
6. (*c*)	**7.** (*a*)	**8.** (*c*)		

LONG ANSWER QUESTIONS

1. What do you understand by nanotechnology? Describe its important features.
2. What are the nanomaterials? How they differ from bulk materials?
3. Describe buckyballs and their uses.
4. What do you understand by nanotubes? How these are produced?
5. Write briefly about nanotubes and their types of classification.
6. Explain the structure of nanotube with chiral vector and write its uses.
7. What are fullerence and its type? Describe any one of them.
8. What are the applications of nanotechnology?
9. How can you express that nanotechnology is need of society?
10. What are the advantages of nanotechnology over present scenerio? Explain with some examples.